AF572692

# *ANNALS OF THE NEW YORK ACADEMY OF SCIENCES*

*Volume 491*

*The New York Academy of Sciences*
*2 East 63rd Street*
*New York, New York 10021*

# REPORTS FROM THE MOSCOW REFUSNIK SEMINAR

ANNALS OF THE NEW YORK ACADEMY OF SCIENCES
*Volume 491*

# REPORTS FROM THE MOSCOW REFUSNIK SEMINAR

*Edited by Joel L. Lebowitz*

*The New York Academy of Sciences*
*New York, New York*
*1987*

**Library of Congress Cataloging-in-Publication Data**

Reports from the Moscow refusnik seminar.

(Annals of the New York Academy of Sciences, ISSN 0077-8923; v. 491)
Bibliography: p.
Includes index.
1. Physics—Congresses. 2. Mathematics—Congresses. 3. Refusniks—Soviet Union—Congresses. I. Lebowitz, Joel Louis, 1930– . II. Title: Reports from the Moscow refusnik seminar. III. Series.
Q11.N5 vol. 491 500 s 87-5555
ISBN 0-89766-377-2 [530]
ISBN 0-89766-378-0 (pbk.)

SP
*Printed in the United States of America*
**ISBN 0-89766-377-2** (cloth)
**ISBN 0-89766-378-0** (paper)
**ISBN 0077-8923**

*Dedicated*

*to*

YACOV AL'PERT

DISTINGUISHED SCIENTIST, HUMANIST, AND HOST OF THE MOSCOW SEMINAR 1981–1986, ON THE OCCASION OF HIS 76TH BIRTHDAY, MARCH 1, 1987

ANNALS OF THE NEW YORK ACADEMY OF SCIENCES

*Volume 491*
*April 15, 1987*

# REPORTS FROM THE MOSCOW REFUSNIK SEMINAR

*Editor*
JOEL L. LEBOWITZ

CONTENTS

# Foreword

JOEL L. LEBOWITZ[a]
*Hill Center for the Mathematical Sciences*
*Rutgers University*
*New Brunswick, New Jersey 08903*

This is the fifth volume of papers by persecuted scientists from the Soviet Union and by Western colleagues concerned with their fate published by the New York Academy of Sciences.[b] As such, they constitute not only a valuable contribution to science, but also a unique chronicle of courage and a cry for help. It is sincerely hoped that the Soviet authorities will heed this cry and permit a humane resolution to the plight of these brave scientists. Time is rapidly running out. Immediate action is required.

The beginning of the Moscow seminars, from which this volume takes its title, dates to 1973. They were and still are organized by Soviet "refusnik" scientists. These are scientists who have applied for and have been denied exit visas to Israel. As a consequence of their application to emigrate—a universally recognized human right encoded also in many treaties to which the Soviet Union is a signatory—these scientists have suffered a variety of punishments. Many of them have lost their jobs, while others have been or are in prison or exile on trumped-up charges—all of them have been demoted and barred from normal participation in the scientific life of their country. To overcome their isolation—tantamount to a scientific death sentence—an unofficial seminar was started in the house of Alexander Voronel. When Voronel was granted permission to leave, Mark Azbel took over. When Azbel was able to leave for Israel in 1978, the seminar moved to the house of Viktor and Irena Brailovsky, where it stayed until the end of 1980.

The departure from the Brailovsky home, unfortunately, was not due to their getting permission to emigrate. Rather, to use a phrase one often hears among the refusniks, Viktor Brailovsky was sent East (to prison and exile), not West. The Brailovsky family, like most others whose papers appear in this volume and hundreds more (counting just scientists), are still being detained against their will; they are prevented from working at their professions and they despair to see their children growing up in a poisoned atmosphere.

After the arrest of Brailovsky, there was a time when the authorities would not permit even this scientific communication among the refusniks. The refusniks, however, did not give up. It was a vital matter to them—scientific survival. Yacov Al'pert, a most distinguished scientist in the field of space plasma physics, managed to reestablish the seminar in his house in 1981, and continued to host it for five years. This required great courage on the part of him and his wife Svetlana. The seminar stayed at the Al'pert's house until March 1986 and this volume is dedicated to them.

During this period, there were 114 sessions of the Seminar and 143 lectures were given: 78 on physics, 44 on mathematics, 18 on biology and medicine, and 3 on chemistry. These seminars were attended by 126 scientists from France, Sweden,

[a]Chairman, Human Rights Committee of The New York Academy of Sciences.
[b]Previous volumes are *Annals* numbers 337, 373, 410, and 452.

Norway, Denmark, England, the United States of America, and Venezuela. Many scientists came especially for the seminar.

Since March, the seminar has been meeting at the houses of Alex Ioffe, Mark Freidlin, and Viktor Brailovsky. The seminar continues to invite participation by Western scientists. Such support is very important to the refusniks both for scientific and human reasons. In addition to needing scientific contacts, they need to feel that they are not totally abandoned. They live on their hopes of someday soon being able to emigrate and resume normal scientific activity.

To understand just how difficult the situations of these colleagues are, let me give just the duration of the plight of the contributors to this volume by indicating the year of their first application for an exit visa. That is when their troubles began: Yacov Al'pert, 1975; Viktor Brailovsky, 1972; Eugene M. Chudnovsky,[c] 1979; L. A. Dickey, 1979; M. I. Freidlin,[c] 1979; Eugene Grechanovsky, 1980; A. D. Ioffe, 1976; A. I. Leonov, 1978; R. Polyak, 1979; D. Soloveichik, 1979.

I urge all scientists reading this to actively support these brave and long suffering colleagues. The New York Academy of Sciences has an active program of "adoptions" described in the letter on the following page. The Committee of Concerned Scientists, 330 7th Avenue, Suite 608, New York, New York, 10001, telephone 212-695-2560, keeps in constant touch with the refusniks and can supply you with information about the seminar. Please contact them if you or a colleague are planning a visit to the Soviet Union.

We must not forget them.

[c]We were very happy to learn just before publication that M. I. Freidlin and E. M. Chudnovsky have received permission to emigrate. Let us hope that the others will also be able to follow soon.

# An Open Letter from Academy Officers

Dear Colleagues:

Scientists the world over are often conspicuous leaders in human-rights activities and have been persecuted by repressive regimes. The protection of their human rights is one of The Academy's ongoing concerns. The Academy's actions in this connection are unrelated to the political persuasion of the government in question—whether it is right wing or left wing is not a deciding factor. Activities directed toward supporting our colleagues include:

- Letters and telegrams of concern and protest to officials of unjust governments that imprison, torture, exile, dismiss from jobs, or deny freedom of inquiry to scientists, doctors, engineers, and educators.
- Formal requests to our own government officials and to appropriate international bodies to intervene on behalf of persecuted colleagues.
- Sponsorship and cosponsorship with other scientific and human-rights groups of conferences, missions to foreign countries, and public demonstrations in support of oppressed scientists.
- Offers of direct assistance, waiving membership fees, and helping with the publication of scientific papers of our colleagues in need.

While the effectiveness of any single action is often in doubt, many of the scientists affected by our efforts say that, over all, our activities are helpful. Appeals for more help confirm this feeling.

There is a great need for scientists to "adopt" a colleague (or colleagues) in trouble and to assist her or him in various ways. To help, you can:

- Send reprints and other scientific information useful in the oppressed scientist's research. (Access to scientific libraries is often denied scientists who are subjected to the policies of repressive regimes.)
- Correspond with the adoptee. This contact can range from occasional postcards to scientific collaboration.
- Send letters of concern to government officials on the scientist's behalf (along the lines of Amnesty International's activities).
- Make visits to the adoptee and encourage traveling friends to do so.
- Help the scientist publish scientific papers in international journals.

The Human Rights Committee of The New York Academy of Sciences, together with The Academy's Committee on Women in Science and in collaboration with the Committee of Concerned Scientists, Inc., has undertaken the task of matching those who want to help with those who need help. Other professional organizations are carrying out similar programs, and we are cooperating with them. As part of these efforts, we strongly urge you to sign up by contacting the Human Rights Committee, The New York Academy of Sciences, Two East Sixty-third Street, New York, N.Y. 10021. In addition, write to us about any preferences or information you might have regarding this activity. We would also be pleased to know of the names and addresses of any oppressed scientists with whom you are now corresponding. We believe that you will find your efforts rewarding despite the fact that at times you may be very frustrated.

"If not, I, who? If not now, when?"

FLEUR L. STRAND, *President*
WILLIAM T. GOLDEN, *President-Elect*
JOEL L. LEBOWITZ, *Chair*
*Human Rights Committee*

# On Fourier Analysis of Geomagnetic Pulsations PC-1, "Twofold" and "Threefold" Pulsations PC-1, and Fundamental Frequencies of the Magnetospheric Cavity[a]

YA. L. AL'PERT

## INTRODUCTION

Some types of geomagnetic pulsations generated in the magnetosphere are isolated packets, $F_s(t)$, of very long, quasi-monochromatic ELF waves whose carrier frequencies, $f_{so}$, vary within very wide limits, namely, $f \approx [(1.5 \times 10^{-3})–7]$ Hz. The durations of these signals, $\tau_s$, vary in different frequency regions from several tens of seconds to hundreds of seconds. On purely formal grounds (and not according to the physical principle), the PC pulsations are divided into five types, PC-1, . . . , 5. Up to now, there is not yet a definite and clear understanding of the process of their formation. A more or less definite frequency range, $f$, corresponds to each of these types of pulsations. Many experimental and theoretical papers deal with the study and description of this interesting and fundamental phenomenon associated with the local and large-scale wave properties of the magnetosphere. (See, for instance, review papers by Jacobs,[1] McPherron *et al.*,[2] Russel *et al.*,[3] Gulyelmi and Troitskaja,[4] and Gulyelmi.[5])

In various experiments, geomagnetic pulsations are recorded as *in situ* in the magnetosphere by artificial Earth satellites and by the extensive network of ground-based observation stations where they are received. These pulsations are guided by magnetic force line plasma ducts that pass through the region of the generation of these wave packets and the magnetically conjugate points on the Earth's surface. Such wave packets are trapped in the ducts due to the dispersion and propagation properties of ULF waves in the magnetoplasma.

Amidst the different types of geomagnetic pulsations, pulsations with the shortest waves, namely, geomagnetic pulsations PC-1, should be singled out. They possess a number of important properties. Discreteness is their interesting, but physically least clear property. The envelope of the packet of these waves often clearly and sufficiently smoothly grows on from the noise level (or, more exactly, from the level of the background of the oscillations of the magnetospheric plasma), reaches a maximum, and then drops to the background level. Usually the trains of such pulsations are observed. They move between the magnetically conjugate points due to their reflection in the appropriate regions of the near-Earth surface. The amplitude of such a sequence,

[a]This paper was written on the basis of three papers by Ya. L. Al'pert and D. S. Fligel. Two of these papers will soon appear in *Planetary and Space Science* (they will be in Russian in *Geomagnetizm i Aeronomiya*); the third one was submitted only recently to these same journals.

$\langle s \rangle$, of signals often initially grows, being amplified apparently in the apogee of their trajectory in the region of their generation, and then decreases and finally vanishes. (See Kenney, Knafich, and Liemohn.[6]) Sometimes, only the train of wave packets growing by the amplitude is observed. This train is suddenly cut off. Some mechanism of the amplification of PC-1 has been discussed (Liemohn,[7] Roux and Solomon,[8] and Gendrin *et al.*[9]). The high degree of the monochromaticity of oscillations is also an essential feature of geomagnetic pulsations PC-1. The copies of original records of $F(t)$ of PC-1 signals given in FIGURES 1 and 2 clearly illustrate their properties.

The gyroresonance instability of the plasma is the main mechanism of the generation of PC-1 pulsations. The excitation of these waves presumably takes place through the action of hot protons in the magnetosphere. This point of view is sufficiently substantiated and generally accepted. Hence, the carrier frequency, $f_{so}$, of each of the packets of waves is a Doppler-shifted frequency with respect to the local value of the gyrofrequency, $\Omega_H = 2\pi F_H$, namely,

$$\omega_{so} = 2\pi f_{so} = \bar{k} \cdot \bar{V}_{b\|} - \rho\Omega_H. \tag{1}$$

Here, $V_{b\|}$ is the component of the velocity of the beam of the hot protons along the direction of the Earth's magnetic field, $H_0$, in the region of the generation of these oscillations, $\bar{k}$ is a wave vector, and $\rho = \pm 1, 2, 3, \ldots$.

Due to the above-mentioned properties of the PC-1 wave packets, $F_s(t)$, their records can be used for a sufficiently complete and detailed Fourier analysis. The results of such a study are described in the following sections of the paper. However, for some reasons pointed out below, it seems proper to discuss briefly the conclusions

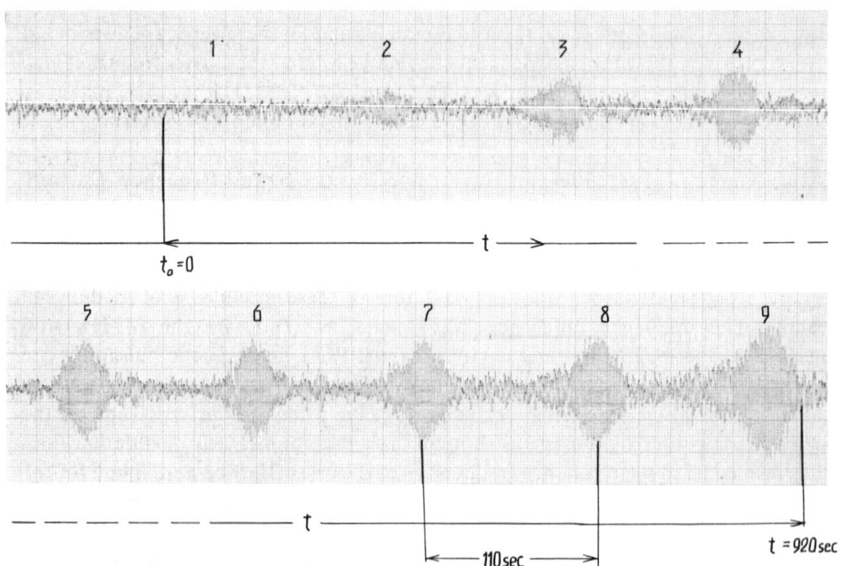


**FIGURE 1.** A copy of original records of $F(t)$. See text.

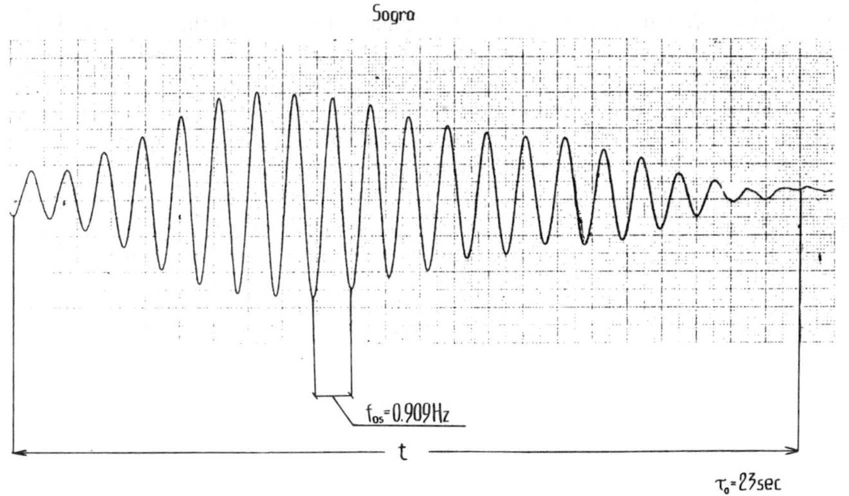


**FIGURE 2.** Another copy of original records of $F(t)$. See text.

drawn by this study and to attach some comments to them for orienting the reader in this introductory section.

Analysis of the properties of the spectra, $S_s(f)$, shows that the bulk of the energy of all wave packets, $F_s(t)$ ($s = 1, 2, 3, \ldots$), is concentrated in the central part of their spectra in the vicinity of the carrier frequencies of signals, $f_{so}$. The spectra have groups of oscillating components (of "satellites"), which, as we assume, are of a different physical nature. The frequencies, $f_{sk}$, of these oscillations of the spectra are distributed more or less randomly within a wide frequency range around the carrier frequency, $f_{so}$. The relative values of their amplitudes, $\delta S_{sk} = [|S_{sk}(f_{sk})|]/[|S_{so}(f_{so})|]$, make up about several percent of the intensity of the central maximum of the spectrum, $|S_{so}|$, namely, $\delta S_{sk} \simeq (1\text{–}6) \times 10^{-2}$.

Also, symmetrical satellites, $f_{sk}^{\mp}$, with respect to the central frequency, $f_{so}$, were found in the spectra. The relative values of their amplitudes are considerably greater than the amplitudes of randomly distributed satellites and of theoretically expected lateral maxima of the spectra.

There are grounds to believe that the nature of these satellites is of modulation character. The frequency differences, $\Delta f_{sk}^{\mp} = |f_{so} - f_{sk}^{\mp}|$, are the fundamental frequencies of the magnetospheric cavity in which PC-1 geomagnetic pulsations are generated. To some extent, an analogy can be drawn between this phenomenon and the so-called combination scattering of light, namely, when in the spectrum of the mercury-vapor lamp's light passing through crystals, symmetrical modulation lines that corresponded to the resonance frequencies of the crystal lattice were found for the first time by Landsberg and Mandelstam[10] (see also Mandelstam[11]). It goes without saying that the character of the excitation of the fundamental oscillations of the magnetosphere and their interaction with the source, that is, with PC-1 wave packets, essentially differs in both cases.

It has been established, which is of particular interest, that frequencies $\Delta f_{sk}^{\mp}$ closely coincide with the frequencies of PC-3 and PC-4 geomagnetic pulsations. Thus, it seems that the source of the latter are the fundamental resonance oscillations of the magnetospheric cavity. Naturally, this idea unites geomagnetic pulsations PC-1 and other types of ELF waves generated in the magnetosphere into a unified process that calls for adequate theoretical consideration.

Definitely expressed maxima on the double frequency, $f_s^{(2)} = 2f_{so}$, and less pronounced maxima on the triple frequency, $f_s^{(3)} \simeq 3f_{so}$, are reliably detected in most spectra of $|S_s(f)|$. This shows that the generation of PC-1 wave packets is accompanied by the generation of "twofold" and "threefold" PC-1 pulsations. The theory of this process must include the nonlinear characteristic of at least an essentially quadratic and cubic form.

The conclusions given here concerning the resonance frequencies of the magnetosphere are of particular interest, and if they prove to be trustworthy, they can play a prominent role in this field of geophysics. Naturally, these conclusions should be approached cautiously and critically. For this purpose, they were described briefly already in the introductory part of this paper.

Theoretically, the problem of the fundamental oscillations of the magnetosphere was considered for the first time apparently thirty years ago (Dungey[12]). Then, it was discussed by many authors.[b] In these studies, a curvilinear "magnetic line of force guide" was regarded as a resonance cavity whose fundamental frequencies are meant. This guide relies by two ends on the base of the ionosphere at the magnetically conjugate points of the lines of force of the Earth's magnetic field, $\overline{H}_0$. This determines the shape of the cavity. Due to mathematical complexity and some unclear points in formulating a number of the physical conditions of the problem, these studies so far have not led to definite quantitative or qualitative results. Earlier, it was already noted that the resonance oscillations of the magnetospheric cavity are possibly the source of long-wave geomagnetic pulsations (see, for example, reference 2). However, there are no experimental data on this score. In this paper, the conclusions concerning the fundamental frequencies of the magnetosphere follow from an argument that is physically very convincing—namely, the symmetricity of the central satellites of the spectrum and their modulation character. This, however, is not yet a convincing proof that these conclusions are sufficiently substantiated. It should be borne in mind that the conditions of such an interaction between PC-1 pulsations and the fundamental oscillations of the region where they are generated can be favorable in certain situations. That is why it is interesting and necessary to analyze experimental data obtained in different latitudes with the different states of the perturbation of the magnetosphere. It should also be noted that the model of the resonance cavity in the form of a magnetic line of force guide, used in theoretical studies of the fundamental oscillations of the magnetosphere (see references 4 and 12–14), to say nothing of the extreme complication of the mathematical problem, should not necessarily be regarded as the only one that is physically adequate to the process under consideration.

[b] See, for instance, review papers by Gulyelmi and Troitskaya,[4] and theoretical surveys by Polyakov *et al.*[13] and Krylov and Lifshits.[14]

However, the discussion of other models that, in our opinion, can be more adequate to the physical formulation of the problem lies beyond the scope of this paper.

## THE INITIAL EXPERIMENTAL DATA: THE TIME FORM $F_s(t)$ OF THE WAVE PACKETS PC-1

The results of this paper were obtained by the treatment of two chains of PC-1 signals $F_s(t)$. One series of six signals was recorded at Novolazarevskaya (Station I),

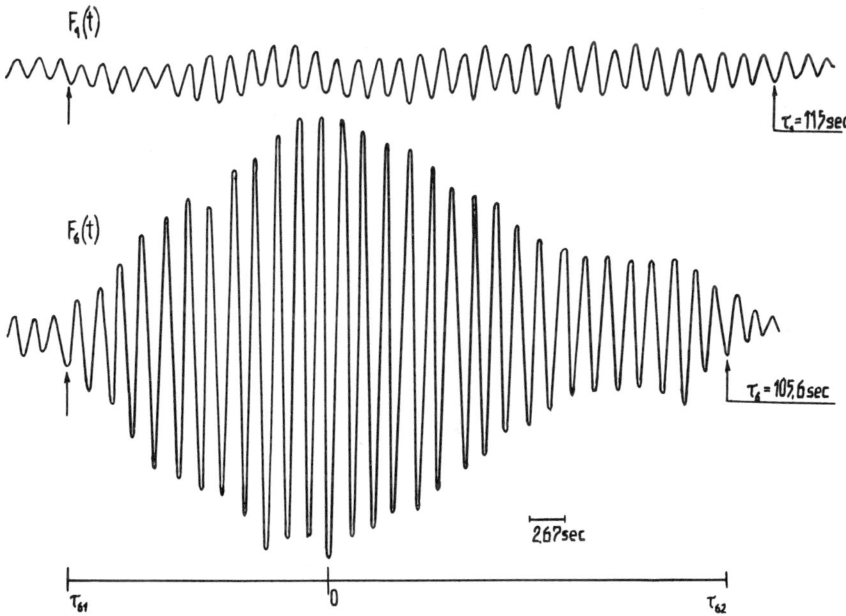


**FIGURE 3.** Copies of two signals ($s = 1$ and $s = 6$) that were recorded at Novolazarevskaya (station I), with geomagnetic latitude $\phi_0 = 66.2°$. The overall duration was 920 seconds. See text.

whose geomagnetic latitude $\phi_0 = 66.2°$. The amplitudes of this sequence of signals increased considerably from $s = 1$ to $s = 6$. The copies of two signals ($s = 1$ and 6) of this series are shown in FIGURE 3. The overall duration of this series was 920 seconds.

The second chain of 12 signals was recorded at Sogra (Station II, $\phi_0 = 57.9°$) and is to a certain extent unique. Practically, the beginning and the end of this process were recorded—namely, the amplitudes of the first and last PC-1 signals on the record are on the order of the amplitudes of the oscillating background $F_0(t)$ of the plasma, that is, of the noise of the region of the generation of PC-1. Unfortunately, the first three signals were recorded with an advanced speed of the tape recorder. This made it

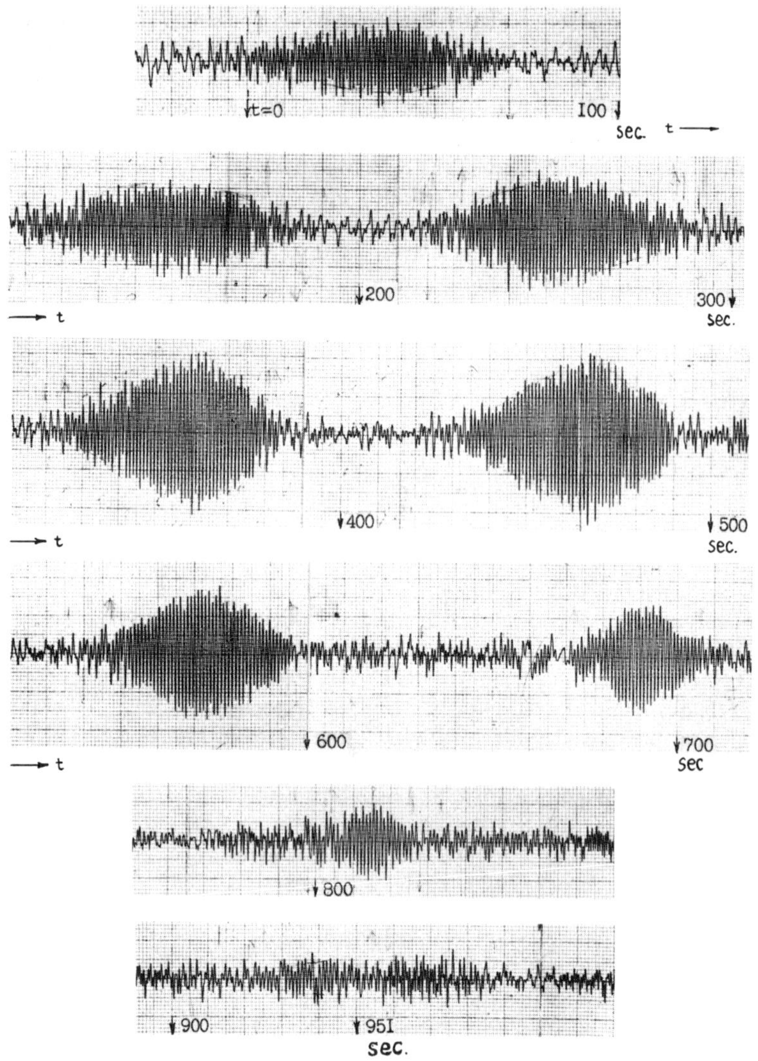


**FIGURE 4.** The analysis of the spectra $s$ of nine signals $F_s(t)$ indicated by indices $s = 1, \ldots, 9$. See text.

impossible to carry out their harmonic analysis with sufficient accuracy. That is why we have performed the analysis of the spectra $s$ of only nine signals $F_s(t)$, which are indicated by indices $s = 1, \ldots, 9$ (see FIGURE 4). The time, $t = 0$, in FIGURE 4 corresponds to the beginning of the signal $s = 1$. The duration of the whole process is 951 s. With the addition of three unexplored signals, the duration of the entire observation session is about 1,310 s (about 22 min). Along with the spectra of $F_s(t)$, we considered the spectra of the oscillating background of the plasma, $F_0(t)$, of the records obtained at Sogra for four time intervals, $\tau_0 = 50$ s: specifically, before the beginning of the signal $s = 1 (t < 0)$, in an interval between some signals, and after the last signal.

It is seen from FIGURES 3 and 4 that PC-1 signals are recognized well by the periodicity of their oscillations even when the intensity of the wave packet is commensurable with the intensity of the background whose oscillations are irregular (see signal $s = 9$ on FIGURE 4). This is revealed most conspicuously by their spectra. It should also be noted that the amplitudes of the chain of PC-1 recorded at Sogra increased slowly up to the signal $s = 5$ and then slowly decreased up to the signal $s = 9$. However, the chain of PC-1 recorded at Station I is cut off when the maximal amplitude $A_s$ ($s = 6$) of $F_s(t)$ is reached.

The main parameters of the wave packets $F_s(t)$—namely, the duration of signals, $\tau_s$, the ratio, $A_s/A_1$, of their amplitudes to the amplitude of the first signal, the values of the carrier frequency of $F_s(t)$, that is, the central frequencies, $f_{so}$, of the spectra, and the values,

$$\delta S_{so} = \frac{|S_{so}(f_{so})|}{|S_{1o}(f_{1o})|} \tag{2}$$

(the ratio of the modules of the central maxima of the spectra to the module of the first signal)—are listed in TABLE 1A.

It seems that the PC-1 signals, as recorded at stations I and II, were generated in the plasmapause close to the geocentric distances of $L = R/R_0 \simeq 6$ and 4, where it can be assumed that the plasma concentration is $N \simeq 15$ and 500 cm$^{-3}$, and the Earth's magnetic field is $H_0 \simeq (1.5 \text{ and } 4) \times 10^{-3}$ oersted, respectively. Then, it follows from TABLE 1A that the velocities of the beams of the hot protons (due to action of which these signals were generated) are $V_{bo} \simeq (6.0 \text{ and } 4.0) \times 10^3$ km/s.

For the complete Fourier analysis of the experimental wave packets $F_s(t)$, it was interesting and necessary to compare them with the spectra $S_s^*(f)$ of some theoretical signals $F_s^*(t)$ that describe their shape well enough. Two models of theoretical signals

TABLE 1A. The Main Parameters of the Wave Packets and of Their Spectra

Station I—Novolazarevskaya

| $\langle S \rangle$ | 1 | 2 | 3 | 4 | 5 | 6 |
|---|---|---|---|---|---|---|
| $\tau_s$, cek | 115 | 141 | 125 | 130 | 148 | 151 |
| $A_s/A_1$ | 1 | 1.33 | 2.42 | 3.63 | 5.09 | 7.75 |
| $f_{so}$(Hz) | 0.2920 | 0.2893 | 0.7893 | 0.2929 | 0.2911 | 0.2829 |
| $\delta S_{so}$ | 1 | 1.78 | 2.56 | 3.81 | 4.85 | 5.92 |

Station II—Sogra

| $\langle S \rangle$ | 1 | 2 | 3 | 4 | 5 | 6 | 7 | 8 | 9 |
|---|---|---|---|---|---|---|---|---|---|
| $\tau_s$, cek | 52 | 55 | 57 | 53 | 53 | 48 | 33 | 26 | 28 |
| $A_s/A_1$ | 1.0 | 1.17 | 1.33 | 1.98 | 2.01 | 1.56 | 1.28 | 0.94 | 0.67 |
| $f_{so}$(Hz) | 1.051 | 1.055 | 1.055 | 1.044 | 1.019 | 1.047 | 1.045 | 1.034 | 1.042 |
| $\delta S_{so}$ | 1.0 | 1.32 | 1.49 | 1.94 | 1.48 | 1.38 | 0.84 | 0.50 | 0.30 |

were used. One type of the signals was given in the form,

$$F_s^*(t) = A_s \cdot \cos \Omega_s t \cdot \cos \omega_{so} t, \qquad \begin{Bmatrix} \Omega_s = \Omega_{s1}, & -\tau_{s1} \leq t \leq 0 \\ \Omega_s = \Omega_{s2}, & 0 \leq t \leq \tau_{s2} \end{Bmatrix}, \tag{3}$$

$(\tau_{s1} + \tau_{s2}) = \tau_s$. It is a sum of two cutoff quarter-periods of cosinusoids with the periods being $T_{s1}$ and $T_{s2}$, respectively, where $T_{s1/4} < -\tau_{s1}$, $T_{s2/4} > \tau_{s2}$. Naturally, the signals of equation 3 were designed so that

$$F_s^*(-\tau_{s1}) = F_s(-\tau_{s1}), \qquad F_s^*(\tau_{s2}) = F_s^*(\tau_{s2}).$$

The other theoretical model represented the sum of two cutoff gaussoids,

$$F_s^*(t) = A_s \cdot \exp\left(-\frac{a_s^2 \cdot t^2}{(\tau_s^*)^2}\right) \cdot \cos \omega_{so} t, \qquad \begin{Bmatrix} a_s = a_{s1}, \tau_s^* = & -\tau_{s1} \leq t \leq 0 \\ a_s = a_{s2}, \tau_s^* = & \tau_{s2} \geq t \geq 0 \end{Bmatrix}, \tag{4}$$

where $a_{s1}$ and $a_{s2}$ were constants determining the shape of the gaussoids.

The spectra of the signals of equation 3 were calculated by means of the program of the rapid Fourier transformation. However, in order to calculate the spectrum of equation 4, it was necessary to set up a new program. What was used was a program expressed through Kramp functions (see Faddeyeva and Terentyev,[15] and Dubovoy and Yaroslavtsev[16]). Practically, it was sufficient to limit ourselves to the theoretical signal of equation 3 because the parameters of the spectra of both signals $F_s^*(t)$ that

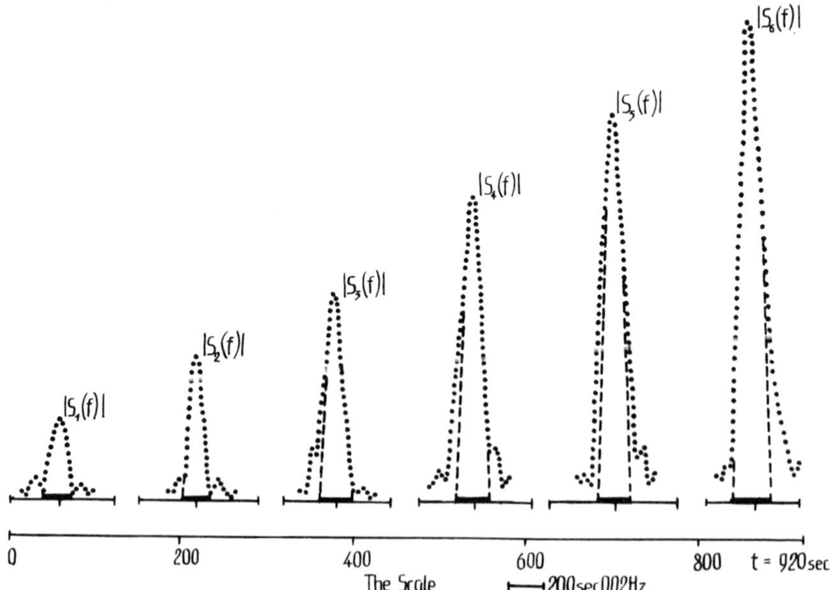


**FIGURE 5.** The evolution of the central parts of all of the six experimental (dots) and theoretical (dotted line) spectra of station I on the time scale $t$. See text.

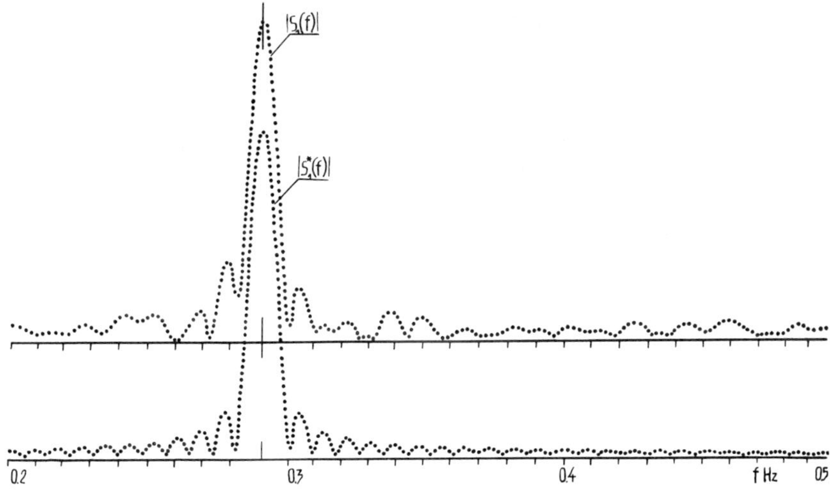


**FIGURE 6.** Moduli of the spectra $|S_s(f)|$ and $|S_s^*(f)|$. See text.

are of interest to us are approximately equal and describe experimental signals $F_s(t)$ well enough.

## THE GENERAL PROPERTIES OF $S_s(f)$ SPECTRA OF $F_s(t)$ WAVE PACKETS

To calculate the spectra, the experimental signals $F_s(t)$ were given numerically pointwise, that is, by the values of $F_s(t)$ in equal time intervals, $\Delta t$. Equal frequency intervals, $\Delta f$, that determine the resolving power of $S_s(f)$ spectra correspond to them. For control and certainty, we used (for station I) two time and frequency intervals, namely, the values of

$$\left.\begin{aligned} \Delta t &= 0.267 \text{ s}, \quad \Delta f = 9.14 \times 10^{-4} \text{ Hz} \\ \Delta t &= 0.801 \text{ s}, \quad \Delta f = 3.048 \times 10^{-4} \text{ Hz} \end{aligned}\right\}, \quad \text{Station I,}$$

$$\Delta t = 0.131 \text{ s}, \quad \Delta f = 1.864 \times 10^{-3} \text{ Hz}, \qquad \text{Station II.}$$

For illustration, in FIGURE 5, the evolution of the central parts of all the six experimental (dots) and theoretical (dotted line) spectra of station I on the time scale $t$ of this process is shown. In detail, different kinds of experimental spectra recorded at I and II are represented in FIGURES 6 and 7. Moduli of the spectra, $|S_s(f)|$ and $|S_s^*(f)|$, are given in FIGURE 6, and their relative values are depicted in FIGURE 7, namely,

$$\delta S_s(f) = \frac{|S_s(f)|}{|S_s(f_{so})|}. \tag{5}$$

Besides for simplicity and for a more graphical representation on this figure, the values of $\delta S_s(f)$ of the central maxima of all the spectra are equal. The consideration of the

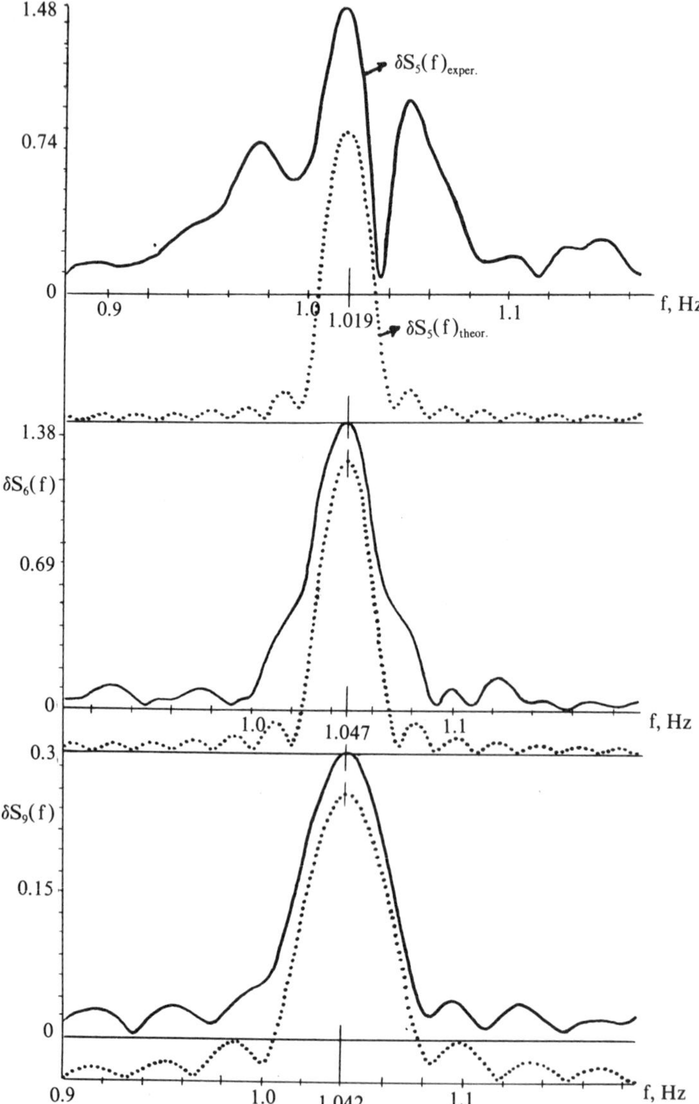


**FIGURE 7.** The relative values, $\delta S_s(f) = |S_s(f)|/|S_s(f_{so})|$, of the moduli in FIGURE 6. See text.

spectra like the spectra shown in FIGURES 6 and 7 allows one to reveal their basic and general features and their evolutions from signal to signal.

Let us note some features of the theoretical spectra $S_s^*(f)$ conditioned by the asymmetry of the envelope of the signal $F_s^*(t)$ and by its cutoff. The moduli of these theoretical spectra have no zeros; they are symmetrical with respect to the central, carrier frequency of the wave packet, $f_{so}$. However, the value of the width of $\Delta f^*_{so}$ of their central maxima is noticeably lower than the values of $\Delta f^{**}_{so}$ of symmetrical signals $F^{**}_s(t)$ that are not cut off, that is, when $\tau_{s1} = \tau_{s2}$ and $F^{**}_s(\tau_1) = F^{**}_s(\tau_2) = 0$. The typical differences of the frequencies, $\Delta f^{**}_{sk} = f^{**}_{sk} - f_{so}$, of the symmetrical signals of equation 3, for example, are:

$$\left.\begin{aligned} \Delta f^{**}_{s1,\min} &= \pm 1.5/\tau_s, \quad \Delta f_{s1,\max} = \pm 1.89/\tau_s, \\ \Delta f^{**}_{s2,\min} &= \pm 2.5/\tau_s, \quad \Delta f_{s2,\max} = \pm 2.93/\tau_s, \\ \Delta f^{**}_{s3,\min} &= \pm 3.5/\tau_s, \quad \Delta f_{s3,\max} = \pm 3.95/\tau_s. \end{aligned}\right\} \tag{6}$$

The values of $\Delta f^*_{sk,\min,\max}$ and of the relative values of their amplitudes,

$$\delta S_{sk,\min} = \frac{|S_s(f^*_{sk,\min})|}{|S_s(f^*_{so})|}, \qquad \delta S_{sk,\max} = \frac{|S_s(f^*_{sk,\max})|}{|S_s(f^*_{so})|}, \tag{7}$$

of the cutoff theoretical signals appreciably depend on the degree of the cutoff of a signal and its asymmetry, especially for the first three lateral maxima of the spectrum ($k = 1$, 2, and 3). It should be noted that these properties of the spectra of the signals can be very useful for diagnosing oscillatory processes of various types.

The spectra $|S_s(f)|$ of experimental signals are, in general, asymmetrical with respect to the central maximum. Naturally, they have an oscillating form. At first glance, all the maxima of the lateral oscillations of $|S_s(f)|$ are randomly distributed with respect to the central frequency, $f_{so}$. However, intensive symmetrical maxima were revealed in the closest vicinity of $f_{so}$. Their behavior differs from the expected theoretical symmetrical maxima. It should also be noted that with the increase of $\langle s \rangle$, that is, with the evolution of the signals, the central maxima of the spectra often become asymmetrical and the widths of the central maxima, $\Delta f_{so} = |f^-_{s1,\min} - f^+_{s1,\min}|$, increase as compared with the theoretical values, $\Delta f^*_{so}$. This has influences on the resolution of the experimental spectra and, as it is seen below, blurs some of the essential details of $|S_s(f)|$.

It follows from TABLE 1A (see equation 2) that the central frequencies, $f_{so}$, of the spectra noticeably vary from signal to signal. It was important to find out whether this change of frequencies was due to the errors of the used method of spectra analysis. For this purpose, spectra $|S_s(f)|$ of signals of cosinusoidal shape, $F(t) = \cos 2\pi f_o t \cdot \cos 2\pi t/T$, in the interval, $\tau = (-T/4 \text{ to } T/4) = 40$ s, for $f_o = 1$ Hz ($f_o$ is the carrier frequency, and $1/T$ is the frequency of the amplitude modulation) were calculated. The values of $\tau$ and $f_o$ are close to the values of $\tau_s$ and $f_s$ of the experimental signals of station II. Spectra $S(f)$ were calculated by two methods—analytically (method I) and by means of a computer (method II)—from the set of the values of $F(t)$ that were given in equal intervals, $\Delta t$, and determined from the graphs of the function $F(t)$. Namely method II was used when the experimental wave packets $F_s(t)$ were processed.

Next, TABLE 1B lists the frequency differences that determine the position of the minima and maxima of the theoretical spectra $|S(f)|$ to the left (−) and right (+) of its central maximum $f_o$, that is, the values, $\Delta f^{\pm}_{k,\min} = |f^{\mp}_{k,\min} - f_o|$, $\Delta f^{\mp}_{k,\max} = |f^{\mp}_{k,\max} - f_o|$, and the relative values of the amplitudes of these maxima, $\delta S_{k,\max} = |S(f_{k,\max})|/|S(f_o)|$, obtained by the two methods.

It follows from TABLE 1B that the errors in determining the frequencies $f_k$ and amplitudes $\delta S_k$ of the lateral maxima of spectra by using the graphical representation of $F(t)$ are small enough. The values of $f_k$ differ from analytical values only in the fourth case of the third sign. Therefore, it seems that the variability of carrier frequencies $f_{so}$ is linked apparently with the fact that the regeneration of each next signal takes place in some cases in conditions that are somewhat different from the conditions of the generation of the previous signal, namely, with other values of plasma parameters. Naturally, the discussion of this problem goes beyond the scope of this paper and requires a special study.

Let us note at the end of this section that the experimental spectra $|S_s(f)|$ enable one to single out a number of their specific features that reveal, to some extent, the structure and the process of the generation of geomagnetic pulsations PC-1. Various oscillations, which can be called satellite components of spectra, are, as it may be assumed, of a different physical nature.

**TABLE 1B.** Frequency Differences That Determine the Position of the Minima and Maxima of the Theoretical Spectra $|S(f)|$ to the Left (−) and Right (+) of Its Central Maximum $f_o$

| | I | II | |
|---|---|---|---|
| $k$ | $\Delta f_{k,\min}^{\mp}$ | $\Delta f_{k,\min}^{-}$ | $\Delta f_{k,\min}^{+}$ |
| 1 | 0.0375 | 0.0370 | 0.0370 |
| 2 | 0.0625 | 0.0620 | 0.0620 |
| 3 | 0.0875 | 0.0870 | 0.0900 |
| 4 | 0.1125 | 0.1120 | 0.1126 |
| $k$ | $\Delta f_{k,\max}^{\mp}$ | $\Delta f_{k,\max}^{-}$ | $\Delta f_{k,\max}^{+}$ |
| 1 | 0.04725 | 0.0470 | 0.0470 |
| 2 | 0.07325 | 0.0730 | 0.0720 |
| 3 | 0.09875 | 0.1000 | 0.0990 |
| $k$ | $\delta S_{k,\max}^{\mp}$ | $\delta S_{k,\max}^{-}$ | $\delta S_{k,\max}^{+}$ |
| 1 | 0.070 | 0.069 | 0.069 |
| 2 | 0.029 | 0.026 | 0.036 |
| 3 | 0.016 | 0.015 | 0.015 |

## THE BACKGROUND OF EXPERIMENTAL SPECTRA

Let us consider, first of all, the properties of the components of the spectra, which may be singled out as its background. Apparently, these oscillations are mainly due to the oscillations $F_o(t)$ of the plasma of the magnetosphere that exist practically constantly. The distribution of the frequencies $f_{sk}$ of the maxima of the satellites and of the relative values of their amplitudes $\delta S_{sk}$ are presented in FIGURE 8 for the spectra recorded at station I. The region of the background

$$\delta S_{sk} = \frac{|S_s(f_{sk})|}{|S_s(f_{so})|} = (\lesssim 1\text{–}3) \times 10^{-2} \tag{8}$$

is distinguished on this figure. One of the criteria of selecting the background components was the following property: From signal to signal, these oscillations

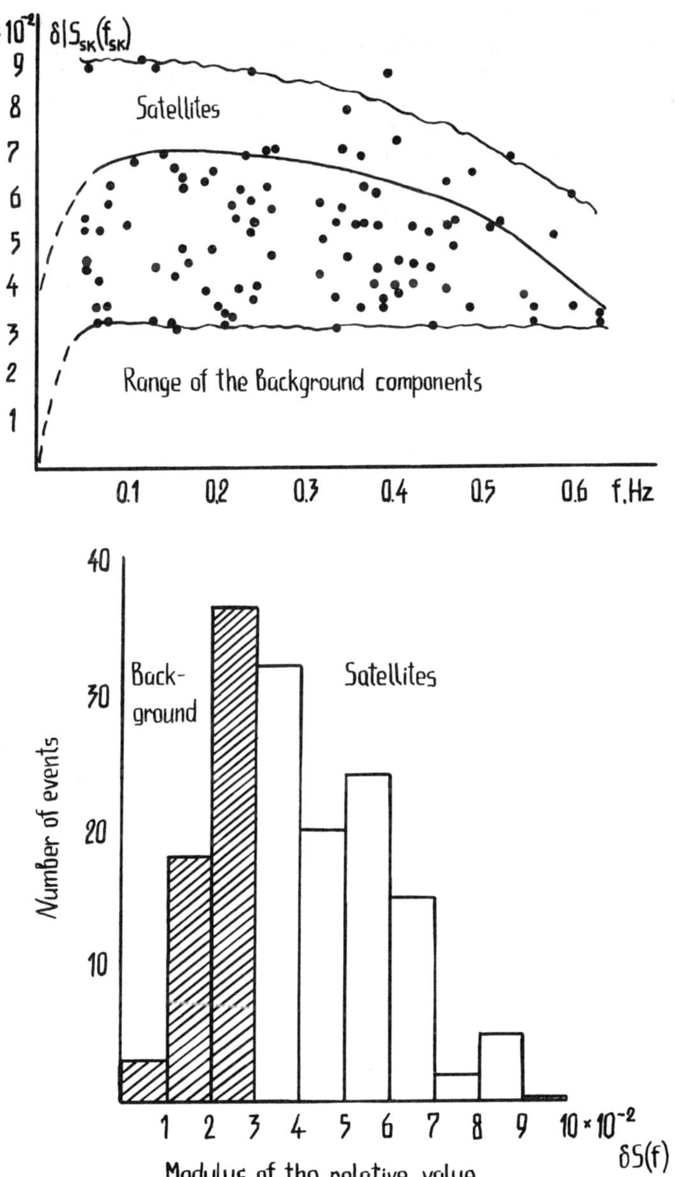


**FIGURE 8.** The distribution of the frequencies $f_{sk}$ of the maxima of the satellites and of the relative values of their amplitudes $\delta S_{sk}$ for the spectra recorded at station I. See text.

comply with equation 8; namely, they are amplified approximately in the same way as the central maximum of the spectrum, that is, proportionally to the value of $|S_s(f_{so})|/|S_1(f_{1o})|$.

In general, there are not too many background components, and their frequencies $f_{sk}$ are distributed more or less randomly. However, it should be borne in mind that some components of the background of the spectrum are covered with other, more intensive oscillations $|S_{sk}(f)|$ that are singled out below as satellites of the spectrum.

It is also interesting and important to find out what is the spectral composition of the background itself in the absence of the generation of the regular signals $F_s(t)$, whether $F_o(t)$ possesses singled out frequency, and how its spectral composition varies

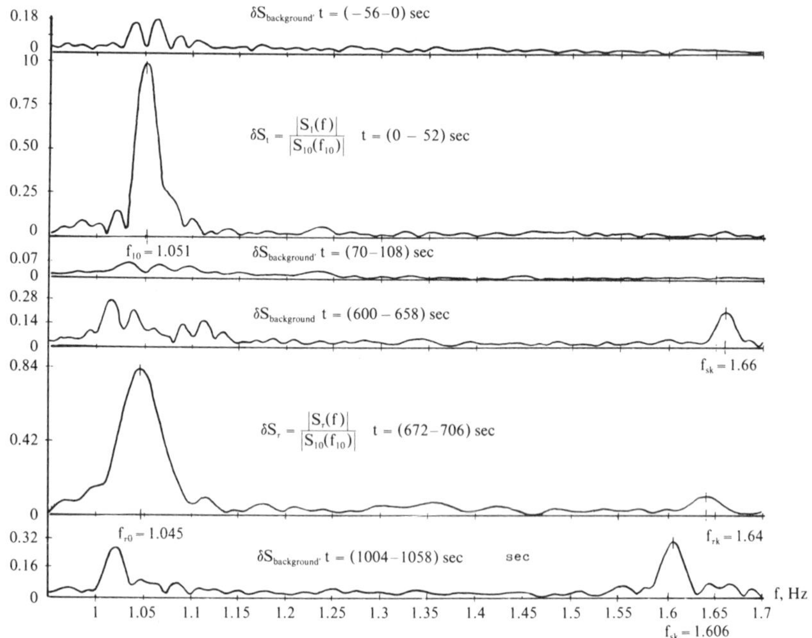


**FIGURE 9.** Spectra $|S_o(f)|$ of the background $F_o(t)$ obtained from the records of $F_s(t)$ of station II for various time intervals. See text.

with time. FIGURE 9 presents spectra $|S_o(f)|$ of the background $F_o(t)$ obtained from the records of $F_s(t)$ of station II for time intervals of $\tau = 50$ s before the generation of the first signal, in intermediate time between the first and seventh signals, and after recording the ninth signal (see FIGURE 4). The spectra of signals $F_1(t)$ and $F_7(t)$ are presented in the same figure for the sake of comparison. Analysis of spectra $|S_o(f)|$ shows that in the neighborhood of the carrier frequencies, $f_{so}$, the background spectra $|S_o(f)|$ have one maximum or several maxima that do not, however, coincide exactly with the values of $f_{so}$ of the neighboring signals. The relative amplitudes of these maxima are $\delta S_o = 0.15$–$0.3$. The amplitudes of other irregularities of spectra $|S_o(f)|$

are of the same order as those of the spectra of signals $|S_s(f)|$, namely, $\delta S_o(f) \gtrsim 0.03$. Of interest is that within the time intervals of $t = (600–658)$ s and $t = (1004–1058)$ s, the background spectra of Sogra have pronounced maxima on frequencies $f_{ok} = 1.66$, 1.61 Hz, with their amplitudes being $\delta S_{ok} = 0.2$ and 0.32. On frequencies close to $f_{ok}$, most of the spectra $|S_s(f)|$ recorded at station II also have similar maxima. The average values of the frequency and the amplitude of this satellite are

$$\left.\begin{aligned} \bar{f}_{sk} &= \left(1.63 \begin{matrix} +0.04 \\ -0.02 \end{matrix}\right) \text{Hz} \simeq 1.5 \bar{f}_{so}, \\ \delta S_{sk} &= \frac{|S_s(f_{sk})|}{|S_s(f_{so})|} \simeq (0.05 \text{ to } 0.22), \overline{\delta S}_{sk} \simeq 0.11. \end{aligned}\right\} \quad (9)$$

The cause of the appearance of this satellite is unclear. Perhaps this is due to the presence of a group of hot particles, $V_{bk} \simeq 1.5\ V_{bo}$, in the generating proton beam. It is worth noting that this satellite manifests itself equally in the spectrum of the background of plasma oscillations $F_o(t)$ and in the spectra of wave packets $F_s(t)$ of PC-1 pulsations.

## THE SATELLITES OF SPECTRA

It seems that some oscillating components, $|S_{sk}(f_{sk})|$, which are singled out here as satellites, appear due to additional oscillations of the plasma generated under the effect of the outbursts of the velocity of the proton flux. The velocities, $V_{bk}$, of these outbursts vary within certain limits in the vicinity of the velocity, $V_{bo}$, of the main part of the beam generating the signal $F_s(t)$. It follows from TABLE 1A and the data given below (see equation 10) that in our case, $V_{bk} \simeq (10^3–1.3 \times 10^4)$ km/s. In all likelihood, the appearance of such outbursts, as well as the distribution of their velocities, is of a more or less random character.

The relative values of the maxima of the amplitude of the satellites $\delta S_{sk} = \delta|S_s(f_{sk})|$ make up only several percent of the intensity of the central maximum of the spectrum $|S_s(f_{so})|$ (see FIGURE 8); however, the values $f_{sk}$ of these components vary within great limits, namely,

$$\delta S_{sk} \simeq (3–9) \times 10^{-2}, \qquad \frac{f_{sk}}{f_{so}} \simeq (0.17–2.24). \quad (10)$$

It is evident from FIGURE 8 that the intensities of the satellites are more or less equally distributed in frequency. Values of $\delta S_{sk} > 7 \times 10^{-2}$ occur rarely, and decreasing with frequency they reach the level of the background $\delta S_{sk} \approx 3 \times 10^{-2}$ on the frequency $f =$ 0.6 Hz for station I. On frequencies $f > 0.6$ Hz up to $f \simeq 1.4$ Hz, the values of $\delta S_{sk}$ continue to decrease slowly up to $\delta S_{sk} \lesssim 2 \times 10^{-2}$. It follows from the histogram given in the lower part of FIGURE 8, in which the distribution of the number of the cases of $\delta S_{sk}$ of background components and satellites within the range of $f \simeq (0.05–0.6)$ Hz is presented, that mainly

$$\begin{aligned} &\delta S_{sk} \text{ of the background} \simeq (2–3) \times 10^{-2}, \\ &\delta S_{sk} \text{ of the satellites} \simeq (3–6) \times 10^{-2}. \end{aligned} \quad (11)$$

The low level of the intensity of the satellites shows that the concentration of the groups of protons ($V_b = V_{bk}$) of the outbursts of the flux is much lower than the concentration of the protons of the main flux ($V_b = V_{bo}$).

According to the division of the oscillations of spectra into the background and satellites, which we have used here, the latter are fully absent, for example, for the data of station I on frequencies of $f >$ (0.6–0.7) Hz up to $f \simeq 1.4$ Hz. It should be noted that the same behavior of the background follows from the spectra of station II. Only the values of $f_{sk}$ and $V_{bk}$ respectively change because of the difference of the values of $f_{so}$ at stations I and II.

## TWOFOLD AND THREEFOLD PC-1 PULSATIONS

The major part of the spectra $|S_s(f)|$ of PC-1 signals has definitely pronounced maxima on the second and third harmonics, that is, on $f_s^{(2)} \simeq 2f_{so}$ and $f_s^{(3)} \simeq 3f_{so}$. The relative amplitudes, $\delta S_s^{(2)}$, of the maxima on the double frequencies, $f_s^{(2)}$, exceed the amplitudes, $\delta S_{sk}$, of the satellites and of the background by two to three times and more. The values of $\delta S_s^{(3)}$ on the triple frequencies also considerably exceed $\delta S_{sk}$ of the satellites in a broad frequency range around $f_s^{(3)}$, even when $\delta S_s^{(3)} \approx 0.03$–$0.05$. The characteristic values of these harmonics are given in TABLE 2. FIGURES 10 and 11 represent narrow regions of the spectra $|S_s(f)|$ around only $f_s^{(2)}$ and $f_s^{(3)}$ of the PC-1 signals recorded at station I.

**TABLE 2.** The Values $f_s^{(2)}$, $\delta S_s^{(2)}$ and $f_s^{(3)}$, $\delta S_s^{(3)}$ of the Twofold and Threefold Pulsations PC-1

| $\langle S \rangle$ | $f_{so}$ (Hz) | $f_s^{(2)}$ (Hz) | $\delta S_s^{(2)}$ | $f_s^{(3)}$ (Hz) | $\delta S_s^{(3)}$ |
|---|---|---|---|---|---|
| Station I—Novolazarevskaya | | | | | |
| 1 | 0.2920 | 0.585 | $7.0 \times 10^{-2}$ | 0.876 | $4.8 \times 10^{-2}$ |
| 2 | 0.2893 | 0.580 | $5.4 \times 10^{-2}$ | — | — |
| 3 | 0.2893 | 0.582 | $6.9 \times 10^{-2}$ | 0.863 | $2.5 \times 10^{-2}$ |
| 4 | 0.2929 | 0.584 | $1.18 \times 10^{-1}$ | — | — |
| 5 | 0.2911 | 0.581 | $1.25 \times 10^{-1}$ | 0.875 | $3.6 \times 10^{-2}$ |
| 6 | 0.2829 | 0.566 | $1.28 \times 10^{-1}$ | 0.846 | $3.4 \times 10^{-2}$ |
| Average values | 0.2896 | 0.580 | $7.2 \times 10^{-2}$ | 0.865 | $3.6 \times 10^{-2}$ |
| Station II—Sogra | | | | | |
| 1 | 1.051 | 2.11 | $1.0 \times 10^{-1}$ | 3.16 | $3 \times 10^{-2}$ |
| 2 | 1.055 | 2.12 | $1.8 \times 10^{-1}$ | 3.17 | $8 \times 10^{-2}$ |
| 3 | 1.055 | 2.12 | $1.3 \times 10^{-1}$ | 3.19 | $6 \times 10^{-2}$ |
| 4 | 1.044 | 2.10 | $1.7 \times 10^{-1}$ | 3.13 | $6 \times 10^{-2}$ |
| 5 | 1.019 | 2.03 | $8 \times 10^{-2}$ | 3.06 | $5 \times 10^{-2}$ |
| 6 | 1.047 | 2.10 | $1.9 \times 10^{-1}$ | 3.16 | $7 \times 10^{-2}$ |
| 7 | 1.045 | 2.10 | $1.1 \times 10^{-1}$ | 3.15 | $7 \times 10^{-2}$ |
| 8 | 1.034 | — | — | — | — |
| 9 | 1.042 | — | — | — | — |
| Average values | 1.044 | 2.097 | $1.37 \times 10^{-1}$ | 3.146 | $6 \times 10^{-2}$ |

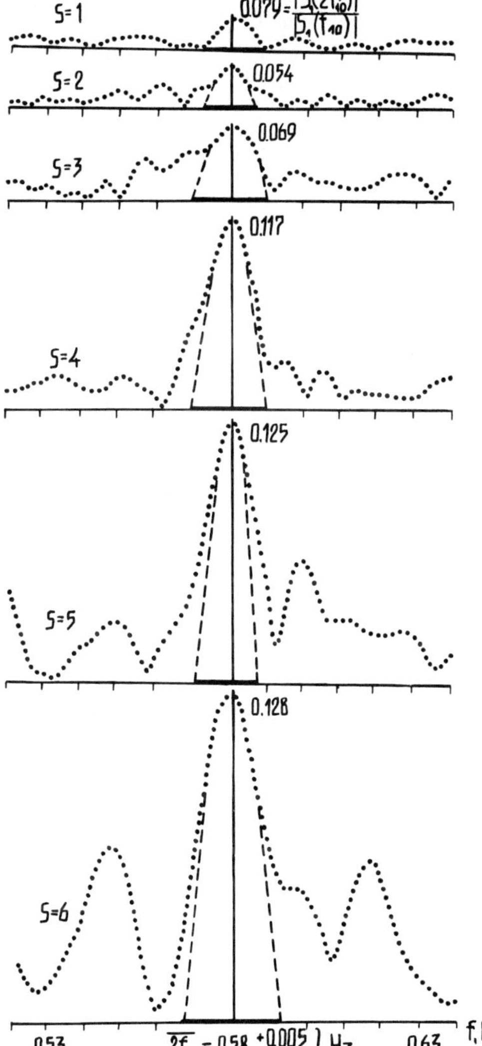


**FIGURE 10.** Narrow regions of the spectra $|S_s(f)|$ around only $f_s^{(2)}$ of the PC-1 signals recorded at station I. See text.

The revealed harmonics of the spectra of PC-1 show that the generation process of these geomagnetic pulsations is accompanied by the generation of "twofold" and "threefold" pulsations PC-1. It is possible that the absence of these harmonics in the spectra of some signals is due to the weak interaction of the beam of the energetic protons with the plasma when the amplitudes of the PC-1 signals $F_s(t)$ decrease. However, it seems that the mechanism of their generation is rather complicated. For example, from the data given in TABLES 1A and 2, it follows that the growth rate of the amplitudes of the second harmonic exceeds the growth rate of the central maxima of the spectra of station I. This means that in some cases during the passage of the region of the amplification of the PC-1 signals, twofold pulsations PC-1 are amplified more intensively. On the other hand, the data obtained from the spectra of station II show that $\delta S_5^{(2)}$ was minimal when, namely, the amplitude of the signal and of its spectra,

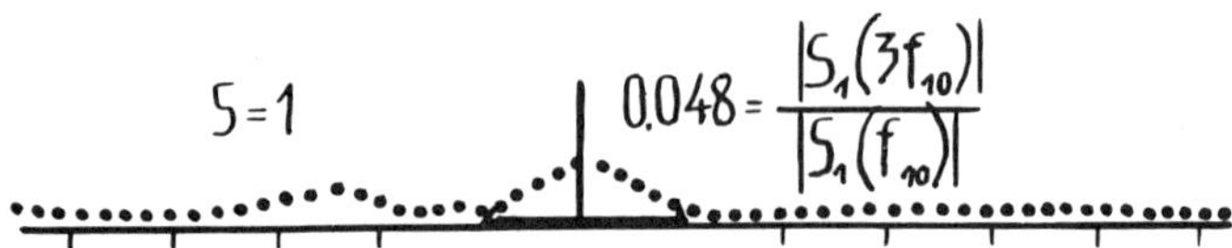


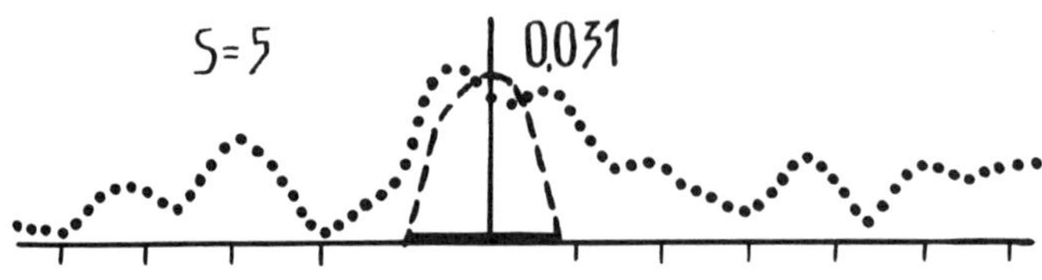


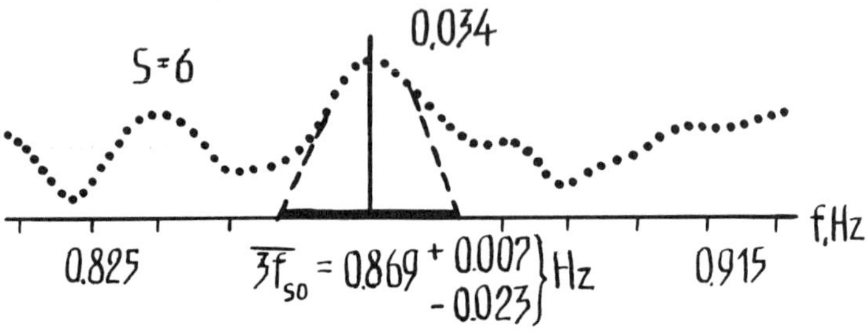


**FIGURE 11.** Narrow regions of the spectra $|S_s(f)|$ around only $f_s^{(3)}$ of the PC-1 signals recorded at station I. See text.

that is, the values of $A_5$ and $\delta S_5(f_{5o})$, were maximal. Conversely, the value of $\delta S_6^{(2)}$ was maximal when the values of $A_6$ and $\delta S_{6o}$ were diminished.

In conclusion of this section, we note that the presence of twofold and threefold pulsations PC-1 shows that the theory of their generation should be essentially nonlinear and should contain a nonlinear characteristic that is at least of both the quadratic and cubic form.

## SYMMETRICAL SATELLITES OF SPECTRA

In the closest vicinity of the central maxima $|S_{so}|$ of the major part of the spectra of PC-1 signals, intensive satellites that are symmetrical with respect to the frequency $f_{so}$ are singled out. Let us denote the frequencies of these satellites by $f_{sk}^-$ and $f_{sk}^+$, and the relative values of their amplitudes by $\delta S_{sk}^-$ and $\delta S_{sk}^+$.

The main property of these satellites is that the moduli of the differences of their frequencies, $\Delta f_{sk}^- = |f_{so} - f_{sk}^-|$, $\Delta f_{sk}^+ = |f_{so} - f_{sk}^+|$, coincide very accurately, while most of the values of their amplitudes, $\delta S_{sk}^-$ and $\delta S_{sk}^+$, surpass several times the level of the amplitudes of the satellites of $|S(f)|$ and of the amplitudes of the theoretical lateral maxima of the spectra $|S^*(f)|$. The modulation character of these satellites enables one to assume that these satellites differ in their physical nature from the satellites considered in the third and fourth sections of this paper. The author assumes that the frequency differences, $\Delta f_{sk}^-$ and $\Delta f_{sk}^+$, are the fundamental resonance frequencies of the region of the magnetosphere in which geomagnetic pulsations PC-1 are generated.

TABLES 3 and 4 list the values of the frequencies $\Delta f_{sk}^{\mp}$ and of the relative amplitudes $\delta S_{sk}^{\mp}$ of the symmetrical satellites revealed in the spectra of both of the stations. It is natural that the pairs of the values $\Delta f_{sk}^{\mp}$ of various signals differ, which, in particular, is due to the variability of the state of the magnetosphere during the experiment. However, the deviations from the average values of $\overline{\Delta f_{sk}^{\mp}}$ are small. It should be borne in mind that in the parts of the spectrum that are occupied by symmetrical satellites, oscillations $|S_s(f)|$ of another type could appear simultaneously. That is why the interference of the oscillations of various types can distort the profile of symmetrical satellites, especially influencing their amplitude $\delta S_{sk}^{\mp}$. This, in particular, can explain the variability of $\delta S_{sk}^{\mp}$ from case to case, especially for the less intensive symmetrical satellites. It is worth noting that the considerable broadening of the central maxima $\Delta f_{so}$ of the spectra greatly hinders the determination of $\Delta f_{sk}^{\mp}$ and $\delta S_{sk}^{\mp}$ because the oscillations of symmetrical satellites lie partly on their slope. This explains why some values of $\Delta f_{sk}^{\mp}$ are given in TABLE 3 with the sign ($\sim$) denoting their certain inaccuracy.

The gaps in this table are also due to some violations of the dependence of $|S_s(f)|$ that are perhaps caused by the interference of plasma oscillations of different types. The irregular character of the emergence of satellites $f_s^{\mp}$ and their absence in some spectra, for example, in $|S_8(f)|$ and $|S_9(f)|$ of the signals recorded at Sogra (see TABLE 4), is apparently due to their small intensity—it is approximately ten times lower than the intensity of the first signal—and hence it is due to the weak interaction of the signals with the cavity of the magnetosphere.

It should also be noted that the first two resonance frequencies ($k = 1, 2$) revealed at station I (TABLE 3) cannot be found at all in the spectra of station II (TABLE 4). This is due to the fact that the duration $\tau_s$ of the wave packets $F_s(t)$ recorded at Sogra are three to four times smaller than the values of the chain of the signals recorded at Novolazarevskaya (station I). Therefore, the half-widths of their central maxima are

**TABLE 3.** Symmetrical Satellites of Experimental Spectra with $\Delta f_{sk}^{\mp} = |f_{so} - f_{sk}^{\mp}|$, $\delta S_{sk}^{\mp} = |S_s(f_{sk}^{\mp})|/|S_s(f_{so})|$[a]

| Station I—Novolazarevskaya | | | | | | | |
|---|---|---|---|---|---|---|---|
| $\langle s \rangle$ | $f_{so}$ (Hz) | $\Delta f_{s1}^-$ (Hz) / $\Delta f_{s1}^+$ (Hz) | $\delta S_{s1}^-$, $\delta S_{s1}^+$ $\times 10^1$ | $\Delta f_{s2}^-$ (Hz) / $\Delta f_{s2}^+$ (Hz) | $\delta S_{s2}^-$, $\delta S_{s2}^+$ $\times 10^2$ | $\Delta f_{s3}^-$ (Hz) / $\Delta f_{s3}^+$ (Hz) | $\delta S_{s3}^-$, $\delta S_{s3}^+$ $\times 10^2$ |
| 1 | 0.2920 | 0.013<br>0.013 | 2.4<br>1.6 | 0.023<br>0.022 | 10.5<br>5.3 | 0.041<br>— | 7.8<br>— |
| 2 | 0.2893 | 0.010<br>0.011 | 1.6<br>1.4 | 0.019<br>0.018 | 12.4<br>5.2 | 0.029<br>0.031 | 7.2<br>5.2 |
| 3 | 0.2893 | 0.012<br>— | 2.5<br>— | 0.022<br>0.027 | 4.6<br>8.7 | 0.036<br>— | 5.6<br>— |
| 4 | 0.2929 | ~0.011<br>0.012 | 2.2<br>1.7 | 0.019<br>0.020 | 9.7<br>7.7 | 0.028<br>0.028 | 5.4<br>7.0 |
| 5 | 0.2911 | 0.009<br>~0.009 | 2.5<br>2.5 | 0.024<br>0.024 | 8.2<br>12.6 | 0.035<br>0.039 | 6.0<br>5.0 |
| 6 | 0.2839 | ~0.012<br>0.011 | 2.2<br>1.8 | 0.026<br>— | 14.6<br>— | 0.039<br>0.033 | 6.0<br>1.9 |
| Average values | $0.2896^{+0.0033}_{-0.0057}$ | $0.011^{+0.002}_{-0.003}$ | 2.2<br>1.8 | $0.022^{+0.005}_{-0.003}$ | 10.0<br>7.9 | $0.035^{+0.004}_{-0.007}$ | 6.3<br>7.0 |

[a] $\langle k \rangle$ = number of the satellite; $\langle s \rangle$ = number of the signal.

**TABLE 4.** Symmetrical Satellites

| | Station II—Sogra | | | | | | | | |
|---|---|---|---|---|---|---|---|---|---|
| $\langle s \rangle$ | $f_{so}$(Hz) | $\Delta f_{s3}^-$(Hz) | $\delta S_{s3}^- \times 10^1$ | $\Delta f_{s3}^+$(Hz) | $\delta S_{s3}^+ \times 10^1$ | $\Delta f_{s4}^-$(Hz) | $\delta S_{s4}^- \times 10^1$ | $\Delta f_{s4}^+$(Hz) | $\delta S_{s4}^+ \times 10^1$ |
| 1 | 1.051 | 0.030 | 1.5 | 0.025 | 2.3 | 0.049 | 0.8 | 0.049 | 1.0 |
| 2 | 1.055 | 0.027 | 1.7 | 0.025 | 1.2 | 0.046 | 1.0 | 0.046 | 1.0 |
| 3 | 1.055 | 0.029 | 2.5 | 0.026 | 1.9 | 0.045 | 1.4 | 0.044 | 0.8 |
| 4 | 1.044 | 0.030 | 1.6 | 0.030 | 1.8 | 0.047 | 1.1 | — | — |
| 5 | 1.019 | 0.042 | 5.4 | 0.033 | 6.8 | — | — | — | — |
| 6 | 1.047 | 0.027 | 3.3 | 0.027 | 3.3 | — | — | — | — |
| 7 | 1.045 | 0.038 | 1.9 | 0.037 | 2.8 | — | — | — | — |
| 8 | 1.034 | — | — | — | — | — | — | 0.046 | 2.8 |
| 9 | 1.042 | — | — | — | — | 0.042 | 2.0 | 0.052 | 1.3 |
| Average values | 1.044 | 0.032 | 2.6 | 0.029 | 2.9 | 0.046 | 1.3 | 0.048 | 1.4 |

much larger, namely, $\Delta f_{so} \simeq (1.15\text{–}1.33)/\tau_s = (0.026\text{–}0.045)$ Hz, and due to the great intensity, the central maxima may cover (absorb) these satellites.

## RESONANCES OF THE MAGNETOSPHERE AND GEOMAGNETIC PULSATIONS PC-3 AND PC-4

A very interesting feature of frequencies $\Delta f_{sk}^{\mp}$ is their close coincidence with the frequencies of the wave packets of another type generated in the magnetosphere. According to the classification established in literature (see references 2 and 4), these are geomagnetic pulsations PC-3 and PC-4. They are observed very often both directly in the magnetosphere and on the Earth's surface.

Results of systematic observations of the pulsations of this type *in situ*—directly in the magnetosphere—on the ATS-I geostationary satellite at the geocentric distance of $R = 6.3\ R_0$ ($R_0$ is the Earth's radius) were described for the first time in reference 17. In these experiments, pulsations were observed on frequencies,

$$f_o = 0.0125\ (\text{PC-4})\ \text{to}\ 0.005\ (\text{PC-5})\ \text{Hz.} \tag{12}$$

In earlier experiments on the Explorer-12 satellite at the distance of $R = (6.8\text{–}10.6)\ R_0$, pulsations of this type were also observed on frequencies,

$$f_o = 0.01\ (\text{PC-4})\ \text{to}\ 0.0055\ (\text{PC-5})\ \text{Hz} \tag{13}$$

(see reference 19). It is easy to notice that the frequencies of pulsations PC-4 closely coincide with the values of the first resonance frequency, $\overline{\Delta f_{s1}^{\mp}} = 0.011$ Hz (see TABLE 3). In the same experiments, individual PC-3 pulsations were also recorded on the frequency,

$$f_o = 0.025\ \text{Hz} \tag{14}$$

(see McPherron *et al.*[20]), which differ little from the second resonance frequency, $\overline{\Delta f_{s2}^{\mp}} = 0.022$. Pulsations PC-3 on

$$f_o = 0.033\ \text{Hz,} \tag{15}$$

which closely coincide with the values of the third resonance frequency, $\overline{\Delta f_{s3}^{\mp}} \simeq 0.033$ Hz, were recorded from the Dodge satellite at the distance of $R = 6.3\ R_0$ (Dwarkin *et al.*[21]).

In various experiments in the magnetosphere, the packets of longitudinal waves were also observed on the same frequencies. An interesting case of such observations is shown in FIGURE 12. In the energy spectra of magnetospheric oscillations recorded from ATS-I, an acute maximum was observed on the frequency close to the first resonance frequency $\Delta f_{s1}^{\mp}$, namely,

$$f_o = 0.0094\ \text{Hz (PC-4)} \tag{16}$$

(Barfield *et al.*[22]). The same event was also studed in papers[23] (Lanzerotti *et al.*, 1971) both from the data of observations on ATS-I and at the magnetically conjugate point on the Earth's surface (College) where the PC-4 wave packet was observed on the same frequency. The results of that experiment are given in the lower part of FIGURE 12. It

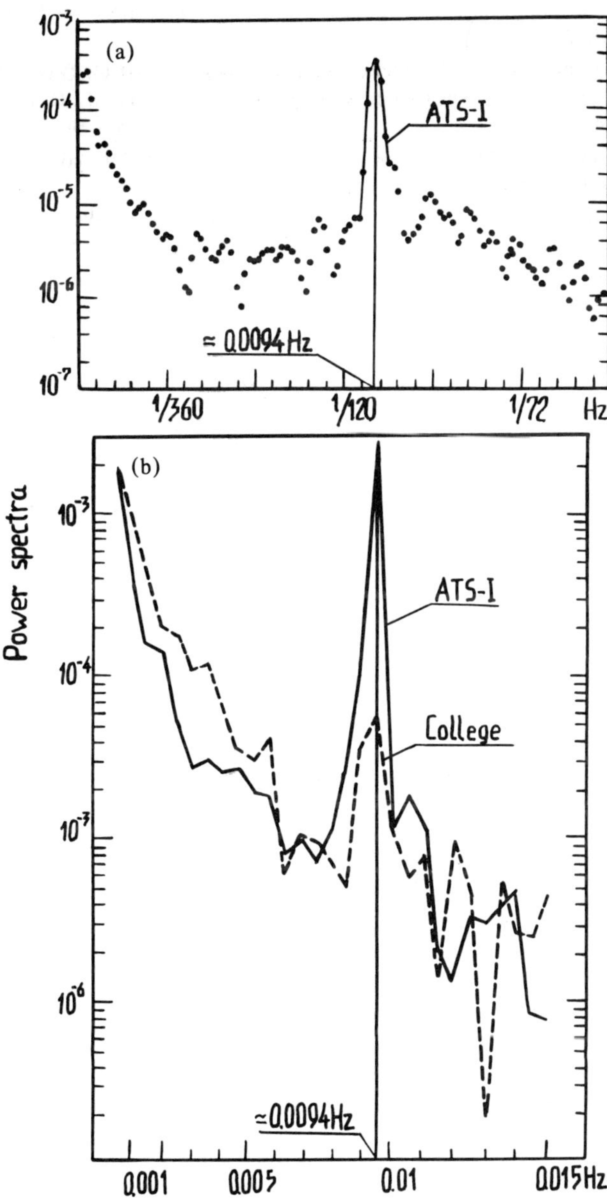


**FIGURE 12.** Packets of longitudinal waves observed in the magnetosphere. See text for descriptions of the top and lower halves of the figure.

goes without saying that at College, the packet of transverse waves PC-4 was recorded, that is, the transformed packet of longitudinal oscillations that was formed in the magnetosphere. The intensity of the wave packet on the Earth was found to be fifty times lower than on ATS-I.

The above-mentioned experimental data testify to the legitimacy of the supposition about the correlations between geomagnetic pulsations PC-3 and PC-4 and resonance oscillations of the magnetospheric region. It was already noted in the introduction that in literature such hypotheses were already put forward (see, for instance, references 2 and 4), but without any proof.

Thus, to some extent, the results given in this section lead to the unification of ULF electromagnetic oscillations of various types and character observed in the magnetosphere into a single process. Of course, these results call for further verification and studies.

## SOME MODIFICATIONS OF THE USED METHOD OF TREATMENT OF EXPERIMENTAL DATA

Various attempts were made to analyze the experimental data considered in this paper by modifying the methods of their treatment. Although these attempts did not lead to other conclusions, it is expedient to point out, at least briefly, the essence of some of them.

First, it seemed that it was most consistent to determine the disturbances and irregularities of the spectrum (the background and satellites) by analyzing the real and imaginary parts of the frequency spectra of wave packets, with the assumption that the experimental signal

$$F_s(t) = F_s^*(t) + \Delta F_s(t), \tag{17}$$

where $F_s^*(t)$ is a model signal described by equations 3 or 4. As it has already been pointed out, the moduli of the spectra of these signals reproduce well the main features of the central, main part of the experimental spectra $|S_s(f)|$ (see, for instance, FIGURES 6 and 7). It follows from equation 17 that

$$\left.\begin{aligned} \mathrm{Re}\,\{\Delta S_s(f)\} &= \mathrm{Re}\,\{S_s(f)\} - \mathrm{Re}\,\{S_s^*(f)\} \\ \mathrm{Im}\,\{\Delta S_s(f)\} &= \mathrm{Im}\,\{S_s(f)\} - \mathrm{Im}\,\{S_s^*(f)\} \end{aligned}\right\}, \tag{18}$$

from which the dependence of the modulus of the spectrum $\Delta S_s(f)$ of the disturbance $\Delta F_s(t)$ on frequency $f$ and, hence, its background and satellites (that is, values $\delta S_{sk}$ and $f_{sk}$) are determined. For this purpose, though, it is necessary to know the phase characteristic of the model signal or, in other words, how and to which extent the beginning of the model signal $F_s^*(t)$ is shifted with respect to $F_s(t)$. Hence, in equation 17, $F_s^*(t)$ must be replaced by the function $F_s^*(t - t_0)$. However, the value $t_0$ is not known and it cannot be determined. The moduli of spectra $F_s^*(t)$ naturally do not depend on it. Despite this difficulty, spectra $\Delta F_s(t)$ were studied for various values of the phase-shifts. These studies led to some contradictions. At the same time, in each version of such calculations, symmetrical satellites $\Delta f_{sk}^{\mp}$ close to those given in TABLES 3 and 4 were singled out. For example, one of the versions of such studies led to six values of resonance frequencies,

$$\overline{\Delta f_{sk}^{\mp}} = 0.011, 0.021, 0.030, 0.038, 0.044, 0.052, \ldots, \quad \textbf{(19)}$$

of the spectra obtained at station I. It is evident that four of them coincide very closely with those obtained above. Certainly, resonances of a higher order are entirely natural. In this connection, it is of interest here to illustrate the variations of the sequences of the relative values of the resonance frequencies of a metal sphere (1), a metal cylinder (2), and a waveguide of the Earth's ionosphere (3), and the frequencies $\Delta f_s^{\mp}$ given in TABLES 3 and 4 and in equation 19. The corresponding data (see, for instance, Al'pert, 1973[24]) lead to the following series of figures:

$$(1)\ f_k/f_{k-1} = 2.14, 1.59, 1.34, 1.22, 1.19, 1.17, \ldots,$$

$$(2)\ f_k/f_{k-1} = 2.18, 1.60, 1.36, 1.28, 1.18, 1.24, \ldots,$$

$$(3)\ f_k/f_{k-1} = 1.81, 1.44, 1.30, 1.22, \ldots,$$

$$(19)\ f_k/f_{k-1} = 2.00, 1.60, 1.27, 1.16, 1.18, \ldots, \quad \textbf{(20)}$$

with $k = 2, 3, 4, \ldots, 7$. It is very instructive that there is a close similarity between the ratios of resonance frequencies in various cases, which evidently follows from equation 20.

The components of the disturbance $\Delta F_s(t)$ were also calculated by a quasi-statistical method from the differences of the moduli $|S_s(f)|$ and $|S_s^*(f)|$. Instead of the values of $\delta S_{sk}$ and $f_{sk}$, their values, averaged statistically to some extent, were calculated—namely, the values of $\delta S_k(f_k)$ and $f_k$ were determined from the dependence $\delta S(f)$ by the following equation,

$$\delta S(f) = \left\{ \frac{1}{S_c} \sum_{S=1}^{S_c} [\delta S_s^2(f) - \delta S_s^{*2}(f)]^{1/2} \right\}; \quad \textbf{(21)}$$

that is, from the sum of the differences of the squares of the current values of the moduli of the spectra of the experimental and theoretical signals. In equation 21,

$$\delta S_s(f) = \frac{|S_s(f)|}{|S_s(f_{so})|}, \qquad \delta S_s^*(f) = \frac{|S_s^*(f)|}{|S_s^*(f_{so})|}. \quad \textbf{(22)}$$

Equation 21 cannot be substantiated strictly. It was interesting, however, to find out whether symmetrical satellites $\Delta f_{sk}^{\mp}$ would be obtained from this equation. The treatment of experimental data led to values that were close to the above values of $\Delta f_{sk}^{-}$ and $\Delta f_{sk}^{+}$ (see equation 19).

It was also reasonable to find out whether symmetrical satellites $|S_s(f_{sk}^{\mp})|$ were linked with the effect of a certain inconstancy of the carrier frequency $f_{so}$ within the limits of one wave packet $F_s(t)$. Because the durations of signals $\tau_s$ were sufficiently long and 30–40 periods of oscillations with the frequency $\sim f_{so}$ took place in them, it was possible to determine approximately the average values of the depth and the index of the frequency modulation of the signal, $\Delta\omega_{so}/\omega_{so}$ and $\Delta\omega_{so}/\Omega_{so}$, assuming that

$$\omega_{so} = 2\pi f_{so}^* \left( 1 + \frac{\Delta\omega_{so}}{\omega_{so}} \cos \Omega_{so} t \right). \quad \textbf{(23)}$$

The values of $\Delta\omega_{so}$ and $\Omega_{so}$ were determined from the smoothed dependence $f_{so}(t)$ constructed from the sequence of the values of $f_{so,l}$ ($l$ = 1, 2, . . . , 30–40). As a result, for instance, values of $\Delta\omega_{so}/\omega_{so} \simeq 0.037$ and 0.051 and $\Delta\omega_{so}/\Omega_{so} \simeq 0.098$ and 0.163 were obtained for the first and the sixth signals of station I, respectively. Because these values were small, it was enough for the calculation of the spectra to use only the first members of the series,

$$\cos\left(\omega_{so}t + \frac{\Delta\omega_{so}}{\Omega_{so}}\sin\Omega_{so}t\right) = J_o\left(\frac{\Delta\omega_{so}}{\Omega_{so}}\right)\cdot\cos\omega_{so}t + \sum_{j=1} J_j\left(\frac{\Delta\omega_{so}}{\Omega_{so}}\right)[\cos(\omega_{so}+j\Omega_{so})t + (-1)^j\cos(\omega_{so}-j\Omega_{so})t], \quad \textbf{(24)}$$

where $J_j(\ldots)$ is the Bessel function and $j$ = 1, 2, 3, . . . . As a result, the moduli of the spectra of the signals of equations 3 and 4, calculated with the replacement of the multiplier $\cos\omega_{so}t$ in them by the series in equation 24, led to the dependencies $S_s(f)$, which practically coincide with the spectra calculated without due account of the effect of the frequency modulation of signals $F_s^*(t)$. Thus, this effect can be ignored.

## BASIC CONCLUSIONS

Let us summarize the conclusions that can be drawn from the study of the frequency spectra $S_s(f)$ of two trains of wave packets of geomagnetic pulsations PC-1. They are listed below.

(1) The energy of PC-1 signals is concentrated chiefly in the vicinity of carrier frequencies $f_{so}$. Hence, the spectrum of velocities, $S(V_b)$, of energetic protons generating these signals has a sharply pronounced maximum. In our cases (I and II), it corresponds to the velocities of the beam, $V_{bo} = (4\text{–}6) \times 10^3$ km per sec, which perhaps changed during the period of measurements by roughly 1.5%.

(2) The frequency spectra of signals have satellites that apparently are due to the fact that the beam of generating hot particles (protons) have outbursts; that is, groups of particles appear in them whose velocities $V_{bk}$ greatly differ from the velocity $V_{bo}$. In the case considered, $V_{bk} = (10^3)\text{–}(1.3 \times 10^4)$ km/s. The energy of each of such outbursts apparently comprises only $(10^{-2}\text{–}10^{-3})$ of the energy of the main proton beam. This can be explained by the fact that the concentration of the particles of outbursts is much lower than the concentration of the main part of the protons of the beam whose velocity is $V_{bo}$.

(3) The moduli of spectra $S_s(f)$ have a definitely pronounced stable maximum in the vicinity of frequencies $f = 2f_{so}$, where the intensities of which exceed several times the intensity of the above-mentioned satellites and the background of the spectrum. Less intensive, but sufficiently clear maxima are also observed in some spectra in the vicinity of frequencies $f = 3f_{so}$. The presence of these maxima shows that in some cases, the process of the generation of geomagnetic pulsations PC-1 is accompanied by the generation of twofold and threefold pulsations PC-1.

(4) In the central parts of frequency spectra $|S_s(f)|$, satellites symmetrical with respect to frequencies $f = f_{so}$ are revealed. The intensity of these satellites is 1.5–8 times higher than the intensity of the bulk of the oscillations of the modulus of the spectrum.

There are some grounds to believe that these satellites differ in their physical nature from the satellites considered above—they are of a modulation character and appear as a result of the interaction between resonance oscillations of the magnetospheric region and PC-1 wave packets generated in it. The following values of four pairs of the supposed fundamental frequencies of the magnetospheric cavity are obtained. The data at Novolazarevskaya (station I, $\phi_0 = 66.2°$) give

$$\overline{\Delta f_{sk}^{\mp}} = (0.011, 0.022, 0.035)\ \text{Hz}, \qquad (k = 1, 2, 3),$$

and the data at Sogra (station II) give

$$\overline{\Delta f_{sk}^{\mp}} = (0.031, 0.047)\ \text{Hz}, \qquad (k = 3, 4).$$

(5) Frequencies $\Delta f_{sk}^{\mp}$ very closely coincide with the frequencies of geomagnetic pulsations PC-3 and PC-4 observed *in situ* from artificial Earth satellites and from the Earth's surface. This leads to the supposition that the process of the generation of PC-3 and PC-4 wave packets is regulated by the resonance properties of the magnetospheric cavity.

For illustration, the results of the analysis of the spectra $|S_s(f)|$ of the chain of signals $F_s(t)$ recorded at Sogra (station II, $\phi_0 = 57.9°$) are summarized in FIGURE 13 for the whole-considered frequency range. This figure indicates the maximum values of the amplitudes $\delta S_s(f)$ of all satellites of nine spectra (dots). Crosses on the figure show the values of $\delta S_{sk}^{\mp}$ for the first symmetrical two maxima of the spectra of the theoretical simulating signals. The central maxima $\delta S_o$ are also indicated on the figure by $s = 1, \ldots, 9$. The values of $\delta S_s(f) > 0.05$ given in FIGURE 13 exceed the level of the background and of the weak satellites of the spectra. The examination of the figure clearly illustrates the above conclusions; that is:

(a) The level of the amplitudes $\delta S_{sk}^{\mp}$ of the central symmetrical satellites of spectra, by which the resonance frequencies of the cavity of the magnetosphere, $\Delta f_{s3}^{\mp} = 0.031$ Hz and $\Delta f_{s4}^{\mp} \simeq 0.047$ Hz, are determined, turns out to be two to four times higher than the level of the regular maxima $\delta S_{sk}^{\mp}$ of the theoretical signals.

(b) In the frequency range of $f > f_{so}$, three regions of maxima can be distinguished: on the second and third harmonics, $\overline{f_s^{(2)}} = 2\bar{f}_{so} \simeq 2.097$ Hz and $f_s^{(3)} = 3\bar{f}_{so} \simeq 3.146$ Hz, along with a steady satellite of unclear origin on the frequency $\bar{f}_{sk} \simeq 1.63$ Hz $\simeq 1.5\bar{f}_{so}$. Near $f_{so}$, irregular satellites $f_{sk}$ appear, which perhaps point to the existence of the group of particles in the generating beam whose velocities differ from the main velocity of the beam.

(c) In the frequency range of $f = (0.41–1)$ Hz $< f_{so}$, a relatively increased number of irregular satellites are observed. Apparently, they also indicate that the beam of protons generating PC-1 pulsations contains groups whose velocities are 1.5–2 times lower.

In conclusion, we should emphasize the following: Some of the results of this paper are of a preliminary character. They require detailed verification and studies in different conditions of the magnetosphere. To understand the wave nature of the magnetosphere, it follows that it is necessary to evolve a general theory that would unite the wave properties of the magnetospheric cavity on ultralow frequencies into a

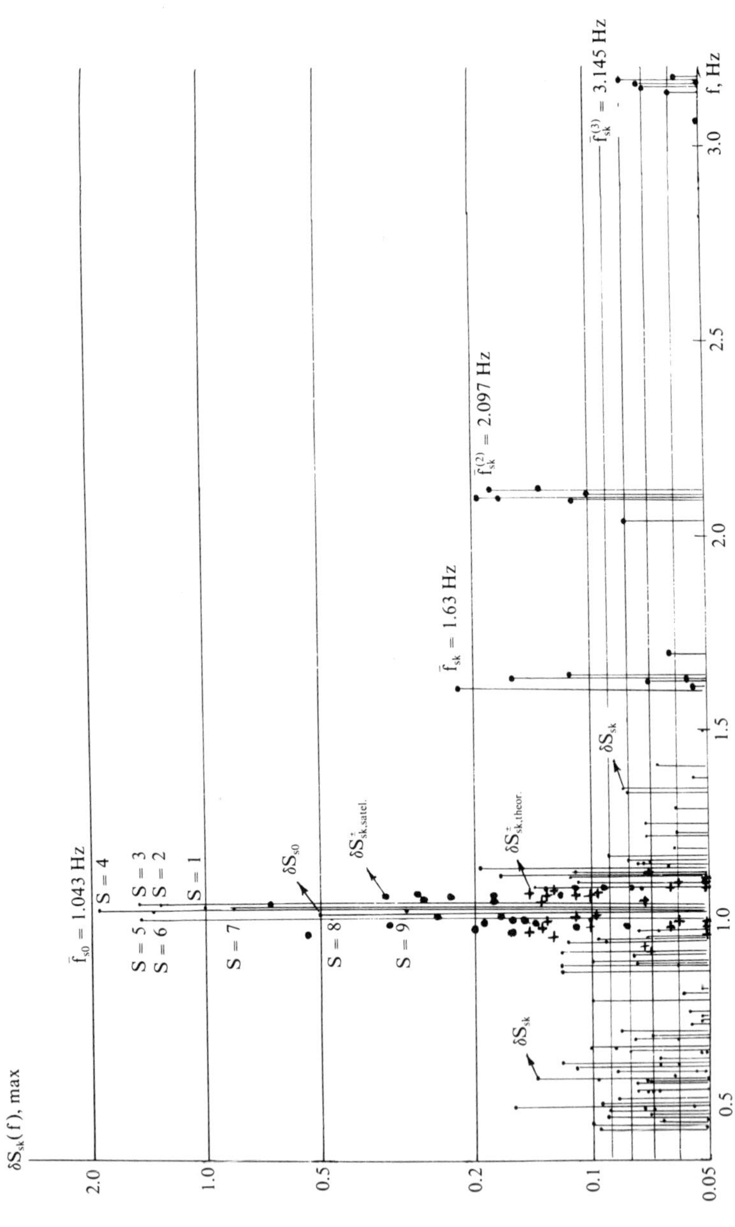


**FIGURE 13.** The maximum values of the amplitudes $\delta S_s(f)$ of all satellites of nine spectra (dots) recorded at Sogra (station II, $\phi_0 = 57.9°$). See text for specifics.

unified process. Such a theory, adequate to this process, should obviously be nonlinear to a great extent.

## ACKNOWLEDGMENTS

I would like to express my gratitude to D. S. Fligel for her help and for the fruitful collaboration in this work.

## REFERENCES

1. JACOBS, J. A. 1970. Geomagnetic Micropulsation. Springer–Verlag. Berlin/Heidelberg.
2. MCPHERRON, R. L., C. T. RUSSEL & P. J. COLEMAN. 1972. Space Sci. Rev. **13:** 411–454.
3. RUSSEL, C. T., R. L. MCPHERRON & P. J. COLEMAN. 1972. Space Sci. Rev. **12:** 810–856.
4. GULYELMI, A. V. & V. A. TROITSKAYA. 1973. Geomagnetic Pulsations and the Diagnostics of the Magnetosphere (in Russian). Nauka. Moscow.
5. GULYELMI, A. V. 1979. MHD Waves in Near-Earth Plasma (in Russian). Nauka. Moscow.
6. KENNEY, J. F., H. B. KNAFICH & H. B. LIEMOHN. 1968. Boeing Sci. Res. Lab. Document no. D1-82-0691; 1969. Boeing Sci. Res. Lab. Document no. D1-82-0890.
7. LIEMOHN, H. B. 1967. J. Geophys. Res. **72:** 39.
8. ROUX, A. & J. SOLOMON. 1970. Ann. Geophys. **26:** 279.
9. GENDRIN, R., S. LACOURLY, A. ROUX, J. SOLOMON, F. Z. FEIGIN, M. V. GOKHBERG, V. A. TROITSKAYA & V. L. YAKIMENKO. 1971. Planet. Space Sci. **19:** 165.
10. LANDSBERG, G. & L. MANDELSTAM. 1928. Naturwissenschaften **16:** 557, 772.
11. MANDELSTAM, L. I. Complete Works (in Russian). Vol. I: 229 and 305.
12. DUNGEY, J. W. 1954. Electrodynamics of the Outer Atmosphere. Pa. State Univ. Sci. Rep. no. 69; 1963. The Earth Environment. Gordon & Breach. New York.
13. POLYAKOV, S. V., V. O. RAPOPORT & V. YU. TRAKHTENGERTS. 1984. Geomagn. Aeron. (in Russian) **24:** 242.
14. KRYLOV, A. L. & A. YE. LIFSHITS. 1983. Fiz. Zemli (Earth Phys.) (in Russian) **11:** 3.
15. FADDEYEVA, V. N. & N. M. TERENTYEV. 1954. Tables of the Values of the Probability Integral with Respect to the Complex Argument (in Russian). State Publishing House of Theoretical and Technical Literature. Moscow.
16. DUBOVOY, A. P. & A. A. YAROSLAVTSEV. 1980. On the calculation of Kramp functions (in Russian). IZMIRAN Collection, pp. 69–75.
17. CUMMINGS, W. D., R. J. O'SULLIVAN & P. J. COLEMAN. 1969. J. Geophys. Res. **74:** 748.
18. CUMMINGS, W. D., F. MASON & P. J. COLEMAN. 1972. J. Geophys. Res. **77:** 748.
19. PATEL, V. L. 1965. Planet. Space Sci. **13:** 485.
20. MCPHERRON, R. L., C. T. RUSSEL & P. J. COLEMAN. 1972. J. Geophys. Res. **13:** 411.
21. DWARKIN, M. L., A. J. ZMUDA & W. D. RADFORD. 1971. J. Geophys. Res. **76:** 3688.
22. BARFIELD, J. N., I. J. LANZEROTTI, C. G. MCLENNAN, G. A. PANLIKAS & M. SHULZ. 1971. J. Geophys. Res. **76:** 5262.
23. LANZEROTTI, I. J. & N. A. TARTAGLIA. 1972. J. Geophys. Res. **77:** 1934.
24. AL'PERT, YA. L. 1973. The Propagation of Electromagnetic Waves and the Ionosphere (Second Edition). Plenum. New York.

# On Generation of Geomagnetic Pulsations PC-1 in the Magnetospheric Cavity

YA. L. AL'PERT

The solution of a system of two connected, ordinary oscillatory differential equations of the second order simulates well the discrete packets of oscillations, $F_s(t)$, of geomagnetic pulsations PC-1 of different form. The excitation of these oscillations in the magnetosphere is due to the cyclotron instability of the plasma.

One of these equations is linear of the "parametric" kind:

$$\ddot{y} + 2\dot{y}\gamma_{i0}[1 - \alpha_0 Z(t)] + \omega^2 y = 0.$$

It determines $Y(t)$. The coefficient, $X(t) = 2\gamma_{i0}[1 - \alpha_0 Z(t)]$, of the term $\dot{y} = dy/dt$ of

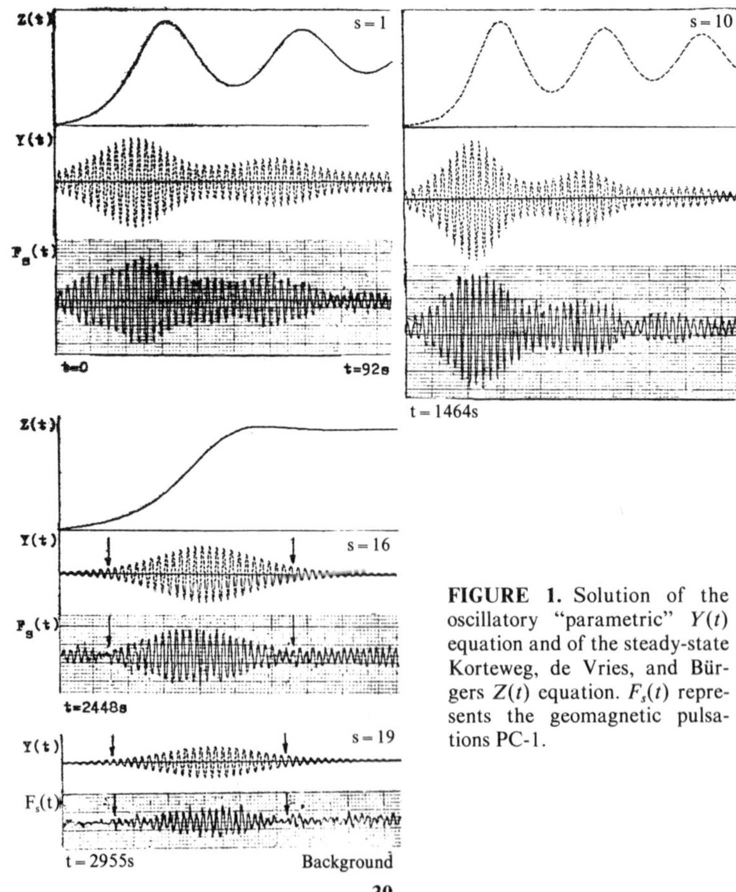


**FIGURE 1.** Solution of the oscillatory "parametric" $Y(t)$ equation and of the steady-state Korteweg, de Vries, and Bürgers $Z(t)$ equation. $F_s(t)$ represents the geomagnetic pulsations PC-1.

this equation is connected through $Z(t)$ with the solution of the second equation. It is nonlinear and follows from the Korteweg, de Vries, and Bürgers equation for stationary waves with a running coordinate, $\xi = (x - V_A t)$ ($V_A$ is the velocity of Alfvén waves); specifically,

$$(\beta/V_A^2)\ddot{Z}(t) - (\mu/V_A^2)\dot{Z}(t) - (\nu_0 - Z/2)Z = 0.$$

Thus, the function $Z(t)$ controls the forming of the envelope of $Y(t)$; it can be of monotonic or oscillatory-solitonic form. Physically, it means that the fast oscillations of the wave packets, $Y(t)$, are modulated by the slow oscillations of $Z(t)$.

An analysis of a chain of $s = 21$ geomagnetic pulsations PC-1 was made. The equations were solved numerically and graphically by the Pocket Computer SHARP. The results of the simulating of four wave packets are given in FIGURE 1. The anisotropy coefficient of the temperature, $T_{i\perp}/T_{i\parallel}$, the concentration, $\mathcal{N}_{ib}$, and the velocity, $v_{0b}$, of the hot protons (the source of the excitation of PC-1) were determined. All these values were found to be in a good agreement with the known data, particularly those obtained on satellites. This allows one to suppose that the used theoretical model is apparently adequate for the process of generation of PC-1 pulsations.

# New Developments in Density Functional Theory[a]

WALTER KOHN

*Institute for Theoretical Physics*
*University of California, Santa Barbara*
*Santa Barbara, California 93106*

In this brief lecture, I shall attempt to accomplish two objectives:

(I) For the benefit of the uninitiated, I shall summarize density functional theory as it stood four or five years ago. This stage may be found described in greater detail in a review by W. Kohn and P. Vashishta.[1]

(II) I shall then describe a number of significant recent developments, mostly within the last two years, that have considerably advanced both the theoretical basis and the practical usefulness of the theory.

This presentation will be limited to a discussion of one-component, nonrelativistic spinless fermions in a nondegenerate ground state, even though density functional theory covers a much broader class of systems (multicomponent, degenerate ground states, thermal ensembles, spin-½ fermions, bosons, relativistic particles, etc.). I shall present and discuss the main results without proof.

## STATUS AS OF 1980/1981

Density functional theory is an in principle exact theory for the quantum mechanical ground state of a many-body system in terms of the particle density, $n(\underset{\sim}{r})$, rather than the $N$-body wave function, $\Psi(\underset{\sim}{r}_1, \underset{\sim}{r}_2, \ldots, \underset{\sim}{r}_N)$, that satisfies the Schrödinger equation,

$$\left[\sum_i \left(-\frac{1}{2}\nabla_i^2 + v(\underset{\sim}{r}_i)\right) + \sum_{i \neq j} \frac{1}{2} u(r_i - r_j) - E\right] \Psi(\underset{\sim}{r}_1, \ldots, \underset{\sim}{r}_N) = 0, \quad (1)$$

with self-evident notation.

Let us fix $N$ and the form of $u(\underset{\sim}{r})$ and let us consider systems corresponding to all possible, sufficiently well-behaved $v(\underset{\sim}{r})$. Clearly, under the assumption of nondegeneracy,

$$v(\underset{\sim}{r}) \rightarrow n(\underset{\sim}{r}) \text{ uniquely.} \quad (2)$$

[Also, of course, $v(\underset{\sim}{r}) \rightarrow H, \Psi(\underset{\sim}{r}_1, \ldots, \underset{\sim}{r}_N)$ and all thc properties of the system.]

[a]This paper was originally submitted for the 1984 meeting, but inadvertently was not published in the proceedings of that meeting.

### *The Lemma of Hohenberg and Kohn*

This states[2] that the correspondence of equation 2 can also be reversed, that is,

$$n(\underset{\sim}{r}) \longrightarrow v(\underset{\sim}{r}) \text{ uniquely} \tag{3}$$

(to within a trivial additive constant). Hence, $n(\underset{\sim}{r})$ also determines $H$, $\Psi(\underset{\sim}{r}_1, \ldots, \underset{\sim}{r}_N)$ and all other properties of the system (which are all governed by $H$).

### *The Variational Principle of Hohenberg and Kohn*

For every $n'(\underset{\sim}{r})$ corresponding to some $v'(\underset{\sim}{r})$ [such $n'(\underset{\sim}{r})$ will be called $v$-representable], we define[2] the functional of $n'(\underset{\sim}{r})$,

$$F[n'(\underset{\sim}{r})] \equiv (\Psi, (T + U)\Psi), \tag{4}$$

where $T$ and $U$ are, respectively, the kinetic and interaction energy operators. With the aid of $F[n]$, we define, for any given $v(r)$, the energy functional,

$$E_v[n'(\underset{\sim}{r})] \equiv \int v(\underset{\sim}{r})n'(\underset{\sim}{r})\, d\underset{\sim}{r} + F[n'(\underset{\sim}{r})]. \tag{5}$$

It could be shown that for each $v(\underset{\sim}{r})$, this functional attains its minimum for the correct ground state density, $n(\underset{\sim}{r})$, corresponding to $v(\underset{\sim}{r})$, and that this minimum is the correct ground state energy:

$$\min_{n'(\underset{\sim}{r})} E_v[n'(\underset{\sim}{r})] = E_v[n(\underset{\sim}{r})] = E. \tag{6}$$

### *The Kohn-Sham Self-Consistent Equations*

Assuming $v$-representability whenever needed (see below in regard to this assumption), the exact ground state density and energy can, in principle, be obtained by the solution of the following set of self-consistent equations:[3]

$$[-\tfrac{1}{2}\nabla^2 + v_{\text{eff}}(\underset{\sim}{r}) - \epsilon_i]\Psi_i(\underset{\sim}{r}) = 0, \tag{7}$$

$$n(\underset{\sim}{r}) = \sum_{1}^{N} |\Psi_i(\underset{\sim}{r})|^2, \tag{8}$$

$$v_{\text{eff}}(\underset{\sim}{r}) = v(\underset{\sim}{r}) + \int n(\underset{\sim}{r} - \underset{\sim}{r}')v(\underset{\sim}{r}')\, d\underset{\sim}{r}' + \frac{\delta E_{\text{xc}}[n(\underset{\sim}{r})]}{\delta n(\underset{\sim}{r})}, \tag{9}$$

where the exchange-correlation functional, $E_{\text{xc}}[n]$, is defined by

$$F[n] = T_s[n] + \tfrac{1}{2}\int n(\underset{\sim}{r})n(\underset{\sim}{r} - \underset{\sim}{r}')n(\underset{\sim}{r}')\, d\underset{\sim}{r}d\underset{\sim}{r}' + E_{\text{xc}}[n]. \tag{10}$$

Here, $T_s[n]$ is the kinetic-energy functional for noninteracting electrons in external potentials $v(\underset{\sim}{r})$. The ground state energy is given by

$$E = \sum \epsilon_i - \frac{1}{2}\int n(\underset{\sim}{r})n(\underset{\sim}{r} - \underset{\sim}{r}')n(\underset{\sim}{r}')\, d\underset{\sim}{r}\, d\underset{\sim}{r}' - \int \frac{\delta E_{xc}[n(\underset{\sim}{r})]}{\delta n(\underset{\sim}{r})} \cdot n(\underset{\sim}{r})\, d\underset{\sim}{r} + E_{xc}[n(\underset{\sim}{r})]. \quad \textbf{(11)}$$

To actually carry out this calculation, an adequate knowledge of $E_{xc}[n]$ is required. The most useful approximation has been the so-called local density approximation,

$$E_{xc}[n(\underset{\sim}{r})] \approx \int \epsilon_{xc}[n(\underset{\sim}{r})]n(\underset{\sim}{r})\, d\underset{\sim}{r}, \quad \textbf{(12)}$$

where $\epsilon_{xc}(n)$ is the exchange-correlation energy per particle of a uniform fermion gas of density $n$.

## NEW DEVELOPMENTS

### *General Theory*

(1) Extension of the lemma of Hohenberg and Kohn to degenerate ground states:[4] Although, in this case, the correspondence $v(\underset{\sim}{r}) \rightarrow n(\underset{\sim}{r})$ is not unique, the inverse correspondence $n(\underset{\sim}{r}) \rightarrow v(\underset{\sim}{r})$ is unique.

(2) Examples of non-$v$-representability:[5,6] Levy and Lieb (in references 5 and 6, respectively) have exhibited perfectly well-behaved functions $n(\underset{\sim}{r})$ that are not ground state densities for any $v(\underset{\sim}{r})$.

(3) A class of $v$-representable densities:[7] Let $n(\underset{\sim}{r})$ be $v$-representable and let $n_1(\underset{\sim}{r})$ be an arbitrary function such that $\int n_1(\underset{\sim}{r})\, d\underset{\sim}{r} = 0$. Then for $\lambda$ small enough, the density

$$\tilde{n}(\underset{\sim}{r}) \neq n(\underset{\sim}{r}) + \lambda n_1(\underset{\sim}{r}) \quad \textbf{(13)}$$

is also $v$-representable (proved only on a finite lattice).

(4) Definition of $F[n]$ for arbitrary $n(\underset{\sim}{r})$, even if not $v$-representable:[5,6]

$$F[n(\underset{\sim}{r})] = \min\, (\Psi_{[n(\underset{\sim}{r})]}, (T + U)\Psi_{[n(\underset{\sim}{r})]}), \quad \textbf{(14)}$$

where for given $n(\underset{\sim}{r})$, the $\Psi_{[n(\underset{\sim}{r})]}$ are the class of determinantal wave functions (not necessarily eigenfunctions) whose density is $n(\underset{\sim}{r})$.

(5) Physical meaning of the highest eigenvalue, $\bar{\epsilon}_i$, of the Kohn-Sham equations:[8]

$$\bar{\epsilon}_i = \text{exact physical ionization energy.} \quad \textbf{(15)}$$

(6) Discontinuity of $v_{eff}$ as a function of $N$:[9–11] For "open" systems, where $N$ becomes a continuous variable, one has for an isolated ground state,

$$\left[\frac{\delta E_{xc}[n(\underset{\sim}{r})]}{\delta n(\underset{\sim}{r})}\right]_{N+\delta N} - \left[\frac{\delta E_{xc}[n(\underset{\sim}{r})]}{\delta n(\underset{\sim}{r})}\right]_{N-\delta N} = C \neq 0, \quad \textbf{(16)}$$

where $C$ is independent of $\underset{\sim}{r}$. (For metals, $C = 0$.)

### *Improvement over the Local Density Approximation*

(7) Whereas straightforward additions of gradient corrections to the local density approximation (equation 12) have not had much success in atomic and solid-state applications, the following form, recently derived by Langreth and Mehl,[12] has given significantly improved results (by factors of 2–5) for atoms:

$$E_{xc}[n(\underset{\sim}{r})] \approx \int \epsilon_{xc}[n(\underset{\sim}{r})]n(\underset{\sim}{r})\, d\underset{\sim}{r} + a \int d\underset{\sim}{r} \frac{|\nabla_n(\underset{\sim}{r})|^2}{[n(\underset{\sim}{r})]^{4/3}} \left(2e^{-F} - \frac{7}{9}\right), \qquad \mathbf{(17)}$$

where

$$F \equiv b|\nabla_n|/n^{7/6}; \qquad a = 2.143, b = 0.2618 \text{ atomic units.} \qquad \mathbf{(18)}$$

Accuracies in the range of 1–4% are obtained.

## REFERENCES

1. KOHN, W. & P. VASHISHTA. 1983. *In* Theory of the Inhomogeneous Electron Gas. S. Lundqvist & N. H. March, Eds.: 79. Plenum. New York.
2. HOHENBERG, P. & W. KOHN. 1964. Phys. Rev. **136:** B864.
3. KOHN, W. & L. J. SHAM. 1965. Phys. Rev. **A140:** 1133.
4. KOHN, W. *In* Highlights of Condensed Matter Theory. International School of Physics, Enrico Fermi, LXXXIX Course (1983). F. Bassani & F. Fumi, Eds. North-Holland. Amsterdam. To be published.
5. LEVY, M. 1982. Phys. Rev. **A26:** 1200.
6. LIEB, E. 1982. *In* Physics as Natural Philosophy: Essays in Honor of Laszlo Tisza on his 75th Birthday. A. Shimony & H. Feshbach, Eds.: 111. MIT Press, Cambridge, Massachusetts.
7. KOHN, W. 1983. Phys. Rev. Lett. **51:** 1596.
8. ALMBLADH, C-O. & U. VON BARTH. To be published.
9. PERDEW, J. P. *et al.* 1982. Phys. Rev. Lett. **49:** 1691.
10. PERDEW, J. P. & M. LEVY. 1983. Phys. Rev. Lett. **51:** 1884.
11. SHAM, L. J. & M. SCHLÜTER. 1983. Phys. Rev. Lett. **51:** 1888.
12. LANGRETH, D. C. & M. J. MEHL. 1983. Phys. Rev. **B28:** 1809.

# Spatial Structure of a Fermion Fluid[a]

J. K. PERCUS

*Courant Institute of Mathematical Sciences*
*and*
*Physics Department*
*New York University*
*New York, New York 10012*

## INTRODUCTION

The understanding of nonuniform fluids constitutes one of the major activities of statistical physics.[1] When several phases are in contact, it appears that new concepts and new tools may be required for this purpose. However, even single phase systems made nonuniform by the action of strong fields can hardly be said to be under effective control. This is especially the case in the quantum domain, where our information on bulk fluids is so meager. Our target in this paper is the electron fluid associated with an atom or molecule with fixed nuclei—a system of clearly overwhelming importance that has been the subject of countless investigations.[2-4] In many ways, though, this system is not exceptional, and, indeed, our discussion will in general encompass the whole class of simple fluids in thermal equilibrium. When we specialize to the electron fluid, we will concentrate on the ground state, $T = 0°$, properties, and for the purpose of assessing approximations and comparing with other techniques, we will trivialize the problem to the case of noninteracting electrons in an external field.

In conformity with the essential similarity of the fluid that we are examining at differential spatial locations, the motif of our analysis will be the comparison of such a fluid with the uniform fluid that it locally resembles. To this end, we first review the density-potential relation for the two extreme cases of very slowly varying and very rapidly varying densities. We then start a systematic expansion of the applied potential in iterated gradients of the density profile, show that it corresponds to an analogous expansion of the free energy, and derive the (modified coefficient) Weiszacker correction[5] to the Thomas-Fermi[6,7] approximation. The inadequacy of this correction is pointed out, and we follow with a nonlocal analysis of the density-potential relation that is valid for both very slow and very rapid density changes. Finally, this analysis is extended to the free energy and tested on model (independent-electron) atoms.

## THERMODYNAMIC BACKGROUND

Our discussion will take place mainly in the context of a grand canonical ensemble that is determined by chemical potential $\mu$ and reciprocal temperature $\beta$. The

[a]This work was supported in part by DOE Contract No. DE-AC02-76ERO3077 and NSF Grant No. CHE-86-07598.

thermodynamic generating function or grand potential is then given by

$$\Omega = -\frac{1}{\beta} \ln \operatorname{tr} e^{-\beta(\hat{H}-\hat{N}\mu)}, \tag{2.1}$$

which reduces to $-PV$ for extended uniform systems. Here, $\hat{H}$ is the system Hamiltonian and $\hat{N}$ is the particle number operator. Much of the development will be modeled after familiar considerations in a classical fluid, in which

$$\Omega = -\frac{1}{\beta} \ln \sum_N \frac{1}{N!} \int \cdots \int e^{-\beta(\Phi_N - N\mu)}\, dr^N, \tag{2.2}$$

with $\Phi_N$ denoting the $N$-particle potential energy alone, and $\mu$ now incorporating the momentum integrations. Our target, however, is the ground state of a Fermion system; in such a case, equation 2.1 reduces to

$$\Omega = E_0(\hat{H} - \hat{N}\mu) = \min_N [E_0(\hat{H}_N) - N\mu], \tag{2.3}$$

where the ground state energy $E_0$ in the latter expression is at fixed particle number $N$. We note that only a single particle number $N$ contributes to $\Omega$, except at the special values

$$\mu_N = E_0(\hat{H}_{N+1}) - E_0(\hat{H}_N), \tag{2.4}$$

where the mean $\langle \hat{N} \rangle$ can be fixed at any value between $N$ and $N + 1$. Thus, $\Omega$ is piecewise linear in $\mu$.

For technical purposes, the reason for choosing a grand ensemble is that it will be extremely convenient to be able to vary the ensemble average particle density $n(r)$ at will by the corresponding variation of an applied external potential $u(r)$. Because $u(r)$ enters into the Hamiltonian in the form $\int \hat{n}(r)u(r)\, d^3r$, the generating function character of $\Omega$,

$$\langle \delta(\hat{H} - \hat{N}\mu) \rangle = \delta\Omega, \tag{2.5}$$

then tells us that in functional derivative notation,

$$n(r) = \delta\Omega/\delta u(r). \tag{2.6}$$

Only the combination $\mu - u(r)$ appears in $\Omega$, and equation 2.6 can also be written as

$$n(r) = -\delta\Omega/\delta[\mu - u(r)]. \tag{2.7}$$

A Legendre transformation,

$$F^B[n] = \Omega + \int [\mu - u(r)]n(r)\, d^3r, \tag{2.8}$$

now allows us to switch the independent variable from $\mu - u(r)$ to $n(r)$, with the consequence that

$$\mu - u(r) = \delta F^B[n]/\delta n(r). \tag{2.9}$$

$F^0[n]$ as defined above is the internal or bulk Helmholtz free energy, from which the full free energy

$$F[n, u] = F^B[n] + \int n(r)u(r)\, d^3r = \Omega + \mu N \tag{2.10}$$

can also be constructed, thereby satisfying

$$\mu = \delta F[n, u]/\delta n(r). \tag{2.11}$$

Note that in the quantum ground state,

$$F[n, u] = E_0(\hat{H}_N) \tag{2.12}$$

numerically, which is consistent with the vanishing of the entropic potential $TS$.

## LIMIT OF SLOW VARIATION

Let us now restrict our attention to some limiting cases, both to get a feeling as to what to expect and to serve as a check on the way that our formalism is developed. To start with, analysis of slowly varying fluids is a routine part of thermodynamics: if the specific Helmholtz free energy at uniform density $n$ is $f(n)$, then in an external potential $u(r)$, the total Helmholtz energy is approximated as

$$F[n, u] = \int n(r)(f(n(r)) + u(r))\, d^3r, \tag{3.1}$$

which is equivalent to

$$F^B[n] = \int n(r)f(n(r))\, d^3r. \tag{3.2}$$

Hence, from equation 2.9 or 2.11, the corresponding density profile is

$$\mu = \mu(n(r)) + u(r), \tag{3.3}$$

where $\mu(n) = dnf(n)/dn$ denotes the chemical potential of a uniform fluid of density $n$.

The tacit assumption in equation 3.1 is that the correlation length for density fluctuations is everywhere zero and, indeed, equation 3.1 may be derived by a scaling process that enforces this condition. This does not quite mean that there are no long-range phenomena present—the kinetic version of equation 3.1 does generate acoustic waves—but it means that they are poorly represented. Some global or collective effects are certainly ignored, and this will remain true, in part, for any correction procedure to equation 3.1 that we may contemplate. However, our objective at this time is rather to concentrate on single particle excitations. A traditional first step in the process of extending the correlation length encompassed applies to the system represented by equation 3.1, in which the interparticle forces are augmented by a slowly varying pair potential $\phi_1(r - r')$. Instead of equation 3.2, the mean field approximation then yields

$$F^B[n] = \int n(r)f(n(r))\, d^3r + \tfrac{1}{2}\iint n(r)\phi_1(r - r')n(r')\, d^3r\, d^3r' \tag{3.4}$$

so that

$$\mu = \mu(n(r)) + u(r) + \int \phi_1(r - r')n(r')\, d^3r'. \tag{3.5}$$

Equation 3.5 is responsible for the classical van der Waals theory[8] of nonuniform fluids, in which $n(r)$ varies on even a slower scale than $\phi_1$. One writes $n(r') = n(r) + (r - r') \cdot \nabla n(r) + \frac{1}{2}(r - r')(r - r'){:}\, \nabla\nabla n(r) + \ldots$ and uses the assumed spherical symmetry of $\phi_1$ to convert equation 3.5 to

$$\mu = \mu(n(r)) + [\textstyle\int \phi_1(R)\, d^3R]n(r) + [\frac{1}{6} \textstyle\int R^2\phi_1(R)\, d^3R]\nabla^2 n(r) + u(r). \quad (3.6)$$

Equation 3.5 is similarly responsible for the Thomas-Fermi approximation of atoms and molecules, where the reference fluid is an ideal electron gas, with $\mu(n) = (\hbar^2/2m)(3\pi^2 n)^{2/3}$, $u(r) = -Ze^2/r$, and $\phi_1(r - r') = e^2/|r - r'|$. Application of $(\frac{1}{4\pi})\nabla^2$ to equation 3.5 then puts it in the desired local form:

$$-\frac{1}{4\pi}\nabla^2\mu(n(r)) + e^2 n(r) = Ze^2\delta(r). \quad (3.7)$$

Thus, equations 3.4 and 3.5 have familiar and desirable consequences that are prototypical of the development that we will pursue.

## LIMIT OF RAPID VARIATION

We proceed next to the opposite regime—that of a region in which the density is controlled by a rapidly changing external potential. The correspondingly very large force effectively decouples the particle being acted upon from the rest of the system, thus reducing the problem to that of one particle. We take advantage of this by using (in the petit ensemble)

$$n(r) = N\langle\delta(r_1 - r)\rangle. \quad (4.1)$$

Suppose now that $R$ is a point at which $u(r)$ is changing rapidly, with, say, a $z$-directed logarithmic gradient of magnitude $\Lambda$. In the case of classical thermal equilibrium, the quantity

$$\rho_N(r_1, \ldots, r_N)e^{\beta u(r_1)}, \quad (4.2)$$

with $\rho_N$ denoting the full $N$-body coordinate distribution, is constant on the scale $1/\Lambda$ in the vicinity of $r_1 = R$. Integrating over $r_2, \ldots, r_N$, so too is $n(r_1)e^{\beta u(r_1)}$, and in the grand ensemble as well. Hence, we can write

$$\mu = u(r) - \frac{1}{\beta}\ln n(r) + \ldots, \quad (4.3)$$

where the remaining terms remain regular as $\Lambda \to \infty$. Moving back one step, this profile comes from a free energy

$$F^B = \frac{1}{\beta}\int [\ln n(r) - 1]n(r)\, d^3r + \ldots, \quad (4.4)$$

which is a local form that is very different from that of equation 3.2 (except for ideal gases).

For a quantum mechanical stationary state (a ground state, in particular), a little more work is required. We first write $u(r_1)$ in the $1/\Lambda$ vicinity of $R$ as $u_R(z_1)$ and replace $r_1$ by $R$ in all other terms. Then, the state $\psi$ is expanded in orthonormal eigenfunctions of the $(N-1)$–body Hamiltonian at fixed $r_1 = R$:

$$H_R(p_2, \ldots, p_N, r_2, \ldots, r_N)\psi_\alpha(r_2, \ldots, r_N; R) = E_\alpha(R)\psi_\alpha(r_2, \ldots, r_N; R). \quad \textbf{(4.5)}$$

Setting

$$\psi(r_1, \ldots, r_N) = \sum_{k,\alpha} \phi_{\alpha k}(z_1; R)\psi_\alpha(r_2, \ldots, r_N; R)e^{ik\cdot x_1}, \quad \textbf{(4.6)}$$

where $x$ signifies the coordinate pair transverse to $z$, the Schrödinger equation $H\psi = E\psi$ is equivalent to

$$\left[\frac{p^2}{2m} + u_R(z)\right]\phi_{\alpha k}(z; R) = [E - E_\alpha(R) - \hbar^2k^2/2m]\phi_{\alpha k}(z; R)$$
$$\equiv \epsilon_{\alpha k}(R)\phi_{\alpha k}(z; R). \quad \textbf{(4.7)}$$

Furthermore,

$$n_R(z) = N\int |\psi(x, z, r_2, \ldots, r_N)|^2\, d^3r_2 \ldots d^3r_N$$
$$= N\,\Sigma\, \phi_{\alpha k}(z; R)\phi^*_{\alpha k'}(z; R)e^{i(k-k')\cdot x}. \quad \textbf{(4.8)}$$

Now, if $\Lambda$ is large compared to any of the wave numbers set by the $\epsilon_{\alpha k}$, all solutions of equation 4.7 are the same in form; that is, linear combinations of two real basic solutions of the second-order equation 4.7, say $\phi^{(1)}$ and $\phi^{(2)}$. Thus, $n(z)$ is bilinear in $\phi^{(1)}$ and $\phi^{(2)}$, and hence factorizable as ($R$ is understood)

$$n(z) = \phi_1(z)\phi_2(z),$$

$$\phi''_{1,2}(z) = \frac{2m}{\hbar^2}\, u(z)\phi_{1,2}(z), \quad \textbf{(4.9)}$$

which is true in a grand ensemble as well. Because the Wronskian

$$\phi'_2\phi_1 - \phi'_1\phi_2 = w \quad \textbf{(4.10)}$$

is a constant, we have $(n/\phi_1)'\phi_1 - \phi'_1(n/\phi_1) = w$, or $\phi'_1/\phi_1 = (n' + w)/2n$. Hence, $(2m/\hbar^2)u = \phi''_1/\phi_1 = (\phi'_1/\phi_1)' + (\phi'_1/\phi_1)^2$ or

$$u = \frac{\hbar^2}{4m}\left(\frac{n''}{n} - \frac{n'^2}{2n^2} + \frac{w^2}{2n^2}\right). \quad \textbf{(4.11)}$$

There are two possibilities. If the local excursion of $u_R(z)$ remains finite, $n(z)$ generally remains nonzero. Thus, the rapidly varying regime is dominated by[9]

$$\mu - u = -\frac{\hbar^2}{4m}\left(\frac{\nabla^2 n}{n} - \frac{|\nabla n|^2}{2n^2}\right) + \cdots, \quad \textbf{(4.12)}$$

which integrates back to

$$F^B = \frac{\hbar^2}{8m}\int \frac{|\nabla n|^2}{n}\, d^3r + \cdots. \quad \textbf{(4.13)}$$

This is recognized as the original Weiszacker inhomogeneity correction. On the other hand, if $u$ goes to infinity, there is no tunneling and hence there is only one solution to equation 4.7. Thus, $w = 0$ in equation 4.11, and equation 4.13 holds even though $n$ falls to zero.

Equations 4.3 and 4.12, it is to be noted, serve to define boundary conditions for (hard) wall-bounded fluids. The wall pressure for a wall to the left is given by

$$P = -\int n(r)\frac{\partial}{\partial z}u(r)\,d^3r, \tag{4.14}$$

where $u(r)$ has the limiting form of an infinite amplitude step function in the $z$-direction normal to the wall, and the volume is a cylinder of unit area normal to $z$. Thus, using equation 4.3 in the classical case,

$$P = -\int n(z)u'(z)\,dz = (1/\beta)\int_{-\infty}^{W} n'(z)\,dz = (1/\beta)\mu_W,$$

where $\mu_W$ is the wall density:

$$P = n_W/\beta. \tag{4.15}$$

In the quantum mechanical version, both $n_W = 0$ and $n'_W = 0$, so

$$n(z) = \tfrac{1}{2}n''_W(z - z_W)^2 + \cdots \tag{4.16}$$

near the wall. However, from equation 4.12,

$$-\int n(z)u'(z)\,dz = (\hbar^2/4m)\int (2n'n''/n - n''' - n'^3/n^2)\,dz$$
$$= (\hbar^2/4m)(n'^2/n - n'')|_W,$$

or simply

$$P = \frac{\hbar^2}{4m}\frac{d^2n}{dz^2}\bigg|_W. \tag{4.17}$$

## GRADIENT EXPANSION

The van der Waals expression (equation 3.6) is an approximation to one in which the fluid density is slowly varying, but without any assumption on the nature of the interparticle interaction. Thus, for the hundredth time or so, let us derive it in a form in which the classical or quantum nature of the fluid is irrelevant. It is based (see FIGURE 1) upon a comparison of the real fluid with a uniform one whose density $n(R)$ coincides with that of the real one at the point of interest.[10] The chemical potential of the reference fluid is now $\mu(n(R))$, with external potential 0. According to equation 2.9, a functional Taylor expansion of the real fluid local chemical potential $\mu - u(r)$ about that of the reference fluid yields

$$\mu - u(r) = \mu(n(R)) + \int \frac{\delta^2 F^B}{\delta n(r)\delta n(r')}\bigg|_R [n(r') - \mu(R)]\,d^3r'$$
$$+ \frac{1}{2}\iint \frac{\delta^3 F^B}{\delta n(r)\delta n(r')\delta n(r'')}\bigg|_R [n(r') - n(R)]$$
$$\cdot [n(r'') - n(R)]\,d^3r'\,d^3r'' + \cdots, \tag{5.1}$$

where the notation $|_R$ means that the evaluation is taken in a uniform fluid of density $n(R)$. Hence, by choosing $R = r$,

$$\mu - u(r) = \mu(n(r)) + \int \frac{\delta^2 F^B}{\delta n(r)\delta n(r')}\bigg|_r [n(r') - n(r)]\, d^3r' + \frac{1}{2}\iint \frac{\delta^3 F^B}{\delta n(r)\delta n(r')\delta n(r'')}\bigg|_r [n(r') - n(r)] \cdot [n(r'') - n(r)]\, d^3r'\, d^3r'' + \cdots. \quad (5.2)$$

Equation 5.2 can be used in many different ways. Here, we will make the weak inhomogeneity gradient expansion:

$$n(r') = n(r) + (r' - r)\cdot\nabla n(r) + \tfrac{1}{2}(r' - r)(r'' - r): \nabla\nabla n(r) + \cdots. \quad (5.3)$$

If we retain in the full series (equation 5.2) terms through two derivatives, only the

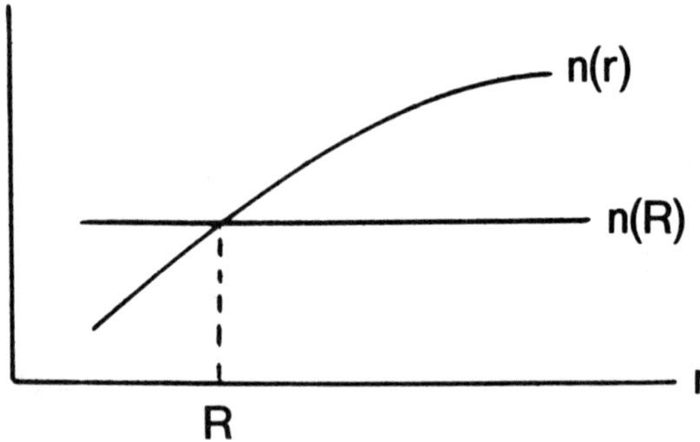


**FIGURE 1.** Local reference fluid.

truncated expression

$$\mu - u(r) = \mu(n(r)) + \int \frac{\delta^2 F^B}{\delta n(r)\delta n(r')}\bigg|_r \left[(r' - r)\cdot\nabla n(r) + \frac{1}{2}(r' - r)(r' - r): \nabla\nabla n(r)\right] d^3r' + \frac{1}{2}\iint \frac{\delta^3 F^B}{\delta n(r)\delta n(r')\delta n(r'')}\bigg|_r \cdot (r' - r)(r'' - r): \nabla n(r)\nabla n(r)\, d^3r'\, d^3r'' \quad (5.4)$$

remains. Equation 5.4 can be simplified considerably. We explicitly assume the isotropy of the pair interaction and hence of the uniform fluid distribution functions. Thus, the integral multiplying the $\nabla n(r)$ term vanishes by parity. Next, we use the identity

$$(r' - r)(r'' - r) + (r'' - r)(r' - r) = (r' - r)(r' - r) + (r'' - r)(r'' - r) - (r'' - r')(r'' - r'), \quad (5.5)$$

together with the symmetry of $\delta^3 F^B/\delta n(r)\delta n(r')\delta n(r'')$ at uniform $n(r)$ with respect to $r' - r$, $r - r''$, and $r'' - r'$, to reduce the second integral. Finally, isotropy allows us to drop all dyadic cross terms. There results

$$\mu - u(r) = \mu(n(r)) + \frac{1}{6}\int \frac{\delta^2 F^B}{\delta n(r)\delta n(r')}\bigg|_r |r' - r|^2\, d^3r'\, \nabla^2 n(r)$$

$$+ \frac{1}{12}\iint \frac{\delta^3 F^B}{\delta n(r)\delta n(r')\delta n(r'')}\bigg|_r |r' - r|^2\, d^3r'\, d^3r''\, |\nabla n(r)|^2. \quad (5.6)$$

Because $\int \delta/\delta n(r'')\, d^3r'' = \partial/\partial n$ for a uniform fluid, equation 5.6 can be further simplified to the form

$$\mu - u(r) = \mu(n(r)) - \tfrac{1}{2}\Lambda(n(r))\nabla^2 n(r) - \tfrac{1}{4}\Lambda'(n(r))|\nabla n(r)|^2, \quad (5.7)$$

where

$$\Lambda(n) = -\frac{1}{3}\int \frac{\delta^2 F^B}{\delta n(r)\delta n(r')}\bigg|_n |r' - r|^2\, d^3r',$$

with the notation $|_n$ now indicating the constant value of the uniform density. Alternatively, we can write

$$\Lambda(n) = -\frac{1}{3}\int \frac{\delta\mu - u(r)}{\delta n(r')}\bigg|_n |r' - r|^2\, d^3r' = -\frac{1}{3}\int R(r) r^2\, d^3r, \quad (5.8)$$

which is in terms of the second moment of the linear response function at uniform density $n$:

$$R(r - r') = \delta(\mu - u(r))/\delta n(r'). \quad (5.9)$$

Finally, equation 5.7 can be integrated back to the corresponding approximate density functional

$$F^B[n] = \int \{n(r) f(n(r)) + \tfrac{1}{4}\Lambda(n(r))|\nabla n(r)|^2\}\, d^3r, \quad (5.10)$$

which is the generalization of the van der Waals free energy alluded to above.

## SPECIAL CASES

Equation 5.10 has been applied, in various guises, to classical thermal equilibrium.[11] In this case, the linear response function (equation 5.9) takes the form,

$$\beta R(r - r') = C(r - r') = \frac{\delta(r - r')}{n} - c_2(r - r'), \quad (6.1)$$

where $c_2$ is the direct correlation function of Ornstein and Zernike. Thus,

$$\Lambda(n) = \frac{1}{3\beta}\int c_2(r) r^2\, d^3r, \quad (6.2)$$

and $\Lambda(n)$ is available to high accuracy numerically and to some extent analytically. There is a difficulty in that both $f(n)$ and $\Lambda(n)$ may be required along the phase transition tie line, and it is not clear just how they should be defined. However, reasonable resolutions of this question have produced what appear to be quite reliable density profiles under a number of circumstances.[12]

In the subject matter of major interest to us, namely, the ground state of a noninteracting but nonuniform Fermi gas, the above difficulties do not occur. We need only various uniform ideal gas properties. To start with, $\mu$ is of course the Fermi energy

$$\mu(n) = \hbar^2 k_F^2/2m, \tag{6.3}$$

where

$$k_F = (6\pi^2 n/g)^{1/3}, \tag{6.4}$$

with $g$ being the spin degeneracy. From $\mu = \partial(nf)/\partial n$, we have

$$f(n) = \frac{3}{5}\mu(n) = \frac{3}{10}\frac{\hbar^2 k_F^2}{m}. \tag{6.5}$$

Next, by switching over to a Fourier transform, that is, $\tilde{h}(k) \equiv \int h(r)e^{ik\cdot r}\, d^3r$, let us write equation 5.8 as

$$\Lambda(n) = \tfrac{1}{3}\nabla_k^2 \tilde{R}(k)|_{k\to 0}. \tag{6.6}$$

Because $\delta(\mu - u(r)) = \int R(r - r')\delta n(r')\, d^3r$, we have for $k \neq 0$, $-\delta u(k) = \tilde{R}(k)\delta\tilde{n}(k)$. Hence, by operating in a system of finite volume $V$, and thus over a discrete $k$-lattice,

$$\tilde{R}(k) = -\partial\tilde{u}(k)/\partial\tilde{n}(k). \tag{6.7}$$

Now, the one-body density matrix for an independent Fermion system in external potential $u(r)$ is

$$\Gamma_1 = \theta\left(\mu - \frac{p^2}{2m} - \frac{1}{V}\sum \tilde{u}(k)e^{-ik\cdot r}\right), \tag{6.8}$$

where $\theta$ is the unit step at the origin, and so

$$\tilde{R}(k)^{-1} = -\frac{\partial\tilde{n}(k)}{\partial\tilde{u}(k)} = -\frac{\partial}{\partial\tilde{u}(k)}\left\langle k\left|\theta\left(\mu - \frac{p^2}{2m} - \frac{1}{V}\sum \tilde{u}(\ell)e^{-i\ell\cdot r}\right)\right|k\right\rangle\Bigg|_{u=0}. \tag{6.9}$$

After some algebra, one therefore finds[13]

$$\tilde{R}(k) = \frac{2}{3}\frac{\hbar^2 k_F^2}{mn}\frac{1}{1 + \dfrac{1-\gamma^2}{2\gamma}\ln\left|\dfrac{1+\gamma}{1-\gamma}\right|}, \tag{6.10}$$

where $\gamma = k/2k_F$.

The limiting forms of equation 6.10, which will be important now and in the sequel, are

$$\tilde{R}(k) \to \frac{\hbar^2 k_F^2}{3mn}\left(1 + \frac{\gamma^2}{3} + \cdots\right) \quad \text{for} \quad \gamma \sim 0$$

$$\to \frac{\hbar^2 k^2}{4mn}\left(1 - \frac{1}{5\gamma^2}\cdots\right) \quad \text{for} \quad 1/\gamma \sim 0. \tag{6.11}$$

In particular, we have at once from equation 6.6,

$$\Lambda(n) = \hbar^2/18mn. \tag{6.12}$$

Insertion of equations 6.5 and 6.12 into equation 5.10 then yields

$$F^B[n] = \frac{3}{10}\frac{\hbar^2}{m}\left(\frac{6\pi^2}{g}\right)^{2/3}\int n(r)^{5/2}\, d^3r + \frac{\hbar^2}{72m}\int \frac{|\nabla n(r)|^2}{n(r)}\, d^3r, \tag{6.13}$$

which is recognized as the standard Thomas-Fermi contribution plus the (corrected) Weiszacker inhomogeneity term.[14]

## NONLOCAL PROFILE EXPANSION

In the fourth and sixth sections of this paper, we have found the explicit effect of density inhomogeneity on the system free energy in two regimes—those of rapid and slow variation. For the two accessible situations of classical thermal equilibrium and noninteracting Fermion ground state, they are (as shown in the table below)

| | *rapid* | *slow* |
|---|---|---|
| *classical* | $\int \frac{1}{\beta}[\ln n(r) - 1]n(r)\, d^3r$ | $\int \frac{1}{12\beta} c^{(2)}(n(r))\lvert\nabla n(r)\rvert^2\, d^3r$ |
| *Fermion* | $\int \frac{\hbar^2}{8m}\frac{\lvert\nabla n(r)\rvert^2}{n(r)}\, d^3r$ | $\int \frac{\hbar^2}{72m}\frac{\lvert\nabla n(r)\rvert^2}{n(r)}\, d^3r$ |

where $c^{(2)}$ is the second moment of the direct correlation. In the Fermion case, it is tempting to say that there is a universal form for the inhomogeneity correction, but that the precise nature of the coefficient must be scrutinized more carefully. However, the classical comparison strongly suggests that the Fermion similarity of form is largely coincidence. Another approach has been used—that of carrying out the gradient expansion found in the fifth section of this paper to higher order (exact results to fourth order are known[15])—but it is clear that the transition to the rapidly varying regime cannot be encompassed in this way.

There is perhaps no overwhelming reason for requiring that both rapid and slow regimes be included when only the weakly singular Coulomb forces of atoms and molecules are of interest, but this requirement is in fact very easy to satisfy if one realizes that the basic expression of equation 5.2 is nonlocal and elects to live with the nonlocality. Let us then return to equation 5.2 and see what happens if a gradient expansion is avoided. We will not even go to second order at this time, and hence, we will use equation 5.2 in the form,

$$\mu - u(r) = \mu(n(r)) + \int R(r - r'; n(r))[n(r') - n(r)]\, d^3r', \tag{7.1}$$

where the uniform density at which the linear response function (equation 5.9) is to be evaluated is indicated explicitly. To see that indeed both regimes can be connected, we first look briefly at the classical case, in which equation 7.1 reduces to

$$\mu - u(r) = \mu(n(r)) - \frac{1}{\beta}\int c_2(r - r'; n(r))[n(r') - n(r)]\, d^3r'. \tag{7.2}$$

We now want to check the profile comparison corresponding to the table given earlier in this section:

$$\mu - u(r) = \begin{cases} \cdots + \dfrac{1}{\beta}\ln n(r) + \cdots & \text{rapid} \\ \cdots + \dfrac{1}{12\beta} c^{(2)\prime}(n(r))\,|\nabla n(r)|^2 & \text{slow} \\ \qquad - \dfrac{1}{6\beta} c^{(2)}(n(r))\nabla^2 n(r) + \cdots . & \end{cases} \tag{7.3}$$

Because the "slow" expression was derived from equation 7.1, albeit to one higher order, it is the "rapid" limit that is in question. In equation 7.2, it is clear that a rapid change in $n(r')$ is masked by the integration, so this term can be dropped. In what remains, we use the well-known expression,

$$\frac{1}{\beta}\int c_2(r'; n(r))\, d^3r' = -\frac{1}{n(r)}\frac{\partial}{\partial n(r)} P^{\text{ex}}(n(r)), \tag{7.4}$$

where superscript "ex" denotes the excess over the ideal gas. Thus, equation 7.2 becomes

$$\mu - u(r) = \frac{1}{\beta}\ln n(r) + \mu^{\text{ex}}(n(r)) - n(r)\frac{\partial \mu^{\text{ex}}(n(r))}{\partial n(r)} + \cdots, \tag{7.5}$$

with the last two terms indeed being the beginning of the low-density expansion of $\mu^{\text{ex}}(0) = 0$.

The classical argument is suggestive, but perhaps not fully convincing. The independent Fermion situation, however, is much clearer. It is best to write $R$ in real space. According to equation 6.11, $\tilde{R}(k)$ diverges as $\hbar^2k^2/4mn$ for large $k$, and so we first make the separation

$$\tilde{R}(k; n) = \frac{\hbar^2k^2}{4mn} - \frac{\hbar^2k_F^2}{5mn} + \Delta\tilde{R}(k; n), \tag{7.6}$$

where

$$\Delta\tilde{R}(k, n) \rightarrow -\frac{32}{175}\frac{\hbar^2k_F^4}{mnk^2} \quad \text{for} \quad k_F/k \rightarrow 0.$$

Thus, equation 7.1 becomes

$$\mu - u(r) = \mu(n(r)) + \frac{\hbar^2}{4m}\frac{\nabla^2 n(r)}{n(r)} + \int \Delta R(r - r'; n(r))[n(r') - n(r)]\, d^3r', \quad (7.7)$$

where

$$\Delta R(r; n) \rightarrow a + \frac{b}{|r|} \quad \text{as} \quad r \rightarrow 0,$$

but

$$\frac{1}{3}\int r^2 \Delta R(r; n)\, d^3r = \frac{4}{9}\frac{\hbar^2}{mn}.$$

Now let us look at the two extremes. For rapidly changing $n(r)$ incited by a rapid change of $u(r)$, the small terms in equation 7.7 can be dropped, thus reducing equation 7.7 to

$$\frac{\hbar^2}{4m}\frac{\nabla^2 n(r)}{n(r)} + u(r) \sim 0, \quad (7.8)$$

which lacks the $|\nabla n|^2/n^2$ term of equation 4.12 due to our first-order truncation of equation 5.2. Therefore, the relation of equation 5.2 near a hard wall will not be recovered at this level of approximation. However, the hardest potentials that one anticipates at the atomic level are Coulomb sources. Suppose, in particular, that

$$u(r) = -Ze^2/r. \quad (7.9)$$

Then, by integrating equation 7.8 over a small spherical volume of radius $a$ about the origin, we have

$$[\hbar^2/4mn(0)] \int a^2 \nabla n(r) \cdot d\Omega = Ze^2 \iint_0^a (1/r) r^2\, dr\, d\Omega = \tfrac{1}{2} Ze^2 a^2 \int d\Omega,$$

so

$$n'(r)/n(r)|_{r=0} = (2m/\hbar^2) Ze^2, \quad (7.10)$$

which is the correct "cusp" boundary condition.

For slowly varying density, on the other hand, the singularity at the origin of $R(r; n)$ that gave rise to the $\nabla^2 n/n$ term in equation 7.7 is joined by the rest of the function. As we know, expansion of $n(r')$ about $r' = r$ then yields

$$\int \Delta R(r - r'; n(r))[n(r) - n(r')]\, d^3r' = -\tfrac{1}{6} \int \Delta R(r', n) r'^2\, d^3r' \nabla^2 n(r),$$

so equation 7.7 becomes

$$\mu - u(r) = \mu(n(r)) + \frac{\hbar^2}{4m}\frac{\nabla^2 n(r)}{n(r)} - \frac{2\hbar^2}{9m}\frac{\nabla^2 n(r)}{n(r)}, \quad (7.11)$$

which, except for the omitted $|\nabla n|^2/n^2$ terms, is the corrected Weiszacker term. In other words, the correction in the Weiszacker term is the full contribution of the tail of the linear response function.[16]

## GENERALIZATION

The form of the decomposition of equation 7.6, with its characteristic high $k$ component, is not limited to an interactionless Fermion fluid. Quite generally, suppose we have a system with Hamiltonian

$$H = H_0 + U,$$

$$H_0 = \sum p_i^2/2m + \frac{1}{2}\sum{}' \phi(r_i - r_j),$$

$$U = \sum u(r_j) = \frac{1}{V}\sum \tilde{u}(k)e^{-ik\cdot r_j}, \tag{8.1}$$

and wave function $\Psi$: $H\Psi = E\Psi$. Then, by first-order perturbation theory (in a petit ensemble—extension to a grand ensemble is automatic),

$$\frac{\partial \Psi}{\partial \tilde{u}(k)} = \frac{1}{V}\frac{1}{E - H + i\eta}\sum e^{-ik\cdot r_j}\Psi, \tag{8.2}$$

which is averaged over $\eta$ and $-\eta$ in the limit $\eta \to 0$. Because

$$\tilde{n}(k) = \langle \Psi | \Sigma e^{ik\cdot r_i} | \Psi \rangle, \tag{8.3}$$

it follows that the uniform system response to a potential change is

$$\left.\frac{\partial \tilde{n}(k)}{\partial \tilde{u}(k)}\right|_{u=0} = \frac{1}{V}\left\langle \Psi \left| \sum_{i,j}\left(e^{ik\cdot r_i}\frac{1}{E - H_0 + i\eta}e^{-ik\cdot r_j} + e^{-ik\cdot r_j}\frac{1}{E - H_0 + i\eta}\right)e^{ik\cdot r_i}\right| \Psi \right\rangle, \tag{8.4}$$

which is transformable at once to

$$\left.\frac{\partial \tilde{n}(k)}{\partial \tilde{u}(k)}\right|_{u=0} = \frac{1}{V}\left\langle \Psi \left| \sum_{i,j} e^{ik\cdot(r_i - r_j)}\left[\frac{1}{E - H_0 + i\eta - \hbar k \cdot p_j/m - \hbar^2k^2/2m} + \frac{1}{E - H_0 + i\eta + \hbar k \cdot p_i/m - \hbar^2k^2/2m}\right]\right| \Psi \right\rangle. \tag{8.5}$$

At high $k$, this reduces to

$$\left.\frac{\partial \tilde{n}(k)}{\partial \tilde{u}(k)}\right|_{u=0} = -\frac{4mn}{\hbar^2k^2}S(k) + O[(1/k)^o], \tag{8.6}$$

where

$$S(k) = \frac{1}{N}\left\langle \sum_{i,j} e^{ik\cdot(r_i - r_j)} \right\rangle$$

is the fluid structure factor. Because $\tilde{R}(k) = -\partial\tilde{u}(k)/\partial\tilde{n}(k)$ and one expects $S(k) =$

$1 + O(1/k^2)$, equation 7.6 thus remains valid in the form,

$$\tilde{R}(k; n) = \frac{\hbar^2 k^2}{4mn} + R_0(n) + \Delta\tilde{R}(k; n), \tag{8.7}$$

where $\Delta\tilde{R}$ is a Laurent series in $k^2$.

Indeed, a similar conclusion follows from our previous general discussion of nonuniform Fermion fluids in the fourth section of this paper. For the high wave number dependence of the linear response, we must apply a very rapidly varying potential perturbation. Hence, according to equation 4.12,

$$R(r, r') = \frac{\delta\mu - u(r)}{\delta n(r')} - \frac{\hbar^2}{4m}\left[\frac{\nabla^2\delta(r - r')}{n(r)} - \frac{\nabla^2 n(r)}{n(r)^2}\delta(r - r') - \frac{\nabla n(r)}{n(r)^2}\cdot\nabla\delta(r - r') + \frac{|\nabla n(r)|^2\delta(r - r')}{n(r)^3}\right]$$

$$= -\frac{\hbar^2}{4m}\frac{1}{n(r)}\nabla\cdot n(r)\nabla\frac{1}{n(r)}\delta(r - r'), \tag{8.8}$$

or in operator form,

$$R \simeq -\frac{\hbar^2}{4m}\frac{1}{n}\nabla\cdot n\nabla\frac{1}{n}. \tag{8.9}$$

In particular, if $n$ is uniform, $R = -(\hbar^2/mn)\nabla^2$, which coincides with the beginning of equation 8.7.

The nonuniform result (equation 8.9) is also of interest in its own right. We recall that in the classical case,

$$R(r, r') = \frac{1}{\beta}\left[\frac{\delta(r - r')}{n(r)} - c_2(r, r')\right], \tag{8.10}$$

which is divided into a singular (zero-range) and regular part, with mean kinetic energy as multiplier. This suggests a quantum mechanical analog of the direct correlation to be defined by

$$R = \frac{1}{4m}\frac{1}{n}p\cdot(n - nc_2n)p\frac{1}{n}. \tag{8.11}$$

However, there is the very general linear response relation

$$\nabla u(r) + \int\frac{\delta - u(r)}{\delta n(r)}\nabla n(r')\, d^3r' = 0 \tag{8.12}$$

for systems with translation-invariant internal forces. Inserting equation 8.9, it now reads

$$n(r)\nabla u(r) - \frac{\hbar^2}{4m}\nabla\cdot[n(r)\nabla]\frac{\nabla n(r)}{n(r)} + \frac{\hbar^2}{4m}\nabla\cdot n(r)\int c_2(r, r')n(r')\nabla'\cdot\left[\frac{\nabla' n(r')}{n(r')}\right]d^3r' = 0, \tag{8.13}$$

or simply

$$n(r)\frac{\partial u(r)}{\partial r_\alpha} + \sum \frac{\partial}{\partial r_\beta} P_{\alpha\beta}(r) = 0, \quad \textbf{(8.14)}$$

where the (symmetric) pressure tensor is defined to within an additive constant by

$$P_{\alpha\beta}(r) = -\frac{\hbar^2}{4m}\left(\nabla_\alpha \nabla_\beta n(r) - \frac{\nabla_\alpha n(r) \nabla_\beta n(r)}{n(r)} + n(r) \int c_2(r, r')\left[\nabla'_\alpha \nabla'_\beta n(r') - \frac{\nabla'_\alpha n(r') \nabla'_\beta n(r')}{n(r')}\right] d^3r'\right). \quad \textbf{(8.15)}$$

The question as to whether reasonable models for $c_2$—for example, that of an appropriate uniform system—give reasonable results has not been investigated.

## NONLOCAL FREE ENERGY EXPANSION: DENSITY ESCALATION METHOD

The expansion in equation 7.1 has two obvious defects. To start with, it represents only a first-order expansion and thus is not correct in the limit of very rapidly changing density. Equally important from our viewpoint, equation 7.1 does not come from a free energy [it is not the result of the operation $\delta/\delta n(r)$] and thus will certainly lack some thermodynamic consistency. In fact, the two are related because the profile equation (equation 5.7) derived from the free energy (equation 5.10) already contains second-order corrections. What we would like is the format of equation 5.7—a decomposition into extended local quantities—for the free energy as well. Because there is no single uniform fluid that can reasonably be used throughout, this is a problem.

A simple formal solution has been suggested by a number of authors,[17] which we will take advantage of. In a way, the uniform reference fluid is taken here as that at zero density, but then the transition between it and the real fluid must be handled intelligently. Suppose that a parameter $\lambda$ is used to turn up the free energy from its zero-density value of 0 to its real fluid value:

$$F^B = F^B(1) = \int_0^1 \partial F^B(\lambda)/\partial\lambda \, d\lambda. \quad \textbf{(9.1)}$$

If the state is described at parameter value $\lambda$ by its density profile $n_\lambda(r)$, we may write instead

$$F^B = \int_0^1 \!\! \int \frac{\partial n_\lambda(r)}{\partial \lambda} \frac{\delta F^B[n_\lambda]}{\delta n_\lambda(r)} d^3r \, d\lambda = \int \left[\int_0^1 \frac{\partial n_\lambda(r)}{\partial \lambda} \frac{\delta F^B[n_\lambda]}{\delta n_\lambda(r)} d\lambda\right] d^3r, \quad \textbf{(9.2)}$$

which, at least formally, has the desired local character. The nature of $n_\lambda(r)$ is totally arbitrary, except that $n_0(r) = 0$ and $n_1(r) = n(r)$, but a particularly convenient path is given by

$$n_\lambda(r) = \lambda n(r), \quad \textbf{(9.3)}$$

so equation 9.2 becomes

$$F^B = \int n(r) \int_0^1 \frac{\delta F^B[n_\lambda]}{\delta n_\lambda(r)}\, d\lambda\, d^3r = \int n(r) \int_0^1 [\mu_\lambda - u_\lambda(r)]\, d\lambda\, d^3r. \quad (9.4)$$

Now, the local quantity $\mu_\lambda - u_\lambda(r)$ can be expanded precisely as in equation 5.2:

$$F^B = \sum \frac{1}{s!} \iint_0^1 \lambda^s R(r, r_1, \ldots, r_s | \lambda n(r))\, d\lambda \prod_1^s [n(r_i) - n(r)] n(r)\, dr^{3s+3}, \quad (9.5)$$

where

$$R(r_1, \ldots, r_t | n) \equiv \left. \frac{\delta^t F^B}{\delta n(r_1) \ldots \delta n(r_t)} \right|_n = \left. \frac{\delta^{t-1}(\mu - u(r_1))}{\delta n(r_2) \ldots \delta n(r_t)} \right|_n$$

is the $t$-th order linear response at uniform density $n$. On carrying out the $\lambda$ integration

$$\overline{R}(r_1, \ldots, r_t | n) \equiv t \int_0^1 \lambda^{t-1} R(r_1, \ldots, r_t | \lambda_n)\, d\lambda, \quad (9.6)$$

we thus have

$$F^B = \sum \frac{1}{s+1!} \int \overline{R}(r, r_1, \ldots, r_s | n(r)) \prod_1^s [n(r_i) - n(r)] n(r)\, dr^{3s+3}. \quad (9.7)$$

In particular, let us truncate at $s = 1$. The $s = 0$ term is trivial to find, and we may symmetrize with respect to $r$ and $r_1 \equiv r'$ in the $s = 1$ term to yield

$$F^B = \int n(r) f(n(r))\, d^3r - \tfrac{1}{4} \iint [n(r) - n(r')][n(r)\overline{R}(r - r' | n(r))] - n(r')\overline{R}(r - r' | n(r'))\, d^3r\, d^3r' + \cdots. \quad (9.8)$$

Equation 9.8 represents the required first-order generalization. It indeed reduces to equation 5.10 under a gradient expansion. For some feeling as to the implications of equation 9.8, we specialize again to the free Fermion case, whose basic content is contained in the expression of equation 6.10. This is most conveniently written as

$$R(k|\gamma) = \frac{8\pi^2\hbar^2}{mg\,k} \frac{\gamma}{1 + \dfrac{1+\gamma^2}{2\gamma} \ln \left| \dfrac{\gamma+1}{\gamma-1} \right|}, \quad (9.9)$$

where $n = (g/4\gamma\pi)k^3/\gamma^3$. Thus, $\overline{R}(k|\gamma) = 2 \int_0^1 \lambda R(k|\gamma\lambda^{-1/3})\, d\lambda$ or

$$\overline{R}(k|\gamma) = 6\gamma^6 \int_\gamma^\infty R(k'|\gamma') \frac{d\gamma'}{\gamma'^7} \quad (9.10)$$

is to be inserted into equation 9.8. At this time, we will be content with a small $n$ or large $\gamma$ limit, in which case, equations 9.9 and 9.10 are readily expanded to read

$$\begin{aligned}\overline{R}(k|\gamma) &= \frac{24\pi^2\hbar^2}{mg\,k}\left(\gamma^3 - \frac{3}{25}\gamma - \frac{24}{1225}\frac{1}{\gamma}\cdots\right)\\ &= \frac{\hbar^2k^2}{2mn}\left(1 - \frac{12}{25}\frac{k_F^2}{k^2} - \frac{384}{1225}\frac{k_F^4}{k^4}\cdots\right),\end{aligned} \tag{9.11}$$

so

$$\overline{R}(r|n) = \frac{\hbar^2}{2mn}\left[-\nabla^2\delta(r) - \frac{12}{25}k_F^2\delta(r) - \frac{96}{1225\pi}k_F^4\frac{1}{r}\cdots\right]. \tag{9.12}$$

Thus, equation 9.8 now becomes

$$\begin{aligned}F^B = \frac{3}{10}\frac{\hbar^2}{m}\left(\frac{6\pi^2}{g}\right)^{2/3}\int n(r)^{5/2}\,d^3r + \frac{24}{1225\pi}\frac{\hbar^2}{m}\left(\frac{6\pi^2}{g}\right)^{4/3}\\ \cdot\iint\frac{[n(r)^{4/3} - n(r')^{4/3}][n(r) - n(r')]}{|r - r'|}\,d^3r\,d^3r',\end{aligned} \tag{9.13}$$

which is significantly different from equation 6.13.

## NONLOCAL FREE ENERGY EXPANSION: DIRECT METHOD

The result of equation 9.7 is, as we have noted, clearly nonunique. Its utility derives from the fact that the input information consists of potentially available uniform system quantities; its disadvantage from the requirement that these be supplied over the full range of densities. This is not the case for profile equation 5.2, and so it would be of interest to know whether such a quasi-local form exists for the free energy as well. The form obtained, however, depends very much on the preferred ordering of the resulting series expansion. For example, one can start with the familiar zero-density functional Taylor expansion,

$$F^B = \sum_1^\infty \frac{1}{s!}\int\cdots\int R(r, r_1, \ldots, r_{s-1}|0)n(r)n(r_1)\ldots n(r_{s-1})\,d^3r\ldots d^3r_{s-1}, \tag{10.1}$$

and process it appropriately. The key step is the observation that led to equation 5.7: because an increase of $dn$ in uniform density is equivalent to an increase of $dn$ at each point in space, then

$$\begin{aligned}\frac{\partial}{\partial n}R(r_1, \ldots, r_t|n) &= \int \frac{\delta R(r_1, \ldots, r_t)}{\delta n(r_{t+1})}\bigg|_n d^3r_{t+1}\\ &= \int R(r_1, r_2, \ldots, r_{t+1}|n)\,d^3r_{t+1}.\end{aligned} \tag{10.2}$$

Thus, we have

$$R(r, r_1, \ldots, r_{s-1}|0) = \sum \frac{1}{t!}(-n)^t \left(\frac{\partial}{\partial n}\right)^t R(r, r_1, \ldots, r_{s-1}|n)$$

$$= \sum \frac{1}{t!}(-n)^t \int \cdots \int R(r, r_1, \ldots, r_{s+t-1}|n)\, d^3r_s \ldots d^3r_{s+t-1}, \quad \textbf{(10.3)}$$

thus transforming equation 10.1 at once to

$$F^B = \sum_{u=1}^{\infty} \int \cdots \int R(r, r_1, \ldots, r_{u-1}|n(r)) \left[\sum_{s=1}^{u} \frac{(-1)^{u-s}}{u - s!\, s!} \right.$$

$$\left. \cdot\, n(r)^{u-s+1} n(r_1) \ldots n(r_{s-1})\right] d^3r \ldots d^3r_{u-1}. \quad \textbf{(10.4)}$$

Equation 10.4 is an expansion in order of linear response starting as

$$F^B = \int \mu(n(r))n(r)\, d^3r - \tfrac{1}{2} \iint R(r, r'|n(r))n(r)[2n(r) - n(r')]\, d^3r\, d^3r'. \quad \textbf{(10.5)}$$

It can readily be rewritten as

$$F^B = \int \{\mu(n(r)) - \tfrac{1}{2}n(r)\mu'(n(r))\}n(r)\, d^3r$$

$$- \tfrac{1}{4} \int \{n(r)R(r, r'|n(r)) - n(r')R(r', r|n(r'))$$

$$\cdot\, [n(r) - n(r')]\}\, d^3r\, d^3r' \ldots, \quad \textbf{(10.6)}$$

which is very similar to equation 9.7. Equation 10.6 is now the prototype for expansions in density differences, for example, $n(r) - n(r')$. To obtain these, we need only replace $n(r_i)$ by $n(r) + [n(r_i) - n(r)]$ and expand, thereby using the symmetry in $n(r_1), \ldots, n(r_{s-1})$ to reduce each term to leading products:

$$n(r_1) \ldots n(r_{s-1}) \rightarrow \sum_{t=0}^{s-1} \binom{s-1}{t} n(r)^{s-1-t}[n(r_1) - n(r)] \ldots [n(r_t) - n(r)]. \quad \textbf{(10.7)}$$

Hence,

$$F^B = \sum_{u=1}^{\infty} \int \cdots \int R(r, r_1, \ldots, r_{u-1}|n(r)) \sum_{t=0}^{u-1} \frac{n(r)^{u-t}}{t!} [n(r_1) - n(r)] \ldots$$

$$\cdot\, [n(r_t) - n(r)] \sum_{s=t+1}^{u} \frac{(-1)^{u-s}}{u - s!\, s - 1 - t!} \frac{1}{s}\, d^3r\, d^3r_1 \ldots d^3r_{u-1}$$

$$= \sum_{u=1}^{\infty} \int \cdots \int \frac{1}{u!} R(r, r_1, \ldots, r_{u-1}|n(r)) \sum_{t=0}^{u-1} (-1)^{u+t+1} n(r)^{u-t}$$

$$\cdot \prod_{1}^{t} [n(r_j) - n(r)]\, d^3r\, d^3r_1 \ldots d^3r_{u-1}$$

$$= \sum_{t=0}^{\infty} \int \cdots \int \left[\sum_{u=t+1}^{\infty} R(r, r_1, \ldots, r_{u-1}|n(r))\, d^3r_{t+1} \ldots d^3r_{u-1}/u! \right.$$

$$\left. \cdot\, (-1)^{u+t+1} n(r)^{u-t}\right] \prod_{1}^{t} [n(r_j) - n(r)]\, d^3r\, d^3r_1 \ldots d^3r_t,$$

or simply

$$F^B = \sum_{t=0}^{\infty} \int \cdots \int \sum_{u=t+1}^{\infty} \frac{1}{u!} (-1)^{u-t+1} n(r)^{n-t} \cdot \left(\frac{\partial}{\partial n(r)}\right)^{u-t-1} R(r, r_1, \ldots, r_t | n(r)) \cdot \prod_1^t [n(r_j) - n(r)] \, d^3r \, d^3r_1 \ldots d^3r_t. \quad \mathbf{(10.8)}$$

If the $u$-summations in equation 10.8 are performed, then by using the relation

$$\frac{\partial}{\partial n}\left[n^t \sum_{u=t+1}^{\infty} \frac{1}{u!} (-1)^{u-t+1} n^{u-t} \left(\frac{\partial}{\partial n}\right)^{u-t-1} R(r, \ldots, r_t | n)\right] = \frac{1}{t!} n^t R(r_1 \ldots r_t | n), \quad \mathbf{(10.9)}$$

equation 10.8 is indeed seen to be identical with equation 9.7.

It is also possible to strike a mean between a density difference expansion (requiring a spread of values of $n$) and a linear response ordered expansion (settling down to the correct uniform value only on carrying out infinite summation) in at least two different ways. The simplest way is to carry out the infinite summation only in the leading term of equation 10.8, thus yielding the approximation

$$F^B = \int f(n(r))n(r) \, d^3r - \tfrac{1}{4} \int \{n(r)R(r, r' | n(r)) - n(r')R(r, r' | n(r'))\} \cdot [n(r) - n(r')] \, d^3r \, d^3r' \ldots . \quad \mathbf{(10.10)}$$

A possible improvement involves a main component $F_0$, which is exactly known (for example, the singular high-gradient or low-density equation 4.13), a separation $F^B = F_0 + F^{ex}$, and then the expansion of the excess $F^{ex}$ as in equation 10.6.

## COULOMB SOURCE—NUMERICAL COMPARISON

We have thus far been concerned primarily with formal expansions. To get some feeling for the adequacy of the resulting approximations, let us check against cases that mimic, in some fashion, the atomic and molecular systems that are our goals. In particular, we consider a closed shell "atom" in which the interelectronic interaction is turned off. Among the various options for comparison, we shall choose that which is numerically and conceptually simplest: we inquire as to how well the exactly known density profile in such a case satisfies the profile equations resulting from various members of our approximation sequence.

Our test then will be for $Z$ noninteracting spin-½ ($g = 2$) electrons in the field of a single Coulomb source

$$u(r) = -\frac{Ze^2}{r}. \quad \mathbf{(11.1)}$$

Restricting our attention to closed shell configurations, the hydrogenic densities are readily computed analytically and one finds for the successive shells:

$$n_K(r) = \left(\frac{Z}{a_0}\right)^3 \frac{2}{\pi} e^{-2R},$$

$$n_L(r) = \left(\frac{Z}{a_0}\right)^3 \frac{1}{8\pi}(2 - 2R + R^2)e^{-R},$$

$$n_M(r) = \left(\frac{Z}{a_0}\right)^3 \frac{2}{243\pi}\left(9 - 12R + 8R^2 - \frac{16}{9}R^2 + \frac{4}{27}R^4\right)e^{-2R/3}, \quad \textbf{(11.2)}$$

where

$$R = Zr/a_0, \qquad a_0 = \hbar^2/me^2. \quad \textbf{(11.3)}$$

It will correspondingly be convenient to scale length, density, wave number, and energy to produce nondimensional quantities:

$$r = \frac{a_0}{Z}R, \qquad n(r) = \left(\frac{Z}{a_0}\right)^3 N(R),$$

$$k_F(r) = \frac{Z}{a_0}K(R) \quad \text{or} \quad K(R) = [3\pi^2 N(R)]^{1/3},$$

$$\mu = (Z^2e^2/a_0)M. \quad \textbf{(11.4)}$$

The first two total densities, "mock-helium" with $Z = 2$ and "mock-neon" with $Z = 10$, are shown in FIGURE 2. They are chosen not because they are analytically simpler (the distinction is irrelevant for subsequent numerical calculations), but because an electron fluid format is most suspect when only a few electrons are involved.

We can start our comparison with the straight Thomas-Fermi approximation, according to which the profile is given (see equations 3.6 and 6.3) by

$$\mu \simeq -\frac{Ze^2}{r} + \frac{\hbar^2}{2m}(3\pi^2 n)^{2/3} \quad \textbf{(11.5)}$$

or, in scaled form,

$$M \simeq -\frac{1}{R} + \frac{1}{2}K(R)^2 \equiv M_{TF}(R). \quad \textbf{(11.6)}$$

A plot of $M_{TF}(R)$ (curve A in FIGURE 3) for our two test cases shows the expected large deviation from constancy. Working our way up slowly, we can then apply the first-order Weiszacker correction (see equation 7.11)

$$M \simeq -\frac{1}{R} + \frac{1}{2}K(R)^2 - \frac{1}{4}\frac{\nabla^2 N(R)}{N(R)} \equiv M_{TFW}(R), \quad \textbf{(11.7)}$$

which results in a substantially better tendency to constancy (curve $B$ in FIGURE 3), but with a large overshoot near the nucleus. It is apparent that the modified Weiszacker correction—the full equation 7.11—would produce an even larger undercompensation.

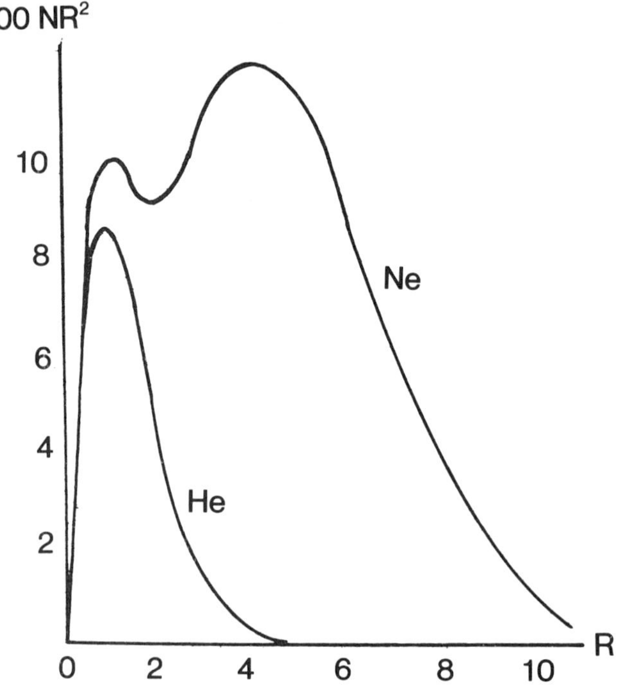


**FIGURE 2.** Profiles for interactionless atoms.

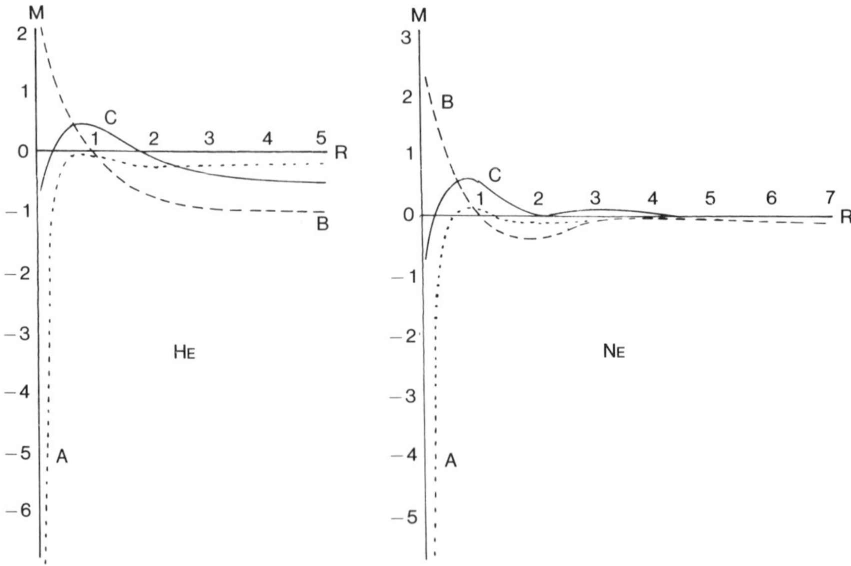


**FIGURE 3.** Scaled chemical potentials for profile approximations: (A) Thomas-Fermi; (B) First-order Weiszacker correction; (C) First-order linear response correction. The left-hand figure is for He; the right-hand one is for Ne.

In order to apply any of the systematic expansions that we have discussed, we must start by writing the linear response in real rather than Fourier space. It is necessary to separate out the singular spatial dependence and so, as in equation 7.6, we set

$$\tilde{R}(k; n) = \frac{\hbar^2 k^2}{4mn} - \frac{\hbar^2 k_F^2}{5mn} + \frac{4\pi\hbar^2}{m}\frac{1}{k}\xi(\gamma), \tag{11.8}$$

where

$$\xi(\gamma) = \frac{\gamma}{1 + \dfrac{1-\gamma^2}{2\gamma}\ln\left|\dfrac{1+\gamma}{1-\gamma}\right|} + \frac{3}{10}\gamma - \frac{3}{2}\gamma^3,$$

$\gamma = k/2k_F$, and $k_F = (3\pi^2 n)^{1/3}$. The spatial non-Weiszacker correction is then

$$\begin{aligned}\Delta R(r; n) &= (1/2\pi)^3 \int \Delta\tilde{R}(k; n)e^{-ik\cdot r}\, d^3k \\ &= (\hbar^2/2\pi m) \int \xi(k/2k_F)e^{-ik\cdot r}/k\, d^3k \end{aligned} \tag{11.9}$$

or

$$\Delta R(r; n) = \frac{8\hbar^2 k_F^2}{m} D(2k_F r),$$

where

$$D(r) = \frac{1}{r}\int_0^\infty \xi(\gamma) \sin \gamma r\, d\gamma.$$

A numerical ploy helps speed the computation of $D(r)$. Because $\xi(\gamma)$ is odd in $\gamma$, we can write instead

$$r\, D(r) = \frac{1}{2i}\int_{-\infty}^{\infty} \xi(\gamma)e^{i\gamma r}\, d\gamma. \tag{11.10}$$

However, $\xi(\gamma)$ is analytic for $|\mathrm{Re}(\gamma)| > 1$ [i.e., for $\mathrm{Re}(\gamma) > 1$, set $(\gamma + 1)/(\gamma - 1) = \rho e^{i\theta}$, where $\rho > 1$, and observe that the denominator of $\xi(\gamma)$ vanishes only if $\ln \rho + i\theta = 1/2(\rho e^{i\theta} + e^{-i\theta}/\rho)$; this implies $\theta/\sin\theta = 1/2(\rho + 1/\rho)$, which cannot be], and decreases as $1/\gamma$ for large $\gamma$. Thus, the contour in equation 11.10 can be distorted to give

$$r\, D(r) = \int_{-1}^{1} \xi(\gamma)e^{i\gamma r}\, d\gamma/2i + 1/2 e^{ir}\int_0^\infty \xi(1 + iz)e^{-zr}\, dz - 1/2 e^{-ir}\int_0^\infty \xi(-1 + iz)e^{-zr}\, dz$$

or

$$r\, D(r) = \int_0^1 \xi(\gamma) \sin \gamma r\, d\gamma + \mathrm{Re}\int_0^\infty e^{ir}\zeta(1 + iz)e^{-zr}\, dz, \tag{11.11}$$

which is readily computed. Note the nontrivial structure with both long-range and characteristic Friedel oscillations (see FIGURE 4).[18]

We can now proceed to the full first-order profile correction and, at no extra cost, include the second-order Weiszacker term (see equation 4.12). Thus,

$$\mu \simeq u(r) + \frac{\hbar^2}{2m} k_F(r)^2 - \frac{\hbar^2}{4m}\frac{\nabla^2 n(r)}{n(r)} + \frac{\hbar^2}{8m}\left|\frac{\nabla n(r)}{n(r)}\right|^2 + \frac{8\hbar^2}{m} k_F(r)^2 \int D(2|r - r'|k_F(r))[n(r') - n(r)]\, d^3r', \quad \textbf{(11.12)}$$

or by scaling via equation 11.4 and by using the usual bipolar reduction of the integral,

$$M \simeq -\frac{1}{R} + \frac{1}{2}K(R)^2 - \frac{1}{4}\frac{\nabla^2 N(R)}{N(R)} + \frac{1}{8}\left|\frac{\nabla N(R)}{N(R)}\right|^2 + \frac{8\pi K(R)}{R} \cdot \iint_\Delta (RD)(2R'K(R))(N(R'') - N(R))R''\, dR'\, dR'' \equiv M(R), \quad \textbf{(11.13)}$$

where $(RD)(x) \equiv x\,D(x)$ and $\Delta$ signifies $|R' - R''| \leq R \leq R' + R''$. The resulting

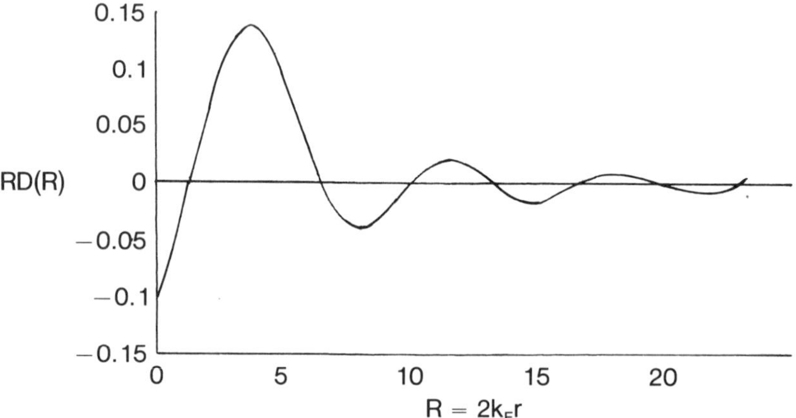


**FIGURE 4.** Linear response function $RD(R)$.

curves (C in FIGURE 3) are a considerable improvement over those of A and B, and by virtue of the second-order Weiszacker term, they even approach the correct constant value of $M$ asymptotically, that is, the energy of the top electron.

It seems clear that the linear response sequence is a promising technique for providing reliable electron densities. It is not clear whether the best direction to follow is that of imposing consistency at each level (in the fashion of equations 9.8 and 10.10) or of proceeding quickly to higher-order approximations. Of course (unless one sticks to mean field-type interaction energies), the reliability of electron plasma response functions is hardly assured. Efforts along all three lines are now being actively pursued.

## ACKNOWLEDGMENTS

The writer would like to thank C. Borzi, A. G. Percus, and G. O. Williams for their help in the numerical computations required for this paper.

## REFERENCES

1. See, for example: PERCUS, J. K. 1982. *In* Liquid State of Matter. E. W. Montroll & J. L. Lebowitz, Eds. North-Holland. Amsterdam.
2. HOHENBERG, P. & W. KOHN. 1964. Phys. Rev. **136:** B864.
3. KOHN, W. & L. J. SHAM. 1965. Phys. Rev. **140:** A1133.
4. See, for example: LANGRETH, D. C. & M. J. MEHL. 1983. Phys. Rev. **B28:** 1809. This reference brings references 2 and 3 up-to-date.
5. VON WEISZACKER, C. V. 1935. Z. Phys. **96:** 431.
6. THOMAS, L. H. 1926. Proc. Cambridge Philos. Soc. **23:** 542.
7. FERMI, E. 1928. Z. Phys. **49:** 73.
8. VAN DER WAALS, J. D. 1944. Z. Phys. Chem. **13:** 657.
9. See also: HOFFMANN-OSTENHOF, M. & T. HOFFMANN-OSTENHOF. 1977. Phys. Rev. **A16:** 1782.
10. LEBOWITZ, J. L. & J. K. PERCUS. 1963. J. Math. Phys. **4:** 248.
11. See, for example: DAVIS, H. T. 1977. J. Chem. Phys. **67:** 3636.
12. EBNER, C., W. F. SAAM & D. STROUD. 1976. Phys. Rev. **A14:** 2264.
13. See, for example: FETTER, A. L. & J. D. WELECKA. 1971. Quantum Theory of Many-Particle Systems. McGraw–Hill. New York.
14. KOMPONEETS, A. S. & E. S. PAVLOVSKI. 1957. Sov. Phys. JETP **4:** 328.
15. HODGES, C. H. 1973. Can. J. Phys. **51:** 1428.
16. See also: JONES, W. & W. H. YOUNG. 1971. J. Phys. **C4:** 1322.
17. See, for example: SAAM, W. F. & C. EBNER. 1977. Phys. Rev. **A15:** 2566.
18. See, for example: MAHAN, G. D. 1981. Many-Particle Physics. Plenum. New York.

# Spin Ordering in a Homogeneous Fermi Liquid

EUGENE M. CHUDNOVSKY

*Prospekt Gagarina 199*
*Kharkov 310080, Union of Soviet Socialist Republics*

## INTRODUCTION

The term "Fermi liquid" is applied to a system of nonlocalized interacting fermions when the average energy of the two-particle interaction is not small in comparison with the average one-particle kinetic energy. Such systems are liquid $^3$He, itinerant electrons in metals, and nuclear matter. Itinerant electrons of pure metals and their compounds represent the widest class of Fermi liquids. Properties of the electron liquid vary when passing from one solid to another because of changes in the electron density and in the effective interaction between the electrons. In other words, the number of different electron Fermi liquids is equal to the number of different metallic compounds. Therefore, one can expect the existence of a great number of different ground states of the electron Fermi liquid in solids.

Fermi liquids can be homogeneous or inhomogeneous depending on whether the probability to find a fermion in the state with given quantum numbers depends on spatial coordinates or not. More exactly, Fermi liquids can be considered as homogeneous if the one-particle density matrix does not depend on coordinates. In solids, it is clear that inhomogeneity of the electron liquid is inevitable because the local electron density correlates with the atomic structure. Itinerant electrons of transition metals originate from the sharing of *s* and *d* (or *f* for rare earth metals) atomic shells. Degrees of sharing for these two groups of electrons are essentially different. The *s*-electrons are approximately homogeneously distributed in space, while the local density of *d*- and *f*-electrons has appreciable maxima at the atomic sites. As a result, in transition metals, the energy band of *s*-electrons is several times wider than the energy band of *d*- or *f*-electrons, and the average kinetic energy of *s*-electrons is greater than the average kinetic energy of *d*- or *f*-electrons. On the other hand, interaction between *d*- or *f*-electrons should be stronger than for *s*-electrons because the number of *d*- or *f*-electrons per atom, and, hence, their density, is greater than the number of *s*-electrons.

However, it is well known that weak attractive interaction between fermions can lead to a superconductivity. Here, we will restrict ourselves to the consideration of nonsuperconducting ground states. In this case, one can hardly expect any ordering in a Fermi liquid if the interaction energy is not large enough in comparison with the kinetic energy of the fermions. According to the above remarks, the ordered state resulting from electron-electron interaction could occur with a greater probability in narrow *d*- or *f*-bands of transition metals.

Using a simplest approach, the electrons of narrow bands are considered as heavy quasiparticles homogeneously distributed in space. Ferromagnetic state,[1] charge density wave (CDW),[2] spin density wave (SDW),[3] and other ground states appear in this model as a result of some definite properties of the two-particle interaction and one-particle energy spectrum. All such approaches are obviously defective in neglecting the effect of the correlation between the local electron density and the atomic structure. Nevertheless, they give correct qualitative description of some important properties of the electron liquid. The reason for this is as follows. The ordered state of a Fermi liquid is characterized by the violation of some symmetry. There are two reasons for the breaking of some symmetry in the ground state of nonrelativistic many-particle systems as well as in relativistic field theory. First, the symmetry can be violated spontaneously. In this case, the symmetry of the ground state is lower than the symmetry of the interaction. Secondly, symmetry of the ground state can be trivially violated by a nonsymmetric interaction. The latter case corresponds to the violation of the translational symmetry of the electron liquid by a crystal field. On the other hand, SDW (which corresponds to the ground state of itinerant electrons in chromium) appears due to a spontaneous symmetry violation in spatially homogeneous Fermi liquid, and its structure parameter does not correlate with an interatomic size. Thus, it is clear that investigation of spontaneous symmetry violation in a homogeneous Fermi liquid becomes important when studying possible states of an electron liquid in transition metals and their compounds.

In the absence of spin ordering, the one-particle density matrix, which defines electron quantum occupation numbers, does not depend on spin variables. When the symmetry about spin rotations is violated, the one-particle density matrix can be represented (as any $2 \times 2$ matrix) as a sum of two terms, $n_{\alpha\beta} = n_1 \delta_{\alpha\beta} + \vec{n}_2 \vec{\sigma}_{\alpha\beta}$, where $\vec{\sigma}_{\alpha\beta}$ are Pauli matrices. Spontaneous violation of the homogeneity of electron charge density, for example, in the case of CDW, would mean that $n_1$ depends on coordinates. Contrary to this case, the electron liquid can be homogeneous in charge density, but inhomogeneous in electron spin density, for example, for SDW, when only $\vec{n}_2$ depends on coordinates. Here, we will consider completely homogeneous ground states of Fermi liquid for which both $n_1$ and $n_2$ do not depend on coordinates.

The most studied ground state of this type is itinerant ferromagnetism. In this case, the one-particle density matrix has the form, $n_{\alpha\beta} = n_1 \delta_{\alpha\beta} + n_2 \vec{\nu} \vec{\sigma}_{\alpha\beta}$, where $\vec{\nu}$ is a unit vector in the direction of the magnetization. The energy band is split into the two bands for the electrons with spins oriented along and opposite $\vec{\nu}$. This model for the first time was applied to the description of ferromagnetism in the iron metal group after it had become clear that the number of ordered spins per atom was fractional for these ferromagnets. The latter phenomenon cannot be understood within the model of localized spins. Nevertheless, the model of localized spins is complementary to the itinerant model when considering neutron scattering or spin waves in the iron metal group and in *f*-metals. Distribution of the local spin density in these metals has maxima at atomic sites that follow maxima in charge density distribution of *d*- or *f*-electrons. Real itinerant ferromagnets, for which neutron scattering does not reveal the existence of localized spins, were discovered in the last two decades. They are the so-called weak itinerant ferromagnets (e.g., $ZrZn_2$, $Sc_3In$, $Ni_3Ga$, and $CrBe_{12}$), which are characterized by small spin splitting of the energy band in comparison with the bandwidth.[4]

One of the possible descriptions of weak itinerant ferromagnetism can be done in a modified theory of Landau's Fermi liquid.[5,6] The nonmagnetic state of Landau's Fermi liquid is unstable if some conditions are fulfilled for spherical harmonics of the amplitude $\vec{\sigma}_1\vec{\sigma}_2\psi(\vec{p}_1, \vec{p}_2)$ of the two-particle forward-exchange scattering ($\vec{p}_1, \vec{p}_2$ are momenta of interacting fermions).[7] In particular, the zero harmonic $\psi_0$ must be negative and large enough in the absolute value for ferromagnetism in a metal. It should be noted, however, that calculation of $\psi(\vec{p}_1, \vec{p}_2)$ for a given alloy and even for a pure metal is a very complicated task. Therefore, one should ad hoc assume negativity and large magnitude of $\psi_0$ when considering ferromagnetism of a metal in Landau's Fermi liquid theory.

An intriguing aspect of the Fermi liquid theory is the possibility of other than ferromagnetic types of spin ordering. These result from the instability of the paramagnetic state with respect to large negative values of $\psi_1$ or some other harmonic of $\psi(\vec{p}_1, \vec{p}_2)$. One of these types of ordering is spontaneous violation of reflectional symmetry,[8] when the one-particle density matrix takes the form, $n_{\alpha\beta} = n_1\delta_{\alpha\beta} + n_2\vec{k}\vec{\sigma}_{\alpha\beta}$, where $\vec{k} = \vec{p}/p$. Spontaneous reflectional symmetry violation (SRSV) is one of the general possibilities for the ground state of the electron liquid. In this case, the energy band is split into two subbands corresponding to different electron helicities. As a result, the majority of electrons have one definite sign of helicity. We strongly believe that SRSV is realized in some transition metal alloys, despite the fact that this state is not yet found. Identification of SRSV in solids is not as easy a task as identification of the ferromagnetism. It should be based on the possibility of proportionality between polar and axial vectors for such a solid; for example, that occurring in some nonequilibrium states of an electric current flowing in the direction of a magnetic field.[9] These symmetry-violating effects are small, however, so their direct observation is not an easy task for experimentalists. Macroscopic magnetic properties of a solid having SRSV can be characterized as itinerant antiferromagnetism or more exactly as itinerant metamagnetism[10] of the type that recently has been discovered in $TiBe_2$.[11]

This paper is organized as follows. Basic relations of Landau's Fermi liquid theory and its modification necessary for the study of spin waves with quadratic dispersion law are discussed in the next section. Ground state, temperature, and field effects in a ferromagnetic Fermi liquid are considered in the third section. Spin waves are then studied in the fourth section. The microscopic analogue for the model of ferromagnetic Fermi liquid based on Green-function method is investigated in the fifth section. The simplest Fermi liquid model of SRSV is considered in the sixth section, while the seventh section contains a discussion of the macroscopic consequences of SRSV in solids. More complicated types of spin ordering in Fermi liquids are introduced and studied in the eighth section. The final section contains some concluding remarks that summarize our results and that may be useful for experimentalists.

Our main purpose is to gather in one place some separate results and to give a common view on spin ordering in a homogeneous Fermi liquid. Thus, only the fifth and eighth sections of this paper fill some gaps and may be considered as new.

## BASIC RELATIONS

The phenomenological theory of Fermi liquids was formulated by Landau, at first for liquid $^3He$.[12] Then, it was spread to electron liquids in metals.[13] In this theory, the equilibrium quasiparticle density matrix has the form of the Fermi distribution,

$$n(\vec{\mathrm{p}}) = \left[1 + \exp\frac{\mathscr{E}(\vec{\mathrm{p}}) - \mu}{T}\right]^{-1}, \tag{2.1}$$

where $\mathscr{E}(\vec{\mathrm{p}})$ is a $2 \times 2$ spin matrix corresponding to the effective quasiparticle Hamiltonian in the momentum representation, $\mu$ is the chemical potential, and $T$ is the temperature. Deviation of Fermi liquids from the thermodynamic equilibrium is described in terms of deviation $\delta n_\sigma(\vec{\mathrm{p}}, \vec{x}, t)$ of the density matrix from the equilibrium matrix in equation 2.1. For small $\delta n$, the variation in the energy density of a Fermi liquid is also small. With the accuracy to the second order in $\delta n$, this variation is given by

$$\delta E(\vec{x}, t) = \mathrm{Sp} \int \frac{d^3\mathrm{p}}{(2\pi\hbar)^3} \mathscr{E}(\vec{\mathrm{p}}) \delta n_\sigma(\vec{\mathrm{p}}, \vec{x}\ t)$$
$$+ \frac{1}{2} \mathrm{Sp}_\sigma \mathrm{Sp}_{\sigma'} \int \frac{d^3\mathrm{p}}{(2\pi\hbar)^3} \int \frac{d^3\mathrm{p}'}{(2\pi\hbar)^3} F_{\sigma\sigma'}(\vec{\mathrm{p}}, \vec{\mathrm{p}}')\, \delta n_\sigma(\vec{\mathrm{p}}, \vec{x}, t)\, \delta n_{\sigma'}(\vec{\mathrm{p}}', \vec{x}, t), \tag{2.2}$$

where $F_{\sigma\sigma'}(\vec{\mathrm{p}}, \vec{\mathrm{p}}')$ describes the interaction between quasiparticles and coincides with the amplitude of forward scattering. Quasiparticle energy can be found as the functional derivative of $\delta E$ with respect to $\delta n$:

$$\mathscr{E}_\sigma(\vec{\mathrm{p}}, \vec{x}, t) = \frac{\delta E}{\delta n} = \mathscr{E}(\vec{\mathrm{p}}) + \mathrm{Sp}_{\sigma'} \int \frac{d^3\mathrm{p}'}{(2\pi\hbar)^3} F_{\sigma\sigma'}(\vec{\mathrm{p}}, \vec{\mathrm{p}}') \delta n_\sigma(\vec{\mathrm{p}}', \vec{x}, t). \tag{2.3}$$

Here, the second term represents the variation in the quasiparticle energy caused by the small deviation of the density matrix from the thermodynamic equilibrium.

Notice that the theory is valid in the limit of small spatial gradients of $\delta n_\sigma(\vec{\mathrm{p}}, \vec{x}, t)$. In this case, the density matrix satisfies the quantum kinetic equation,[13]

$$\frac{\partial n}{\partial t} - \frac{i}{\hbar}\{\mathscr{E}, n\}_- + \frac{1}{2}\left\{\frac{\partial \mathscr{E}}{\partial \vec{\mathrm{p}}}, \frac{\partial n}{\partial \vec{x}}\right\}_+ - \frac{1}{2}\left\{\frac{\partial n}{\partial \vec{\mathrm{p}}}, \frac{\partial \mathscr{E}}{\partial \vec{x}}\right\}_+ = \left[\frac{dn}{dt}\right]_c, \tag{2.4}$$

which allows one to find the spectrum of collective excitations. Here, $\{\cdot\cdot\cdot\}_\mp$ are the commutators and anticommutators, and $[dn/dt]_c$ is the collision integral usually considered in $\tau$-approximation. The smallness of the spatial gradients of $n_\sigma(\vec{\mathrm{p}}, \vec{x}, t)$ corresponds to the excitations with momenta small in comparison with the Fermi momentum $\mathrm{p}_F$. In this limit, the validity of the phenomenological Landau's theory can be confirmed using the Green-function method.[12,14]

The electron liquid of transition metals is created due to the collectivization of electrons of two different atomic shells. It is easy to generalize Landau's theory to the case of two components.[15] In this case, the density matrix is not only a $2 \times 2$ matrix on spin variables, but it is also the nondiagonal matrix on band variables. Its elements give occupation numbers for $s$- and $d$- (or $f$-) electrons and interband interference. The scattering function $F$ also becomes a complicated matrix on spin and band variables. The spectrum of collective excitations in a nonferromagnetic two-component Fermi liquid is rather complicated. It consists of eight branches, including charge density oscillations, correlated oscillations of occupation numbers in different energy bands, two branches associated with interband tunneling, and four branches of spin density oscillations. As well as for one-component Fermi liquid, the phenomenological model of that type can be derived from the Green-function method.[15] It can also be applied to the case of a two-component ferromagnetic Fermi liquid. For the iron metal group, the two-component model is in a qualitative agreement with experimental data for iron and cobalt, and in a quantitative agreement for nickel. The latter has a spin splitting of the energy band that is small compared to the bandwidth.[16] Only in this case can Landau's theory be applied to consideration of itinerant ferromagnetism. The reason for this will be discussed in the fifth section. In this paper, we would like to give more attention to the possible spin symmetries of the ground state of a homogeneous Fermi liquid. Thus, we will not consider rather awkward two-component models[15,16] in the remainder of this paper.

We are now interested in ground states with a spin symmetry lower than the symmetry of the interaction between quasiparticles. Several assumptions that do not follow from the theory are used below for simplicity. First, a quasiparticle Fermi surface is assumed to be spherically symmetric. Secondly, we assume that the scattering function $F$ is invariant under space and spin rotations. The general form of $F$, then, is given by

$$F = \varphi(\vec{p}, \vec{p}') + \vec{\sigma}\vec{\sigma}'\psi(\vec{p}, \vec{p}'), \tag{2.5}$$

where the second term corresponds to spherically symmetric exchange interaction. Thirdly, we assume that $F_{\sigma\sigma'}(\vec{p}, \vec{p}')$ remains unchanged for weak violation of symmetry of the ground state.

Near the Fermi surface, both functions $\varphi$ and $\psi$ in equation 2.5 may be expanded into the series of Legendre polynomials depending on $\cos\vartheta = \vec{k}\vec{k}'$ (where $\vec{k} = \vec{p}/p$):

$$\varphi(\vec{p}, \vec{p}') = \sum_{\ell} \varphi_{\ell} P_{\ell}(\cos\vartheta), \qquad \psi_{\ell}(\vec{p}, \vec{p}') = \sum_{\ell} \psi_{\ell} P_{\ell}(\cos\vartheta). \tag{2.6}$$

As is known,[7,14] for stability of the Fermi surface, the following conditions must be fulfilled:

$$1 + (2\ell + 1)^{-1} g\varphi_{\ell} > 0, \qquad 1 + (2\ell + 1)^{-1} g\psi_{\ell} > 0, \qquad \ell = 0, 1, 2, \ldots, \tag{2.7}$$

where $g$ is the density of electron states at the Fermi level. We will consider the situation where the conditions in equation 2.7 are violated for some coefficients $\psi_{\ell}$, and hence where the ground state symmetry about spin rotations is broken.

For $1 + g\psi_0 < 0$, the Fermi liquid is unstable against ferromagnetism at $T = 0$. Landau's theory is valid when $|1 + g\psi_0| \ll 1$, which corresponds to small spin splitting of the energy band. In this case, the Fermi liquid approach for dependence of the magnetization on temperature and magnetic field[17] coincides with Edwards and Wohlfarth's theory for weak itinerant ferromagnets.[4] For $1 + \frac{1}{3}g\psi_1 < 0$, the Fermi liquid is unstable against SRSV.[8] In this case, the energy band is split into two subbands for different electron helicities. A more complicated situation may occur when both stability conditions of equation 2.7 for $\ell = 0$ and for $\ell = 1$ are broken. This case will be considered in the eighth section of this paper.

Considerations of spin waves in ferromagnetic and in nonferromagnetic Fermi liquids essentially differ from each other. Spin waves in the nonmagnetic Fermi liquid are correlated oscillations of two Fermi surfaces corresponding to different electron spin projections. They are characterized by a linear dispersion law. On the other hand, spin waves in the ferromagnetic Fermi liquid correspond to spatial oscillations of the order parameter and have a quadratic dispersion law. To describe these oscillations, let us note that Landau's theory deals with a contact interaction between quasiparticles. This local formulation of the theory is sufficient for consideration of the magnitude of local magnetization in a ferromagnetic Fermi liquid. It becomes insufficient, however, when considering correlation between orientations of the magnetization in different space points. Moreover, as it will be shown in the fourth section, the local formulation of the theory can be applied only to the consideration of collective excitations with a linear dispersion law. The modification, which is necessary to consider spin waves in a ferromagnetic Fermi liquid, consists of the introduction of a nonlocal interaction between fermions.[6] In this case, the deviation of the total energy $\delta E_t$ from its equilibrium value due to an arbitrary small perturbation of the one-particle density matrix has the form,

$$\delta E_t = \mathrm{Sp}_\sigma \int d\tau \mathscr{E} \delta n_\tau + \tfrac{1}{2}\, \mathrm{Sp}_\sigma \mathrm{Sp}_{\sigma'} \int d\tau \int d\tau' F_{\tau\tau'} \delta n_\tau \delta n_{\tau'}, \tag{2.8}$$

where $d\tau = (2\pi\hbar)^{-3} d^3\mathrm{p}\, d^3x$, and $F_{\tau\tau'}$ depends on $|\vec{x} - \vec{x}'|$. In the local limit, $F_{\tau\tau'} = F_{\sigma\sigma'}(\vec{\mathrm{p}}, \vec{\mathrm{p}}')\delta(\vec{x} - \vec{x}')$, and we return to equation 2.2.

## FERROMAGNETIC FERMI LIQUID

It is easy to see that the spin-independent part of the scattering function $F$ does not play any role and, hence, may be omitted when considering spin effects only. Thus, we can replace $F_{\tau\tau'}$ in equation 2.8 by the short-range exchange potential,

$$F_{\tau\tau'} = \sigma\sigma' F_{\mathrm{pp}'}(|\vec{x} - \vec{x}'|). \tag{3.1}$$

The applicability of the phenomenological theory is restricted to the small gradients of the density matrix,

$$rq \ll 1, \tag{3.2}$$

where $r$ is the interaction radius and $q^{-1}$ is the typical dimension of nonuniformness. By expanding $\delta n(\vec{x}')$ into a series in $(\vec{x} - \vec{x}')$, this inequality permits us to rewrite equation 2.8 in the form,[6]

$$\delta E_t = \int d^3x \left\{ \mathrm{Sp}_\sigma \int \frac{d^3\mathrm{p}}{(2\pi\hbar)^3} \mathscr{E}(\vec{\mathrm{p}}) \delta n_\mathrm{p} \right.$$
$$\left. + \frac{1}{2} \mathrm{Sp}_\sigma \mathrm{Sp}_{\sigma'} \int \frac{d^3\mathrm{p}}{(2\pi\hbar)^3} \int \frac{d^3\mathrm{p}'}{(2\pi\hbar)^3} \vec{\sigma}\vec{\sigma}' \psi_{\mathrm{pp}'} [\delta n_\mathrm{p} \delta n_{\mathrm{p}'} - r^2_{\mathrm{pp}'} \nabla \delta n_\mathrm{p} \nabla \delta n_{\mathrm{p}'}] \right\}, \tag{3.3}$$

where

$$\psi_{\mathrm{pp}'} = \int d^3x F_{\mathrm{pp}'}(x), \tag{3.4}$$

$$r^2_{\mathrm{pp}'} = \frac{1}{2} \int d^3x \cdot x^2 F_{\mathrm{pp}'}(x) \Big/ \int d^3x F_{\mathrm{pp}'}(x). \tag{3.5}$$

The nonlocal character of the interaction results in necessity of expanding $\delta E_t$ in powers not only of the density matrix perturbation, but also of the gradients of the latter.

Quasiparticle energy is determined as the variational derivative of the functional of equation 3.3 with respect to $\delta n$. As a result, perturbation of the quasiparticle energy due to a small perturbation of the density matrix is given by

$$\delta \mathscr{E}_{\mathrm{p}} = \mathrm{Sp}_{\sigma'} \int \frac{d^3\mathrm{p}'}{(2\pi\hbar)^3} \vec{\sigma}\vec{\sigma}' \psi_{\mathrm{pp}'}(\delta n_{\mathrm{p}'} + r^2_{\mathrm{pp}'}\Delta\delta n_{\mathrm{p}'}) \tag{3.6}$$

($\Delta$ is the Laplacian).

The invariance of the Fermi liquid under spin rotations is spontaneously violated in the ferromagnetic state. A noninvariant term must then be introduced in the quasiparticle energy,

$$\mathscr{E}(\vec{\mathrm{p}}) = \mathscr{E}_0(\vec{\mathrm{p}}) - a(\vec{\mathrm{p}})\vec{\nu}\vec{\sigma}, \tag{3.7}$$

where $\vec{\nu}$ is a unit vector in the direction of the magnetization and $a\vec{\nu}$ can be considered to be an effective internal field acting in the system. In this case, the quasiparticle density matrix (equation 2.1) takes the form,

$$n(\vec{\mathrm{p}}) = \tfrac{1}{2}(n_+ + n_-) + \tfrac{1}{2}(n_+ - n_-)\vec{\nu}\vec{\sigma}, \tag{3.8}$$

$$n_\pm = \left[1 + \exp\frac{\mathscr{E}_\pm - \mu}{T}\right]^{-1}, \quad \mathscr{E}_\pm(\vec{\mathrm{p}}) = \mathscr{E}_0(\vec{\mathrm{p}}) \mp a(\vec{\mathrm{p}}), \tag{3.9}$$

where we have chosen the axis of spin quantization directed along $\vec{\nu}$.

Let us consider now a small uniform rotation of the magnetization, $\delta\vec{M} = M\delta\vec{\nu}$. Corresponding perturbations of the quasiparticle energy and of the density matrix are given by

$$\delta\mathscr{E}(\vec{\mathrm{p}}) = -a(\vec{\mathrm{p}})\vec{\sigma}\cdot\delta\vec{\nu}, \tag{3.10}$$

$$\delta n(\vec{\mathrm{p}}) = \tfrac{1}{2}(n_+ - n_-)\vec{\sigma}\cdot\delta\vec{\nu}. \tag{3.11}$$

Substituting them into equation 3.6, we obtain the integral equation for $a(\vec{\mathrm{p}})$:

$$a(\vec{\mathrm{p}}) = -\int \frac{d^3\mathrm{p}'}{(2\pi\hbar)^3} \psi(\vec{\mathrm{p}}, \vec{\mathrm{p}}')(n_+ - n_-). \tag{3.12}$$

If only $\psi_0$ in the series of equation 2.6 has a nonzero negative value, equation 3.12 gives[17]

$$a = \frac{|\psi_0|}{\mu_e} M, \tag{3.13}$$

where $M$ is the magnetization,

$$M = \mu_e \int \frac{d^3\mathrm{p}}{(2\pi\hbar)^3}(n_+ - n_-), \tag{3.14}$$

and $\mu_e$ is the electron magnetic moment. Equation 3.14 makes it possible to calculate the temperature dependence of the magnetization and the Curie temperature, $T_c$, in terms of the energy band structure.[17] Note that the chemical potential in $n_\pm$ is also a function of $M$ and $T$ that must be determined from the condition of constant electron density. For weak itinerant ferromagnets, it should be emphasized that the dependence,

$$M(T) = M(0)\left[1 - \frac{T^2}{T_c^2}\right]^{1/2}, \tag{3.15}$$

calculated in this way can correspond to reality only for low temperatures, when the contribution of the Fermi excitations into the $M(T)$ dependence is the main one. At low temperatures, it is a good approximation even for the iron metal group.[17] Notice also that equation 3.14 always has a trivial solution, $M(0) = 0$. The nontrivial solution of lower energy, $M(0) \neq 0$, exists only if the condition $1 + g\psi_0 < 0$ is fulfilled.

The consideration can be easily generalized to the case of external magnetic field and magnetic anisotropy acting in the system.[18] In this case, the effective internal field is proved to be

$$\vec{a} = \mu_e \vec{B} - \frac{\psi_0}{\mu_e}\vec{M} - \frac{\psi_a}{\mu_e}(\vec{M}\vec{\eta})\vec{\eta}, \tag{3.16}$$

where $\psi_a$ corresponds to the small anisotropic term, $\psi_a(\vec{\sigma}\vec{\eta})(\vec{\sigma}'\vec{\eta})$, introduced into the two-particle interaction, and $\vec{\eta}$ is a unit vector in the direction of the anisotropy axis. Within this model, magnetization curves at different temperatures, $\vec{M}(\vec{B}, T)$, can be obtained for ferromagnetic electron liquid. Let us assume uniaxial ($\psi_a < 0$) anisotropy acting in the $z$ direction and a magnetic field applied in the $xz$-plane. In this case, the magnetization also is directed in the $xz$-plane. We are interested in the absolute value of the magnetization, $M(B_x, B_z, T)$, and in the angle, $\theta(B_x, B_z, T)$, that the magnetization makes with the anisotropy axis. Using the Fermi liquid approach (for details, see the eighth section), one can obtain the following equations for relative variables, $\vec{m} = \vec{M}(\vec{B}, T)/M(0, 0)$ and $\vec{b} = \vec{B}/M(0, 0)$:[18]

$$m^3 + [k\chi_0 \sin^2\theta - \tau]m = \chi_0(b_x \sin\theta + b_z \cos\theta), \tag{3.17}$$

$$\frac{b_x}{\sin\theta} - \frac{b_z}{\cos\theta} = km, \tag{3.18}$$

where $k = \mu_e^{-2}|\psi_a|$ is the dimensionless anisotropy strength, $\chi_0$ is the zero-field longitudinal magnetic susceptibility at $T = 0$,

$$\chi_0 = -\frac{\mu_e g}{1 + g(\psi_0 + \psi_a)}, \tag{3.19}$$

$\tau = (1 - T^2/T_c^2)$, and the Curie temperature is given by

$$T_c^2 = \frac{6}{\pi^2}[1 + g(\psi_0 + \psi_a)] \cdot \frac{g^2}{(gg'' - g'^2)}, \tag{3.20}$$

where $g'$ and $g''$ are derivatives of the paramagnetic density-of-state function with respect to the Fermi energy. Fulfillment of two conditions is necessary for the occurrence of the ferromagnetic ground state, that is, $M(0, 0) \neq 0$, in the anisotropic Fermi liquid:

$$1 + g(\psi_0 + \psi_a) < 0, \quad g'' - \frac{g'^2}{g} < 0. \tag{3.21}$$

Notice also that the spin splitting of the energy band is of the order of $T_c$. It is considered to be small in comparison with the Fermi energy, which requires $|1 + g[\psi_0 + \psi_a]| \ll 1$.

Equations 3.17 and 3.18 give several branches for $\theta(b_x, b_z)$. One of them corresponds to the absolute minimum of the thermodynamic potential, while others correspond to local minima. There is a curve in the $(b_x, b_z)$-plane bounding the region of the magnetic field where metastable magnetic states and hysteresis phenomena exist in an anisotropic ferromagnetic electron liquid. This curve is defined by the equation,

$$\tau(b_x^{2/3} + b_z^{2/3}) - k\chi_0 b_z^{2/3} = k^{-2}(b_x^{2/3} + b_z^{2/3})^4. \tag{3.22}$$

For isotropic exchange interaction ($\psi_a = 0$) or at high magnetic field, where irreversible effects are negligible, the magnetization law (equation 3.17) follows Edwards and Wohlfarth's theory for weak itinerant ferromagnetism.[4]

In considering the effect of the magnetic field, we have neglected Landau diamagnetism. This can be justified in the case of a narrow energy band, which is of major interest when considering spin ordering in transition metals (see the introduction). Indeed, in a narrow band, the effective electron mass $m^*$ is large in comparison with the bare mass $m$. Therefore, the energy quantum for orbital quantization, $\Delta\mathscr{E}_{\text{orb}} = e\hbar/m^*c$, is small in comparison with the energy quantum for spin quantization, $\Delta\mathscr{E}_{\text{spin}} = e\hbar/mc$. Neglecting orbital quantization is thus a good approximation for consideration of spin effects in narrow bands.

## SPIN WAVES IN FERROMAGNETIC FERMI LIQUID

As is clear from the above consideration, the nonlocal term in the quasiparticle interaction does not contribute to the ground state in the case of uniform magnetization. It plays an important role, however, when considering spin waves in a ferromagnetic Fermi liquid. We will restrict ourselves to the case of isotropic exchange interaction and will assume for simplicity that $\psi_{pp'}$ and $r_{pp'}^2$ in equation 3.6 are constants, $\psi_{pp'} = \psi_0$, $r_{pp'}^2 = r^2$. Notice that the taking account of weak magnetic anisotropy is important for the calculation of spin wave spectrum only at very small spin wave momenta, that is, at the wavelengths corresponding to ferromagnetic resonance frequencies.

Let us consider a small perturbation of a one-particle density matrix,

$$\delta n = \vec{\sigma}\vec{\zeta}_p \exp(i\vec{q}\vec{x} - i\omega t), \tag{4.1}$$

caused by the propagation of a spin wave in the direction of the equilibrium magnetization, $z$. According to equation 3.6, the related variation in the one-particle energy is given by

$$\delta\mathscr{E} = \psi_0(1 - r^2q^2) \int \frac{d^3\mathrm{p}}{(2\pi\hbar)^3} \vec{\sigma}\vec{\zeta}_\mathrm{p} \exp(i\vec{q}\vec{x} - i\omega t). \tag{4.2}$$

The kinetic equation (equation 2.4) linearized in $\tau$-approximation with respect to $\delta n$ and $\delta\mathscr{E}$ has the form,

$$-i\omega\delta n + \frac{1}{2} i\vec{q} \left\{\frac{\partial\mathscr{E}}{\partial\vec{\mathrm{p}}}, \delta n\right\}_+ - \frac{1}{2} i\vec{q} \left\{\delta\mathscr{E}, \frac{\partial n_0}{\partial\vec{\mathrm{p}}}\right\}_+ - \frac{i}{\hbar}\{\delta\mathscr{E}, n_0\}_- - \frac{i}{\hbar}\{\mathscr{E}, \delta n\}_- = -\frac{\delta n}{\tau}, \tag{4.3}$$

where $\mathscr{E}$ and $n_0$ are the equilibrium energy and the density matrix given by equations 3.7 and 3.8, respectively. Spin waves in ferromagnetic Fermi liquids are oscillations of circular components of $\vec{\zeta}_\mathrm{p}$:

$$\zeta_\mathrm{p}^\pm = \zeta_\mathrm{p}^{(x)} \pm \zeta_\mathrm{p}^{(y)}. \tag{4.4}$$

With the help of equations 3.7, 3.8, 4.1, and 4.2, it is not difficult to find out that the kinetic equation (equation 4.3) gives two independent dispersion relations corresponding to right and left polarizations of a spin wave:[5,16]

$$D_\pm = 1 - \psi_0(1 - r^2q^2) \int \frac{d^3\mathrm{p}}{(2\pi\hbar)^3} \frac{\vec{q}\vec{u}\left(\dfrac{\partial n_+}{\partial\mathscr{E}_+} + \dfrac{\partial n_-}{\partial\mathscr{E}_-}\right) \mp \dfrac{2}{\hbar}(n_+ - n_-)}{\vec{q}\vec{u} \pm \dfrac{2a}{\hbar} - \omega - \dfrac{i}{\tau}} = 0, \tag{4.5}$$

where $\vec{u} = \partial\mathscr{E}_0/\partial\vec{\mathrm{p}}$, and $n_\pm$, $\mathscr{E}_\pm$, and $a$ are given by equations 3.9 and 3.13.

Integration in equation 3.5 over the angle variables gives, in the limit of $\tau \to \infty$, the following expressions for the real $D'_\pm$ and the imaginary $D''_\pm$ parts of the functions:

$$D'_\pm = 1 - \frac{1}{2}\psi_0(1 - r^2q^2) \int d\mathscr{E}_0 g(\mathscr{E}_0) \cdot \left\{I_\pm\left(\frac{\partial n_+}{\partial\mathscr{E}_+} + \frac{\partial n_-}{\partial\mathscr{E}_-}\right) \pm 2(I_\pm - 1)\frac{n_+ - n_-}{\hbar\lambda_\pm}\right\}, \tag{4.6}$$

$$D''_\pm = -\frac{\pi}{2}\psi_0(1 - r^2q^2) \int d\mathscr{E}_0 \frac{g(\mathscr{E}_0)}{qu} \theta\left(1 - \frac{|\lambda_\pm|}{qu}\right) \cdot \left\{\lambda_\pm\left(\frac{\partial n_+}{\partial\mathscr{E}_+} + \frac{\partial n_-}{\partial\mathscr{E}_-}\right) \pm \frac{2}{\hbar}(n_+ - n_-)\right\}, \tag{4.7}$$

where we have introduced

$$I_\pm = 1 + \frac{1}{2}\frac{\lambda_\pm}{qu} \ln\left|\frac{\lambda_\pm - qu}{\lambda_\pm + qu}\right|, \qquad \lambda_\pm = \omega \pm \frac{2a}{\hbar}, \tag{4.8}$$

$$\theta(x) = \begin{cases} 1, x > 0 \\ 0, x < 0. \end{cases} \quad \textbf{(4.9)}$$

Notice that equation 4.6 is exactly fulfilled in the limit of $q = 0$, $\omega = 0$ due to equation 3.13 for the equilibrium spin splitting of the energy band.

The nonzero value of $D''_{\pm}$ corresponds to the collisionless damping of spin waves. For long waves,

$$\hbar q u_F \ll 2a \quad \textbf{(4.10)}$$

($u_F$ is the Fermi velocity for $a = 0$), and the collisionless damping is absent. Thus, according to equation 4.6, the spectrum of spin waves is defined as[6]

$$\omega = \frac{2a}{\hbar}\left(r^2 + \frac{\hbar^2}{36} g^{-1}(gu^2)''_{\mathscr{E}_F}\right) q^2. \quad \textbf{(4.11)}$$

It should be noted that for $a = 0$, equations 4.6 and 4.7 define the spin wave spectrum in a nonmagnetic Fermi liquid.[12] In this case, spin waves have a linear dispersion law, which is contrary to the quadratic dispersion law for a ferromagnetic Fermi liquid. Thus, for a nonmagnetic Fermi liquid, the nonlocal part of the two-particle interaction ($r \neq 0$) gives a negligible $q^2$-correction to the linear dispersion law. On the contrary, for a ferromagnetic Fermi liquid, the nonlocal interaction can define the main contribution to the spin wave spectrum and, hence, cannot be neglected.

The dispersion law for ferromagnons (equation 4.11) found in the collisionless limit is also valid for $\omega\tau \gg 1$, where $\tau$ is the time of electron spin relaxation. In this case, the spin wave decrement is equal to $\tau^{-1}$. It is also of interest to obtain the spin wave spectrum in the hydrodynamic limit, when $\omega\tau \ll 1$. Neglecting again the magnetic anisotropy effects, that is, considering large spin wave frequencies in comparison with the frequency of uniform ferromagnetic resonance, one can show that, in the hydrodynamic limit, the kinetic equation reduces to the Landau-Lifshitz equation for the magnetization,

$$\frac{\partial \vec{M}}{\partial t} = \frac{2ar^2}{\hbar M}[\vec{M} \times \Delta \vec{M}], \quad \textbf{(4.12)}$$

which gives

$$\omega = \frac{2a}{\hbar} r^2 q^2. \quad \textbf{(4.13)}$$

The spin wave decrement in the hydrodynamic limit is given by

$$\Gamma = \omega^2 \tau. \quad \textbf{(4.14)}$$

Notice that ferromagnons in collisionless and hydrodynamic limits are analogous to the zero and first sounds in a nonmagnetic Fermi liquid.[14]

The second term in equation 4.11 (determined by the local exchange and the band structure) can be small in comparison with the first term. Comparison of dispersion laws of low-frequency and high-frequency ferromagnons in Ni[6,19] shows that the second part of the ferromagnon energy does not exceed 15%. Nevertheless, precise identification of this term in spin wave spectrum of metallic ferromagnets is of great interest because of its purely Fermi liquid origin.

## GREEN-FUNCTION METHOD

In this section, we will show how the results of the phenomenological model studied in previous sections can be obtained within a microscopic approach, that is, using the Green-function method.

When the invariance of a Fermi liquid under spin rotations is violated, the one-fermion Green-function becomes a nondiagonal matrix on spin variables. As a result, it can be represented in the form,

$$(G^{-1})_{\alpha\beta} = (G_0^{-1})\delta_{\alpha\beta} - \Sigma_{\alpha\beta}, \tag{5.1}$$

where $G_0$ is a Green-function for zero magnetization,

$$G_0(P) = [\mathscr{E} - \mathscr{E}_0(\vec{\mathrm{p}}) + i\delta_{\mathrm{p}}]^{-1}, \tag{5.2}$$

$P = (\vec{\mathrm{p}}, \mathscr{E})$ is the four-momentum, and $\Sigma$ is the mass operator responsible for symmetry violation,

$$\Sigma = a(\vec{\mathrm{p}})\vec{\nu}\vec{\sigma} \tag{5.3}$$

($\vec{\nu}$ is a unit vector in the direction of the magnetization). It is evident that equation 5.1 is equivalent to

$$G_{\alpha\beta} = \tfrac{1}{2}(G_+ + G_-)\delta_{\alpha\beta} + \tfrac{1}{2}(G_+ - G_-)\vec{\nu}\vec{\sigma}, \tag{5.4}$$

where

$$G_\pm = [\mathscr{E} - \mathscr{E}_0(\vec{\mathrm{p}}) \mp a(\vec{\mathrm{p}}) + i\delta_{\mathrm{p}}]^{-1}. \tag{5.5}$$

Let us consider the series of loop diagrams for the vertex function $\Gamma_{\alpha\beta,\delta\gamma}(P_1, P_2; Q)$ represented in FIGURE 1. Here, the black square corresponds to the total vertex function $\Gamma_{+-,-+}(P_1, P_2; Q)$, the white square corresponds to the irreducible vertex function $\tilde{\Gamma}_{+-,-+}(P_1, P_2; Q)$, the plus and minus signs correspond to positive and negative projections of fermion spins on the direction of the magnetization, and $Q = (\hbar\vec{q}, \hbar\omega)$ is the four-momentum transfer. FIGURE 1 is equivalent to the following equation for $\Gamma$:

$$\Gamma_{+-,-+}(P_1, P_2; Q) - \tilde{\Gamma}_{+-,-+}(P_1, P_2; Q) - i\hbar \int \frac{d^4P}{(2\pi\hbar)^4}\tilde{\Gamma}_{+-,-+}\left(P_1, P + \frac{Q}{2}; Q\right)$$
$$\cdot\, \Gamma_{+-,-+}\left(P - \frac{Q}{2}, P_2; Q\right) \cdot G_+\left(P - \frac{Q}{2}\right) G_-\left(P + \frac{Q}{2}\right). \tag{5.6}$$

For a nonferromagnetic Fermi liquid, $G_+(P) = G_-(P)$ and the poles of the two Green-functions corresponding to the upper and lower lines of every loop come together in the limit of $Q \to 0$. That is why the diagram series shown in FIGURE 1 defines the main contribution to the vertex function for small momentum transfer. In this limit, equation 5.6 is exact for a nonferromagnetic Fermi liquid.[12,20]

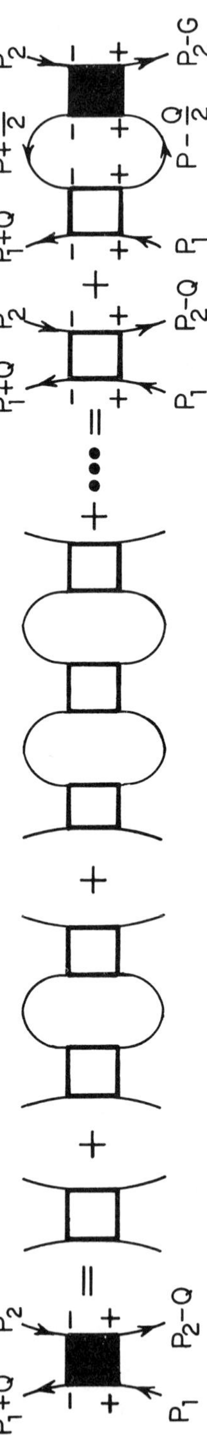


**FIGURE 1.** Diagram series for the vertex function.

For a ferromagnetic Fermi liquid, the poles of $G_+$ and $G_-$ do not coincide in the limit of $Q \to 0$. The diagram series shown in FIGURE 1 represents now the main contribution to the vertex function only for $Q \to 0$ and $a \to 0$, that is, when $G_+ \to G_-$. Thus, at small values, equation 5.6 is approximately valid also for weak itinerant ferromagnetism when we have $a \ll \mathscr{E}_F$.

As is known,[20] poles of $\Gamma$ define the spectrum of collective excitations in a system. To solve the integral equation (equation 5.6), some simplifying assumptions should be made with respect to its kernel $\tilde{\Gamma}(P_1, P_2; Q)$. For a nonferromagnetic Fermi liquid, the dependence of $\tilde{\Gamma}$ on $Q$ leads to a negligible $q^2$-correction of the linear dispersion law of spin waves. In this case, the simplest approach corresponds to a constant $\tilde{\Gamma}$. On the contrary, as will be shown below, $Q$-dependence of $\tilde{\Gamma}$ is essential when considering spin waves in a ferromagnetic Fermi liquid. Thus, we will keep dependence of $\tilde{\Gamma}$ on $Q$ and will neglect for simplicity dependence of $\tilde{\Gamma}$ on $P_1$ and $P_2$. In this case, the explicit form of the vertex function can be found from equation 5.6:

$$\Gamma(Q) = \frac{\tilde{\Gamma}(Q)}{1 + \tilde{\Gamma}(Q)\pi(Q)}, \tag{5.7}$$

where the polarization operator $\pi(Q)$ is given by

$$\pi(Q) = i\hbar \int \frac{d^4P}{(2\pi\hbar)^4} G_+\left(P - \frac{G}{2}\right) G_-\left(P + \frac{G}{2}\right). \tag{5.8}$$

It is easy to perform the integration over the fourth component of $P$ in equation 5.8. As a result, we have

$$\pi(Q) = -\int \frac{d^3\mathrm{p}}{(2\pi\hbar)^3} \frac{n_+\left(\vec{\mathrm{p}} - \frac{\hbar\vec{q}}{2}\right) - n_-\left(\vec{\mathrm{p}} + \frac{\hbar\vec{q}}{2}\right)}{\hbar\omega + \mathscr{E}_+\left(\vec{\mathrm{p}} - \frac{\hbar\vec{q}}{2}\right) - \mathscr{E}_-\left(\vec{\mathrm{p}} + \frac{\hbar\vec{q}}{2}\right) + i\delta}. \tag{5.9}$$

The spin wave spectrum is defined by the poles of $\Gamma(Q)$:

$$1 + \tilde{\Gamma}(Q)\pi(Q) = 0. \tag{5.10}$$

We are interested in the $\omega(q)$ solutions of this equation for $\hbar q \ll \mathrm{p}_F$ when equation 5.6 is valid. In this case, $\tilde{\Gamma}(Q)$ can be represented, for symmetry reasons, in the form,

$$\tilde{\Gamma}(Q) = \tilde{\Gamma}(0) + \tfrac{1}{2}\tilde{\Gamma}''_q \cdot q^2 + \tfrac{1}{2}\tilde{\Gamma}''_\omega \cdot \omega^2, \tag{5.11}$$

where $\tilde{\Gamma}''_{q,\omega}(0)$ are second derivatives of $\tilde{\Gamma}$ at $Q - 0$. Functions $n_\pm$ and $\mathscr{E}_\pm$ in equation 5.9 also can be developed as a series in powers of $q$. As a result, we can rewrite equation 5.10 in the form,

$$1 - [\tilde{\Gamma}(0) + \frac{1}{2}\tilde{\Gamma}''_q(0) \cdot q^2 + \frac{1}{2}\tilde{\Gamma}''_\omega(0) \cdot \omega^2] \cdot \int \frac{d^3\mathrm{p}}{(2\pi\hbar)^3} \frac{n_+(\vec{\mathrm{p}}) - n_-(\vec{\mathrm{p}}) - \frac{1}{2}\hbar\vec{q}\left(\frac{\partial n_+}{\partial\vec{\mathrm{p}}} + \frac{\partial n_-}{\partial\vec{\mathrm{p}}}\right)}{\hbar\omega - 2a - \frac{1}{2}\hbar\vec{q}\left(\frac{\partial\mathscr{E}_+}{\partial\vec{\mathrm{p}}} + \frac{\partial\mathscr{E}_-}{\partial\vec{\mathrm{p}}}\right)} = 0. \tag{5.12}$$

For $q = 0$, $\omega = 0$, this equation determines the spin splitting of the energy band:

$$2a = -\tilde{\Gamma}(0) \int \frac{d^3\mathrm{p}}{(2\pi\hbar)^3} (n_+ - n_-). \tag{5.13}$$

A comparison of this formula with equations 3.13 and 3.14 gives the relation between $\tilde{\Gamma}(0)$ and the parameter $\psi_0$ of the phenomenological theory:

$$\tilde{\Gamma}(0) = 2\psi_0. \tag{5.14}$$

For nonzero $q$ and $\omega$, equation 5.12 gives the quadratic dispersion law for spin waves, $\omega \propto q^2$. In this case, the last term of the expansion of $\tilde{\Gamma}$ given by equation 5.11 is proportional to $q^4$ and can be omitted in equation 5.12 when considering the spin wave spectrum. Thus, it can be easily seen that equation 5.12 coincides with one of those in equation 4.5 (namely, $D_+ = 0$) to an accuracy of the main terms in spin wave dispersion law. Correspondence between $\tilde{\Gamma}''_q(0)$ and the parameters of the phenomenological theory is given by

$$\tilde{\Gamma}''_q(0) = -4\psi_0 r^2. \tag{5.15}$$

Equation 5.12 then leads to the same results for the spin wave spectrum as the phenomenological approach in the collisionless limit studied in the previous section.

The Green-function method confirms the results of the phenomenological theory for ferromagnetic Fermi liquid studied in previous sections. Besides, the microscopic approach allows us to establish the bounds of applicability of the phenomenological theory. The results of this section show that the Green-function method leads to the phenomenological formulae for the ground state and for spin waves when a summation of the loop diagrams only is sufficient. As it was shown above, the smaller that the ferromagnetic spin splitting of the energy band is, the better is the validity of the loop expansion for the vertex function. Thus, our phenomenological model can be valid only for weak itinerant ferromagnetism.

## SPONTANEOUS REFLECTIONAL SYMMETRY VIOLATION

In previous sections, we have studied the ferromagnetic state of a Fermi liquid that is characterized by the order parameter $\langle\vec{\sigma}\rangle$. This state is realized as a result of an instability of a nonferromagnetic Fermi liquid for large negative values of $\psi_0$. Here, $\psi_0$ is the zeroth coefficient of the expansion of the spin-dependent two-particle scattering function $\psi(\vec{\mathrm{p}}, \vec{\mathrm{p}}')$ into a series of Legendre polynomials. Now, we will consider another type of spin ordering in a Fermi liquid that is realized when the stability conditions (equation 2.7) are broken for the first coefficient $\psi_1$, that is, $1 + \frac{1}{3}g\psi_1 < 0$. To single out the instability caused by a large negative value of $\psi_1$, we assume in this section that only the $\psi_1$ coefficient of the expansion of $\psi(\vec{\mathrm{p}}, \vec{\mathrm{p}}')$ has a nonzero value. Realization of the corresponding instability is seen in spontaneous reflectional symmetry violation (SRSV). SRSV is characterized by a correlation between the spin and the momentum of an electron. The order parameter for SRSV is the average helicity $\langle\vec{\sigma}\vec{k}\rangle$ over the electron ensemble, $\vec{k} = \vec{\mathrm{p}}/\mathrm{p}$. In this state, the quasiparticle energy and the density matrix are noninvariant under space reflections:[8]

$$\mathscr{E}(\vec{p}) = \mathscr{E}_0(\vec{p}) - b(\vec{p})\vec{k}\vec{\sigma}, \tag{6.1}$$

$$n(\vec{p}) = \tfrac{1}{2}(n_+ + n_-) + \tfrac{1}{2}(n_+ - n_-)\vec{k}\vec{\sigma}, \tag{6.2}$$

$$n_\pm = \left[1 + \exp\frac{\mathscr{E}_\pm - \mu}{T}\right]^{-1}, \qquad \mathscr{E}_\pm = \mathscr{E}_0(\vec{p}) \mp b(\vec{p}), \tag{6.3}$$

where $2b(\vec{p})$ is the energy gap between the electron states with different helicities.

A consideration analogous to that made in the third section gives the following equation for $b$:

$$b = -\frac{\psi_1}{3}\int \frac{d^3p}{(2\pi\hbar)^3}(n_+ - n_-). \tag{6.4}$$

In our approximation, $b$ does not depend on $\vec{p}$. It is obvious that equation 6.4 always has the trivial solution of $b = 0$. Nonzero solutions corresponding to the lower energy exist when $1 + \tfrac{1}{3}g\psi_1 < 0$. Before their investigation, note that the right-hand side of equation 6.4 is odd with respect to the parameter $b$. Therefore, the existence of a nonzero solution $b$ of equation 6.4 means also the existence of the solution with an opposite sign, $-b$. For $b > 0$, the majority of electrons have positive helicity, while for $b < 0$, the average electron helicity $\langle\vec{\sigma}\vec{k}\rangle$ is negative. With the help of the density matrix of equation 6.2, it is easy to obtain for the order parameter,

$$\langle\vec{\sigma}\vec{k}\rangle = \frac{\displaystyle\int \frac{d^3p}{(2\pi\hbar)^3}(n_+ - n_-)}{\displaystyle\int \frac{d^3p}{(2\pi\hbar)^3}(n_+ + n_-)}. \tag{6.5}$$

SRSV represents the new type of spin ordering in a homogeneous Fermi liquid. The majority of spins of electrons is ordered along (for $b > 0$) or against (for $b < 0$) their momenta. Both states have equal energies and, hence, equal probabilities when SRSV occurs in a given metallic sample. This fact reflects the initial unbroken symmetry of the interaction between quasiparticles and is common for all cases of spontaneous symmetry violation.

For better understanding of SRSV in solids, let us compare it with the ferromagnetic ordering in a system characterized by the uniaxial magnetic anisotropy. When cooling below the Curie temperature in the absence of an external magnetic field, spin orderings along and opposite to the easy axis $\vec{\nu}$ have equal probabilities. As a result, the nuclei of ferromagnetic phase arising in different regions of a solid have different signs of $\langle\vec{\sigma}\vec{\nu}\rangle$. Then, ferromagnetic regions dilate and interact with each other until the magnetic state with the lowest energy is established. In zero magnetic field, this state is characterized by zero net magnetization due to the creation of stable magnetic domains with different signs of $\langle\vec{\sigma}\vec{\nu}\rangle$. The domains appear to decrease the magnetodipole energy caused by nonzero $\langle\vec{\sigma}\rangle$ in a space surrounding the ferromagnet. The decrease of the magnetodipole energy compensates the positive energy of the domain walls dividing neighboring magnetic domains. Nonzero net magnetization can be obtained in an external magnetic field that creates preferential direction of the magnetization. The latter appears due to the interaction of the magnetic field with the ferromagnetic order parameter $\langle \vec{\sigma} \rangle$.

Another situation takes place for SRSV. In this case, we have no long-range field that appears around the sample due to nonzero average helicity. Hence, we have no reason to expect the appearance of domains with different signs of $\langle \vec{\sigma}\vec{k} \rangle$ in a homogeneous solid (like magnetic domains in a ferromagnet). There would be nothing to compensate the positive energy of domain walls in the case of SRSV. Thus, for a monocrystalline solid, SRSV should be characterized by a nonzero average helicity over the solid. Of course, the sign of SRSV should be random for different monocrystalline samples. On the other hand, in a polycrystalline solid, SRSV would be disguised due to the randomness in a sign of $\langle \vec{\sigma}\vec{k} \rangle$ in different crystallites and, hence, zero average helicity over the solid. In this case, we would have no possibility to create a preferential sign of average helicity because we would have no long-range field interacting with the order parameter $\langle \vec{\sigma}\vec{k} \rangle$.

Our statement regarding the exact equivalence of two signs of SRSV is valid without taking into account short-range weak interactions. As is known, weak interactions violate the invariance between left and right and make one definite sign of $\langle \vec{\sigma}\vec{k} \rangle$ preferable to the other. For the electron liquid, this effect gives a substantial value of $\langle \vec{\sigma}\vec{k} \rangle$ only at a very high density, such as when one is studying neutron stars or the early universe.[21] For densities of electron liquid in metals, the distances between fermions are much greater than the radius of weak interactions. As a result, the effect of weak interactions is extremely small. Nevertheless, the question arises whether the weak interactions can bias the system to take a definite sign of $\langle \vec{\sigma}\vec{k} \rangle$ when the phase transition into the SRSV state occurs. In this case, the role of the weak interaction is similar to the role of a very weak magnetic field for a ferromagnetic phase transition. In the Salam-Weinberg model, the energy gap between electron states of different helicities in metals is of the order of $\Delta \sim 10^{-27}$ erg. Due to the uncertainty principle, the time interval that is necessary for the revealing of the weak interaction is of the order of $\tau \sim \hbar/\Delta \sim 1$ s. For all possible experiments, it is much greater than the characteristic times of random processes leading to the definite sign of $\langle \vec{\sigma}\vec{k} \rangle$ at SRSV phase transition. The effect of the weak interaction thus seems to be unimportant for SRSV.

It should be emphasized that SRSV in solids should be as strong an effect as the ferromagnetism. It means that in the general case, the helicity splitting, $2b$, of the energy band should be of the order of the bandwidth, which gives $\langle \vec{\sigma}\vec{k} \rangle \sim 1$. Here, we will consider the case of weak SRSV, when $\langle \vec{\sigma}\vec{k} \rangle \ll 1$ and $b \ll \mathscr{E}_F$.

Notice that SRSV can also be considered using the Green-function method. This consideration is analogous to that for a ferromagnetic Fermi liquid. The one-particle Green-function corresponding to the quasiparticle Hamiltonian (equation 6.1) is given by

$$G_{\alpha\beta}(P) = [\mathscr{E} - \mathscr{E}_0(\vec{\mathrm{p}}) + b(\vec{\mathrm{p}})\vec{k}\vec{\sigma}_{\alpha\beta} + i\delta_{\mathrm{p}}]^{-1}. \quad \textbf{(6.6)}$$

Using the Green-function method, it can be shown that the phenomenological Fermi liquid theory is valid only for weak SRSV. It requires the condition $|1 + \tfrac{1}{3}g\psi_1| \ll 1$ analogous to the condition $|1 + g\psi_0| \ll 1$ in the theory of weak itinerant ferromagnetism.

According to equations 6.4 and 6.5, we have for the band splitting,

$$b = -\frac{\psi_1}{3} n_0 \langle \vec{\sigma} \vec{k} \rangle, \tag{6.7}$$

where $n_0$ is the spatial density of electrons. The average helicity in the case of a weak SRSV is given by

$$\langle \vec{\sigma} \vec{k} \rangle^2 = \frac{6g^3}{n_0^2} \left| \frac{1 + \frac{1}{3} g\psi_1}{g'' - 3\frac{(g')^2}{g}} \right| \left(1 - \frac{T^2}{T_R^2}\right). \tag{6.8}$$

Here, $T_R \sim |1 + \frac{1}{3} g\psi_1|^{1/2}$. $\mathscr{E}_F$ is the temperature of the SRSV phase transition caused by Stoner excitations only. The contribution of the collective spin-dependent fluctuations to the temperature dependence of the order parameter can be unessential, however, only at low temperature, $T \ll T_R$. In this limit, $\langle \vec{\sigma} \vec{k} \rangle$ is of the order of

$$\langle \vec{\sigma} \vec{k} \rangle \sim \pm |1 + \tfrac{1}{3} g\psi_1|^{1/2}. \tag{6.9}$$

## MACROSCOPIC CONSEQUENCES OF SRSV

How can SRSV reveal itself in solids? Uniqueness of this type of symmetry violation consists in the existence of the pseudoscalar order parameter $\langle \vec{\sigma} \vec{k} \rangle$, which gives the possibility of proportionality between polar and axial vectors for such a solid.[9] At first, it might seem that the magnetic field applied to a solid with SRSV could cause the electric current flowing in the direction of the field due to the correlation between the spin and the momentum of an electron. We will show, however, that this effect is absent in thermodynamic equilibrium and that SRSV reveals itself in a more subtle way.

In the presence of the magnetic field, the quasiparticle energy is given by

$$\mathscr{E}(\vec{p}) = \mathscr{E}_0(\vec{p}) - b\vec{\sigma}\vec{k} - \mu_e \vec{\sigma} \vec{B}. \tag{7.1}$$

Choosing the axis of spin quantization directed along the vector

$$\vec{\ell} = b\vec{k} + \mu_e \vec{B}, \tag{7.2}$$

the density matrix may be presented in the form,

$$n = \frac{1}{2}(n_+ + n_-) + \frac{1}{2}(n_+ - n_-)\frac{\vec{\ell}\vec{\sigma}}{\ell}, \tag{7.3}$$

$$n_\pm = \left[1 + \exp\frac{\mathscr{E}_\pm - \mu}{T}\right]^{-1}, \qquad \mathscr{E}_\pm = \mathscr{E}_0(\vec{p}) \mp \ell(\vec{p}). \tag{7.4}$$

Now, the energy gap between two subbands corresponding to different spin polarizations is $2\vec{\mathcal{L}}(\vec{p})$, where

$$\mathcal{L} = (b^2 + \mu_e^2 B^2 + 2\mu_e b\vec{B}\vec{k})^{1/2}. \tag{7.5}$$

The unusual symmetry of the electron spectrum given by equations 7.4 and 7.5 is obvious. Below, we will consider the macroscopic consequences of the lack of reflectional symmetry of the electron spectrum. Note that the latter has the simplest form in a weak magnetic field, $\mu_e B \ll b$. In this case, we have (with the accuracy to the linear term in $\vec{B}$)

$$\mathcal{E}_\pm = \mathcal{E}_0(\vec{p}) \mp b \mp \mu_e \vec{B}\vec{k}. \tag{7.6}$$

The electric current is given by

$$\vec{j} = e \cdot \mathrm{Sp} \int \frac{d^3p}{(2\pi\hbar)^3} \frac{\partial \mathcal{E}}{\partial \vec{p}} n, \tag{7.7}$$

where $\mathcal{E}$ and $n$ are defined by equations 7.1 and 7.3. In the SRSV state, the velocity of the quasiparticle $\partial\mathcal{E}/\partial\vec{p}$ becomes a matrix in spin variables. Substituting equations 7.1 and 7.3 into equation 7.7, it is easy to obtain by direct calculation that $\vec{j} = 0$. Hence, in a thermodynamic equilibrium, the magnetic field does not induce the electric current flowing in the direction of the field, despite the correlation between the spin and the momentum of an electron. This result becomes obvious after the following qualitative consideration. The existence of a nonzero dissipative electric current would mean that the work is being done by the constant magnetic field on the electrons to move them along the field. Thus, the presence of a magnetic field in a conducting sample would be inconsistent with the condition of a thermodynamic equilibrium. This evidently would lead to a contradiction for a normal metal.

One can consider the SRSV in a superconducting phase, which may originate from a normal SRSV state as a result of a superconducting phase transition.[22] Let us consider a long thin superconducting strip of a thickness smaller than the penetration depth for the magnetic field. It can be easily seen that the magnetic field parallel to the strip cannot induce even a nondissipative electric current. Such an effect would contradict the relativity principle. Indeed, passing to the coordinate system that slowly moves together with the superconducting liquid, we would reduce the problem to the initial one.

An electric current parallel to $\vec{B}$ does occur in a number of nonequilibrium situations;[9] for example, in a time-varying magnetic field, $\vec{B}(t) = \vec{B}\exp(-i\omega t)$. Nonequilibrium corrections to the density matrix can be found from the kinetic equation (equation 2.4) in the $\tau$-approximation. For $\omega\tau \ll 1$, we have

$$\delta n = \tau\vec{\sigma}\mu_e \dot{\vec{B}} \frac{\partial n}{\partial \mathcal{E}_0}. \tag{7.8}$$

With the accuracy to the linear term in $\dot{\vec{B}}$, this gives for the current,

$$\vec{j} = -\frac{1}{2} e n_0 v_F \frac{b}{\mathcal{E}_F} \frac{d\tau}{d\mathcal{E}_F} \mu_e \dot{\vec{B}}. \tag{7.9}$$

Although equation 7.9 was derived for $b \ll \mathcal{E}_F$, it should give a correct order of magnitude of the current for strong SRSV, that is, when $b \sim \mathscr{E}_F$. Even in this case, an experimental observation of the symmetry-violating current (equation 7.9) may turn out to be very difficult because of the presence of much greater transverse currents induced by a time-varying magnetic field.

Another symmetry-violating nonequilibrium effect is a magnetization directed along the electric current, that is, a magnetization induced by an electric field. For an arbitrary density matrix, the magnetization is defined as

$$\vec{M} = \mu_e \mathrm{Sp} \int \frac{d^3\mathrm{p}}{(2\pi\hbar)^3} \sigma n. \tag{7.10}$$

Using the kinetic equation in the presence of the electric field for $n$, it is possible to obtain[23]

$$\vec{M} = \frac{1}{2} \mu_e n_0 \frac{b}{\mathscr{E}_F} \frac{d\tau}{d\mathscr{E}_F} v_F e \vec{E}. \tag{7.11}$$

Even in the case of a strong electric field (large current), it is also a very weak effect.

Recently, Vilenkin has proposed to use positron annihilation for revealing SRSV in metals.[24] Vilenkin has shown that the photon pair momentum distribution in such metals is asymmetric in the direction of positron polarization. This asymmetry (of the order of $b/\mathscr{E}_F$) should be easily detectable by present experimental techniques. We think this method could become the basic one for direct identification of SRSV in metals.

To narrow down a class of solids that can be possible candidates for SRSV, it is desirable to point out other effects that give indirect evidence of SRSV. In particular, metamagnetic behavior of itinerant electrons in some metal can be an indirect indication of SRSV.[10] In the presence of the magnetic field, the electrons can be divided into two groups moving inside the hemispheres with spins directed along and against $\vec{B}$. Due to the strong correlation between the spin and the momentum of an electron, the behavior of these two groups of electrons in a magnetic field is similar to the behavior of antiferromagnetic subnetworks. One subnetwork consists of electrons having a positive momentum projection on the direction of the magnetic field. Another subnetwork corresponds to the negative one.

Metamagnetism reveals itself in a typical nonlinear behavior of the magnetization in an external magnetic field. This behavior in our case is determined by the molecular field (equation 7.2) and deviates from the paramagnetic behavior. Calculation of the magnetization (equation 7.10) for the density matrix (equation 7.3) gives for the susceptibility, $\chi = M/B$, the relation,

$$\chi - \chi_0 \approx -\chi_0 \left(\frac{b}{\mathscr{E}_F}\right)^2, \tag{7.12}$$

where $\chi_0$ does not depend on $b$, and $b$ depends on the temperature and the magnetic field.

The temperature dependence of $b$ was studied in the previous section. Let us now investigate the dependence of $b$ on the magnetic field. In zero magnetic field, a self-consistent equation for $b$ is given by equation 6.4. In the presence of the magnetic field, equation 6.4 is replaced by

$$b(\vec{k}) + \psi_1 \int \frac{d^3p'}{(2\pi\hbar)^3} \vec{k}\vec{k}'(n_+ - n_-) \frac{\vec{k}\vec{\ell}(\vec{k}')}{\ell(\vec{k}')} = 0, \quad (7.13)$$

where $\ell$ and $n_\pm$ are given by equations 7.2 and 7.4. Equation 7.13 evidently has a trivial solution $b = 0$ like in the case of zero magnetic field. Another solution corresponding to SRSV can be obtained from equation 7.13 under the assumption that the band splitting is small compared with $\mu$. In this case, by expanding $n_\pm$ in a series of $\ell$ and performing the integration, we obtain

$$b^2 = b_0^2(T) - (\mu_e B)^2, \quad (7.14)$$

where $b_0(T)$ is the zero-field value of the parameter $b$ given by equation 6.7. Notice that the above calculation of $b$ is valid due to the assumption $|1 + \frac{1}{3}g\psi_1| \ll 1$. The dependence of $b$ on $\vec{k}$ appears in terms of higher order in $|1 + \frac{1}{3}g\psi_1|$.

Equation 7.14 gives the restoration of the reflectional symmetry at the field $B_R(T) = \mu_e^{-1}b_0(T)$. For $B > B_R$, the electron liquid turns into a paramagnetic state with $b = 0$. Combining field calculation of $b$ with results of the previous section, we find that SRSV exists in the region of temperatures and magnetic fields restricted by the inequality

$$\frac{T^2}{T_R^2(0)} + \frac{B^2}{B_R^2(0)} < 1, \quad (7.15)$$

where $T_R(0)$ is the critical temperature at zero field, $B_R(0) = \mu_e^{-1}b_0(0)$.

Equations 7.12 and 7.14 show that SRSV should manifest itself as a tendency to an antiferromagnetic behavior in the temperature and field dependence of the susceptibility. It should be emphasized that we obtain a tendency to antiferromagnetic or, more exactly, to metamagnetic behavior for a magnetically homogeneous electron liquid, which differs essentially from the case of SDW. The latter manifests itself in neutron diffraction measurements as spatially inhomogeneous spin ordering, while SRSV corresponds to homogeneous spin ordering. Note that one can consider SRSV as spin ordering in the momentum space.

A compound that has recently been proposed as a candidate for itinerant electron metamagnetism is $TiBe_2$.[11] It reveals the deviation from linearity in the magnetization versus field relation, and it also has phase transitions on temperature, $T_R(0) \approx 10$ K,[25] and magnetic field, $B_R(0) \approx 6T$.[26] Detailed comparison of the theory with experimental data on $TiBe_2$[10] shows that unusual properties of this compound could result from weak SRSV. Unfortunately, direct observation of weak SRSV (e.g., with the use of positron annihilation) is impeded by a small value of $\langle\vec{\sigma}\vec{k}\rangle$.

There is also another model of antiferromagnetism in homogeneous Fermi liquid.[27] It is spontaneous violation of the symmetry with respect to arbitrary spin rotations and rotations in the space of the orbital momentum of an electron. In transition metals with cubic symmetry of a crystalline lattice, the energy band of $d$-electrons is split by the crystal field into a twice degenerate $e_g$-band and a three times degenerate $t_{2g}$-band. This allows one to introduce the isotopic spin $\vec{\tau}$($\tau = \frac{1}{2}$ for $e_g$-band and $\tau = 1$ for $t_{2g}$-band) of an electron and to consider the $(\vec{\tau}\vec{\tau}')(\vec{\sigma}\vec{\sigma}')\psi(\vec{p}, \vec{p}')$ term in the two-particle scattering function. Quasiparticle energy in the antiferromagnetic state is then given by

$$\mathscr{E}(\vec{p}) = \mathscr{E}_0(\vec{p}) + b(\vec{p})(\vec{m}\vec{\sigma})(\vec{\nu}\vec{\tau}), \tag{7.16}$$

where $\vec{m}$ and $\vec{\nu}$ are unit vectors in spin and isospin spaces, respectively, and $\tau$-projections numerate antiferromagnetic subnetworks. It should be noted, however, that one can hardly expect the exact isospin degeneracy of the energy band. In the absence of the exact degeneracy, the model does not give the exact compensation of the magnetization by different subnetworks. This corresponds to the case of itinerant ferromagnetism. For that reason, the SRSV model seems more attractive for the description of itinerant antiferromagnetism than the isospin model.

## OTHER TYPES OF SPIN ORDERING

In previous sections, we have considered instabilities of a Fermi liquid with respect to either large negative values of $\psi_0$ or large negative values of $\psi_1$. In the first case, the itinerant ferromagnetism occurs as a result of the instability, while for the second case, we obtain SRSV. It is clear, however, that in the general case, taking into account both $\psi_0$ and $\psi_1$, as well as higher coefficients in the expansion of $\psi(\vec{p}, \vec{p}')$, can be important. For instance, it can be important when considering magnetic properties of $TiBe_2$ within the Fermi liquid theory. It is known that a small addition of copper to pure $TiBe_2$ easily turns this compound into a ferromagnetic state. This could mean that the inequality $1 + g\psi_0 > 0$ for pure $TiBe_2$ easily turns to $1 + g\psi_0 < 0$, which gives $|1 + g\psi_0| \ll 1$.

In this section, we will study the situation where stability conditions (equation 2.7) are broken for both $\psi_0$ and $\psi_1$. For simplicity, we will restrict ourselves to the consideration of the two first harmonics of the scattering function,

$$\psi_{\sigma\sigma'}(\vec{p}, \vec{p}') = (\psi_0 + \psi_1\vec{k}\vec{k}')\vec{\sigma}\vec{\sigma}', \tag{8.1}$$

and will seek the quasiparticle energy in the form,

$$\mathscr{E}(\vec{p}) = \mathscr{E}_0(\vec{p}) - \vec{\ell}(\vec{p})\vec{\sigma}, \tag{8.2}$$

$$\vec{\ell}(\vec{p}) = a\vec{\nu} + b(\vec{\nu}\vec{k})\vec{\nu}, \tag{8.3}$$

where $\vec{\nu}$ is some unit vector. Then, the quasiparticle density matrix is given by

$$n = \tfrac{1}{2}(n_+ + n_-) + \tfrac{1}{2}(n_+ - n_-)\vec{\nu}\vec{\sigma}, \tag{8.5}$$

$$n_\pm = \left[1 + \exp\frac{\mathscr{E}_\pm - \mu}{T}\right]^{-1}, \qquad \mathscr{E}_\pm = \mathscr{E}_0(\vec{p}) \mp \ell(\vec{p}), \tag{8.6}$$

where $\ell = a + b\vec{\nu}\vec{k}$.

The effective field $\vec{\ell}(\vec{p})$ consists of two terms. The first one is invariant under space reflection. This term is responsible for a spontaneous magnetization. The second term in equation 8.3 is noninvariant under space reflection. It is responsible for SRSV. Note that the effective field in equation 8.3 is the simplest one that satisfies the equation of self-consistency. A simpler sum $a\vec{\nu} + b\vec{k}$ does not satisfy this equation.

To obtain this equation, let us perform a small rotation of $\vec{\nu}$: $\vec{\nu} \rightarrow \vec{\nu}'$. Perturbations of the quasiparticle energy and the density matrix are

$$\delta\mathscr{E} = \mathscr{E}(\vec{p}, \vec{\nu}') - \mathscr{E}(\vec{p}, \vec{\nu}), \qquad \delta n = n(\vec{p}, \vec{\nu}') - n(\vec{p}, \vec{\nu}). \tag{8.7}$$

After substitution of $\delta\mathscr{E}$ and $\delta n$ into equation 3.6 and some simple transformations, this equation can be divided into two parts with respect to their symmetry under space reflection. This gives two equations for parameters $a$ and $b$:

$$a + \psi_0 \int \frac{d^3p}{(2\pi\hbar)^3} (n_+ - n_-) = 0, \tag{8.8}$$

$$b + \psi_1 \int \frac{d^3p}{(2\pi\hbar)^3} (\vec{k}\vec{\nu})(n_+ - n_-) = 0. \tag{8.9}$$

We will study zero temperature solutions of these equations for a weak symmetry violation, that is, for $\ell \ll \mu$. Developing $n_\pm$ as a series in $\ell$, we have (with the accuracy up to terms of the third order in $\ell$)

$$n_+ - n_- = -2\frac{\partial n_F}{\partial \mathscr{E}_0}\ell - \frac{1}{3}\frac{d^3 n_F}{\partial \mathscr{E}_0^3}\ell^3, \tag{8.10}$$

where $n_F$ is the Fermi distribution function, $n_F = [1 + \exp(\mathscr{E}_0 - \mu/T)]^{-1}$. After substitution of equation 8.10 into equations 8.8 and 8.9, the integration over momenta can be replaced by the integration over the energy and angle variables:

$$\int \frac{2 \cdot d^3p}{(2\pi\hbar)^3} \to \int d\mathscr{E}_0 g(\mathscr{E}_0) \int \frac{d\Omega}{4\pi}. \tag{8.11}$$

Integrating by parts, we obtain

$$a + \psi_0 g(\mu) \int \frac{d\Omega}{4\pi} \ell(\vec{k}) + \frac{1}{6}\psi_0 g''(\mu) \int \frac{d\Omega}{4\pi} \ell^3(\vec{k}) = 0, \tag{8.12}$$

$$b + \psi_1 g(\mu) \int \frac{d\Omega}{4\pi} (\vec{k}\vec{\nu})\ell(\vec{k}) + \frac{1}{6}\psi_1 g''(\mu) \int \frac{d\Omega}{4\pi} (\vec{k}\vec{\nu})\ell^3(\vec{k}) = 0. \tag{8.13}$$

Integration over angle variables, then gives

$$a\left[1 + \psi_0 g(\mu) - \frac{1}{6}\frac{g''(\mu)}{g(\mu)}(a^2 + b^2)\right] = 0, \tag{8.14}$$

$$b\left[1 + \frac{1}{3}\psi_1 g(\mu) - \frac{1}{2}\frac{g''(\mu)}{g(\mu)}\left(a^2 + \frac{b^2}{5}\right)\right] = 0, \tag{8.15}$$

where we have taken into account the conditions

$$|1 + \psi_0 g| \ll 1, \qquad |1 + \tfrac{1}{3} g\psi_1| \ll 1. \tag{8.16}$$

Note that the chemical potential $\mu$ in equations 8.14 and 8.15 is also a function of parameters $a$ and $b$. This function can be found from the condition that the electron spatial density $n_0$ is fixed due to the stiffness of a crystalline lattice and does not depend on parameters $a$ and $b$:

$$Sp \int \frac{d^3p}{(2\pi\hbar)^3} n = n_0. \tag{8.17}$$

Calculation with the density matrix in equation 8.5 gives, at zero temperature,

$$\mu = \mu_0 - \frac{1}{2}\frac{g'(\mu_0)}{g(\mu_0)}\left(a^2 + \frac{b^2}{3}\right). \tag{8.18}$$

Equations 8.14 and 8.15 can also be obtained by the minimization of the symmetry-violating part of the thermodynamic potential,

$$\delta\Omega = Sp_\sigma \int \frac{d^3p}{(2\pi\hbar)^3} (\mathscr{E}_0 - \mu)\delta n_\sigma(\vec{p}) + \frac{1}{2} Sp_\sigma Sp_{\sigma'} \int \frac{d^3p}{(2\pi\hbar)^3} \int \frac{d^3p'}{(2\pi\hbar)^3} \cdot (\psi_0 + \psi_1 \vec{k}\vec{k}')\vec{\sigma}\vec{\sigma}'\delta n_\sigma(\vec{p})\delta n_{\sigma'}(\vec{p}'), \tag{8.19}$$

where $\delta n$ is the symmetry-violating part of the density matrix. Rather awkward calculations give (with the accuracy up to terms of the fourth order in $a$ and $b$)

$$\delta\Omega = \frac{1}{2} g(\mu)\left[[1 + \psi_0 g(\mu)]a^2 + \frac{1}{3}\left(1 + \frac{1}{3}\psi_1 g(\mu)\right)b^2\right] - \frac{1}{12} g''(\mu)\left(\frac{a^4}{2} + \frac{b^4}{10} + a^2b^2\right), \tag{8.20}$$

where the conditions of equation 8.16 are also taken into account. Minimization of $\delta\Omega$ with respect to parameters $a$ and $b$ leads to equations 8.14 and 8.15.

To investigate these equations, it is convenient to introduce the dimensionless variables,

$$1 + \psi_0 g(\mu) = -f_0^2, \qquad 1 + {}^1\!/\!_3\psi_1 g(\mu) = -f_1^2, \tag{8.21}$$

$$\alpha = \frac{a}{\mu}, \qquad \beta = \frac{b}{\mu}, \qquad \omega = \frac{\delta\Omega}{\mu^2 \cdot g(\mu)}, \qquad \frac{g''(\mu)}{g(\mu)} = -\frac{6\gamma^2}{\mu^2}. \tag{8.22}$$

(Note that for free electrons, $g(\mu) \propto \sqrt{\mu}$, $\gamma^2 = 1/24$.) In terms of new variables, equations 8.14, 8.15, and 8.20 can be rewritten as

$$\alpha[-f_0^2 + \gamma^2(\alpha^2 + \beta^2)] = 0, \tag{8.23}$$

$$\beta\left[-\frac{1}{3}f_1^2 + \gamma^2\left(\alpha^2 + \frac{\beta^2}{5}\right)\right] = 0, \tag{8.24}$$

$$\omega = -\frac{1}{2}f_0^2\alpha^2 - \frac{1}{6}f_1^2\beta^2 + \frac{1}{2}\gamma^2\left(\frac{\alpha^4}{2} + \frac{\beta^4}{10} + \alpha^2\beta^2\right). \tag{8.25}$$

The conditions $g\psi_0 < -1$, $g\psi_1 < -3$, which are necessary for spin ordering in a Fermi liquid, can be fulfilled for a large enough density of states at the Fermi level

$g(\mu_0)$. For itinerant ferromagnets, the Fermi level $\mu_0$ is often situated at the maximum of the density-of-states function $g(\mu_0)$. For simplicity, we will use this assumption in our calculation. According to equation 8.18, it gives $\mu \approx \mu_0$.

It is easily seen that equations 8.23 and 8.24 have four different classes of solutions. We list them together with the corresponding values of the changes in reduced thermodynamic potential:

$$\alpha = 0, \qquad \beta = 0, \qquad \omega = 0, \tag{8.26a}$$

$$\alpha^2 = \frac{f_0^2}{\gamma^2}, \qquad \beta = 0, \qquad \omega = -\frac{f_0^4}{4\gamma^2}, \tag{8.26b}$$

$$\alpha = 0, \qquad \beta^2 = \frac{5}{3}\cdot\frac{f_1^2}{\gamma^2}, \qquad \omega = -\frac{5f_1^4}{36\gamma^2}, \tag{8.26c}$$

$$\begin{cases} \alpha^2 = (5f_1^2 - 3f_0^2)/12\gamma^2, \qquad \beta^2 = 5(3f_0^2 - f_1^2)/12\gamma^2 \\ \omega = -[f_0^2(5f_1^2 - 3f_0^2) + 5f_1^2(3f_0^2 - f_1^2)]/48\gamma^2. \end{cases} \tag{8.26d}$$

The solution in equation 8.26a corresponds to a minimum of the thermodynamic potential when the stability conditions $1 + g\psi_0 > 0$ and $1 + \frac{1}{3}g\psi_1 > 0$ are fulfilled. In other cases, it is a metastable solution. The solution in equation 8.26d exists for the range of parameters $f_1^2/3 < f_0^2 < 5f_1^2/3$. For this range, comparison of the value of $\omega$ given by equation 8.26d with $\omega$ of equations 8.26b and 8.26c shows that the solution in equation 8.26d is metastable. Hence, for $f_0^2 > \sqrt{5/3}\,f_1^2$, the ground state corresponds to the case of equation 8.26b, while for $f_0^2 < \sqrt{5/3}\,f_1^2$, the case of equation 8.26c is realized. Note that our calculations using the expansion of the density matrix into a power series of order parameters is verified only when the conditions $f_0^2 \ll 1$ and $f_1^2 \ll 1$ are fulfilled.

The anisotropic form of SRSV given by equations 8.2 and 8.3 was introduced in this section with the hope of obtaining the state of a Fermi liquid when both ferromagnetism and SRSV are presented. Our investigation shows, however, that such a state would be metastable. Either the ferromagnetic state or the SRSV is realized depending on the relation between Fermi liquid parameters. Thus, it becomes necessary to compare the thermodynamic potential of the ferromagnetic state with the thermodynamic potential of the isotropic SRSV considered in previous sections. An analogous calculation gives for isotropic SRSV,

$$\omega = -\frac{f_1^4}{4\gamma^2}, \tag{8.27}$$

which is lower than $-5f_1^4/36\gamma^2$ for anisotropic SRSV. However, note that the latter can be energetically favorable in anisotropic Fermi liquid, for example, for a strongly anisotropic Fermi surface. In this case, the direction of $\vec{\nu}$ in equation 8.3 would coincide with one of the anisotropy axes of a solid.

By comparing equation 8.26b with equation 8.26c or with equation 8.27, we obtain the phase diagram shown in FIGURE 2. It represents stable states of a Fermi liquid depending on the relation between $\psi_0$ and $\psi_1$. Phase transitions between paramagnetic, ferromagnetic, and SRSV states become possible after taking the temperature and the magnetic field effects into consideration. Changes in pressure or in impurity concentration can also cause phase transitions due to the changing of Fermi liquid parameters.

In this section, we have approximated the scattering function $\psi(\vec{p}, \vec{p}')$ by two coefficients, $\psi_0$ and $\psi_1$. Note that even such a simple form of $\psi(\vec{p}, \vec{p}')$ permits the choosing of a more general form of the quasiparticle energy than that given by

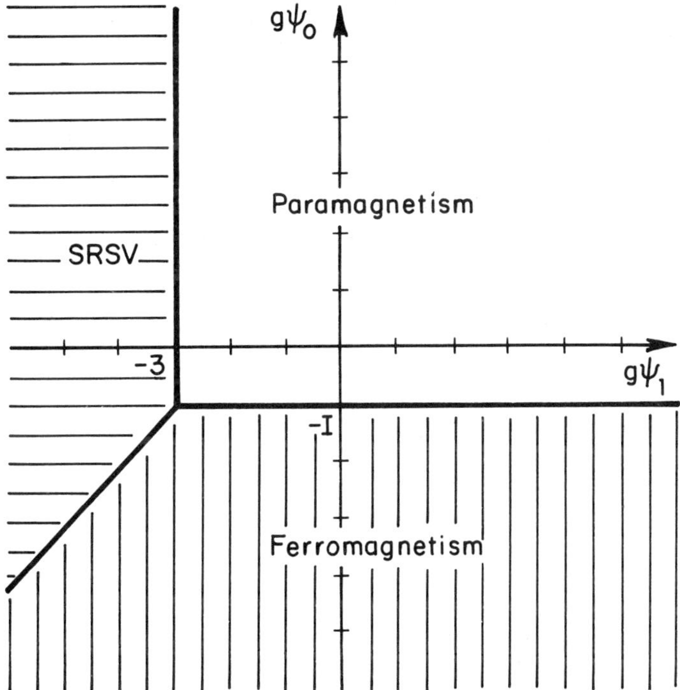


**FIGURE 2.** Phase diagram at $T = 0$ for a Fermi liquid with a scattering function approximated by coefficients $\psi_0$, $\psi_1$.

equations 8.2 and 8.3; for example,

$$\mathcal{E} = \mathcal{E}_0(\vec{p}) - a\vec{\nu}\vec{\sigma} - b\vec{k}\vec{\sigma} - c(\vec{k}\vec{\nu})(\vec{\nu}\vec{\sigma}) - d[\vec{k} \times \vec{\nu}]\vec{\sigma}. \tag{8.28}$$

For more complicated $\psi(\vec{p}, \vec{p}')$, the ground state of a Fermi liquid can also be more complicated. Investigation of a general case, however, as well as of the case of an anisotropic Fermi surface, is much more difficult.

## CONCLUSIONS

We have described different types of spin ordering in a homogeneous Fermi liquid. For this study, we used simplified forms of two-particle interaction and of quasiparticle energy spectrum. Our consideration, however, allows us to formulate some general statements regarding spin symmetry of the ground state of a Fermi liquid.

In the absence of spin ordering, the quasiparticle energy $\mathcal{E}_0(\vec{p})$ does not depend on spin variables; that is, in a given energy band, the quasiparticle energy is a function of the quasiparticle momentum only and does not depend on quasiparticle spin projection. The spin ordering in a Fermi liquid means that for a given quasiparticle momentum, the quasiparticle energy depends on the quasiparticle spin projection on some axis. The direction of this axis can also be a function of the momentum. The general form of the effective quasiparticle Hamiltonian is therefore given by

$$\mathcal{E}(\vec{p}) = \mathcal{E}_0(\vec{p}) - \vec{\ell}(\vec{p})\vec{\sigma}, \tag{9.1}$$

where $\vec{\ell}(\vec{p})$ can be considered to be an effective field acting on the quasiparticle spin. The effective field $\vec{\ell}(\vec{p})$ is the order parameter arising as a result of a spontaneous symmetry violation in a system of interacting fermions.

Equation 9.1 represents a general form of the effective electron Hamiltonian in metals. The latter defines magnetic and other properties of a solid. For nonzero $\vec{\ell}(\vec{p})$, these properties depend essentially on the symmetry of the effective field with respect to space reflection. The ground state of electron liquid is characterized by a spontaneous magnetization when $\vec{\ell}(\vec{p})$ is even with respect to the transformation $\vec{p} \rightarrow -\vec{p}$. The simplest situation of that type corresponds to a constant $\vec{\ell}$. This case is usually used for the consideration of itinerant ferromagnetism. In the ferromagnetic state, the ensemble average electron spin $\langle\vec{\sigma}\rangle$ has a nonzero value, and its direction defines the direction of the effective field $\vec{\ell}$. It should be noted that the ferromagnetic electron spectrum is even also with respect to time inversion because both $\vec{\sigma}$ and $\langle\vec{\sigma}\rangle$ change their signs under the transformation $t \rightarrow -t$. This is a fundamental property of stationary states in quantum mechanics.

Another general situation is realized when the effective field $\vec{\ell}(\vec{p})$ is odd with respect to the transformation $\vec{p} \rightarrow -\vec{p}$. This situation corresponds to the spontaneous reflectional symmetry violation (SRSV) in a Fermi liquid. SRSV is characterized by zero $\langle\vec{\sigma}\rangle$, but nonzero average helicity $\langle\vec{\sigma}\vec{k}\rangle$ over the electron ensemble. In the simplest case, when $\vec{\ell}$ is directed along $\vec{p}$, we can say that spins of electrons are ordered along their momenta. Note that the electron spectrum for SRSV is still invariant under time inversion because both $\vec{p}$ and $\vec{\sigma}$ change their signs under the transformation $t \rightarrow -t$. The $T$-invariance of the SRSV state of a Fermi liquid is possible for the spontaneous violation of $P$-invariance because the latter is accompanied by the spontaneous violation of $SU(2)$-invariance of the ground state with respect to spin rotations. Thus, the SRSV state more exactly corresponds to the spontaneous violation of $P \otimes SU(2)$ symmetry, while the ferromagnetic state corresponds to the spontaneous violation of $SU(2)$ symmetry.

By analogy with the itinerant ferromagnetism, the SRSV state of an electron liquid could also be called itinerant antiferromagnetism. The role of mirror antiferromagnetic subnetworks is played by the electrons moving along and against some axis. Indeed, due to the correlation between the spin and the momentum of an electron, these two groups of electrons can be considered as two ferromagnetic liquids. Their magnetizations exactly compensate each other. There is, however, an essential difference between this model and other models of antiferromagnetism.

Two other models of antiferromagnetism are considered in the literature. One of them is supplied to spins localized at sites of a crystalline lattice. This model explains antiferromagnetic behavior in terms of several compensating antiferromagnetic sub-networks put into each other. Another model of antiferromagnetism, namely, a spin density wave (SDW), is applied to itinerant electrons. SDW is characterized by a spontaneous magnetization depending on the coordinates, for example, by a local magnetization that rotates in the perpendicular plane when moving along some direction through a solid. Both these models of antiferromagnetism assume the existence of a spatial spin structure that reveals itself for neutron scattering in a solid.

On the other hand, the SRSV model of antiferromagnetism does not lead to any spatial spin structure of the ground state because it is characterized by zero $\langle \vec{\sigma} \rangle$. Therefore, neutron diffraction measurements cannot reveal any spatial spin order for itinerant antiferromagnetism of that type. Note that such a situation was established for itinerant antiferromagnet $TiBe_2$. There also exist alternative explanations of the unusual magnetic behavior of this compound. Thus, we need some kind of an experiment that can directly prove SRSV for a given solid.

There are two different groups of such experiments. The first group includes observation of macroscopic symmetry-violating effects, such as those occurring in some nonequilibrium states of an electric current flowing in the direction of the magnetic field. In such experiments, the existence of proportionality between polar and axial vectors would prove SRSV. These effects are weak, however, and their observation requires precision experimental techniques.

Another group of experiments could be based on direct observation of a preferential sign of electron helicity in a solid. Positron annihilation techniques can be most useful for that purpose.[24] Using a polarized electron beam, the effect of SRSV can be easily observed in photon pair momentum distribution. The asymmetry in the momentum distribution should be proportional to $\langle \vec{\sigma}\vec{k} \rangle$. Detection of this asymmetry would be a direct experimental proof of SRSV. This test should be applied to all compounds that reveal antiferromagnetic behavior in the absence of a spatial spin order detected by neutron diffraction measurements. However, note that for a strong SRSV ($\langle \vec{\sigma}\vec{k} \rangle \sim 1$), when the effect of SRSV could be easily detectable by present positron annihilation techniques, observation of the antiferromagnetic behavior would be, on the contrary, impeded because it would be clearly expressed at very high magnetic fields only. For weak SRSV, when antiferromagnetic behavior could be observed in lower fields, the anisotropy in photon pair momentum distribution for polarized positron annihilation would be as weak as $\langle \vec{\sigma}\vec{k} \rangle$.

Ferromagnetic ordering in a solid is accompanied by the creation of domains having opposite directions of $\langle \vec{\sigma} \rangle$. In the absence of the magnetic field, the average $\langle \vec{\sigma} \rangle$ over the sample may prove to be zero due to the creation of the domains. SRSV in a polycrystalline solid can also be characterized by random signs of $\langle \vec{\sigma}\vec{k} \rangle$ in different crystallites and, thus, by zero average $\langle \vec{\sigma}\vec{k} \rangle$ over the solid. This does not remove the antiferromagnetic behavior of a solid, but impedes the direct observation of nonzero $\langle \vec{\sigma}\vec{k} \rangle$ in such experiments as positron annihilation. For a monocrystalline solid, we see no reason for the creation of SRSV domains with different signs of $\langle \vec{\sigma}\vec{k} \rangle$. No long-range field is induced by a nonzero $\langle \vec{\sigma}\vec{k} \rangle$, contrary to the ferromagnetic case in which the magnetic field induced by $\langle \vec{\sigma} \rangle$ is the reason for the creation of ferromagnetic domains. Note that the absence of a long-range field associated with the order parameter $\langle \vec{\sigma}\vec{k} \rangle$ is also the main reason for the difficulties in the experimental detection of SRSV.

In this paper, using simplified forms of the two-particle interaction, we obtained only such stable states of a Fermi liquid that had definite parity, namely, the even parity for ferromagnetism and the odd one for SRSV. We see no reasons, however, why spin-ordered states with nondefinite parity could not exist. In the case of nondefinite parity, the effective field $\vec{\mathfrak{L}}(\vec{p})$ can be divided into two parts—one of which is even and another which is odd with respect to space reflection. This state represents a superposition of ferromagnetism and antiferromagnetism and could be called itinerant ferromagnetism. Detecting the antiferromagnetic component in this case could also be much more difficult than the ferromagnetic one. It is possible that some well-known itinerant ferromagnets have SRSV components of the effective field.

SRSV is a macroscopic parity-violating effect. It has nothing to do with extremely weak parity violation by the weak interaction. SRSV belongs to the class of phenomena where the symmetry of the ground state is lower than the symmetry of interaction. It occurs, as a lowest energy state, due to the effective exchange interaction of electrons. How exotic is SRSV in a comparison with itinerant ferromagnetism? In terms of Landau's Fermi liquid theory, the itinerant ferromagnetism is realized when $g\psi_0 < -1$, where $g$ is the density of electron states at the Fermi level and $\psi_0$ is the zero spherical harmonic of two-particle interaction. The SRSV state is realized when $g\psi_1 < -3$, where $\psi_1$ is the first harmonic. For alkali metals, the parameters, $g\psi_0$ and $g\psi_1$, calculated from spin wave spectra,[28] are small, for example, $g\psi_0 \approx -0.3$ and $g\psi_1 \approx -0.2$ for potassium. In transition metals, these parameters are greater due to a large value of the density of states. In particular, for a large number of itinerant ferromagnets, the inequality $g\psi_0 < -1$ is fulfilled. Fulfillment of the condition $g\psi_1 < -3$ is, of course, more difficult than $g\psi_0 < -1$. Therefore, SRSV should be a rarer effect in solids than in itinerant ferromagnetism. Nevertheless, we see no reasons why the conditions of SRSV could not be realized in some metals, even taking into account a great number of different transition metal compounds. Physical effects associated with SRSV are rather subtle. For this reason, SRSV may have passed unnoticed even in well-studied materials, and its experimental detection may prove to be a challenging task for experimentalists.

## ACKNOWLEDGMENTS

I thank Alex Vilenkin for many helpful discussions of the problem. I also thank Anthony Leggett for a discussion on the role of weak interaction for SRSV in solids.

## REFERENCES

1. FRENKEL, YA. I. 1928. Z. Phys. **49:** 31.
2. WIGNER, E. 1934. Phys. Rev. **46:** 1002.
3. OVERHAUZER, A. W. 1960. Phys. Rev. Lett. **4:** 462.
4. WOHLFARTH, E. P. 1968. J. Appl. Phys. **39:** 1061; EDWARDS, D. M. & E. P. WOHLFARTH. 1968. Proc. R. Soc. **A303:** 127.
5. ABRIKOSOV, A. A. & I. E. DZYALOSHINSKY. 1958. Zh. Eksp. Teor. Fiz. **35:** 771.
6. AKHIEZER, I. A. & E. M. CHUDNOVSKY. 1976. Solid State Commun. **19:** 5.

7. Pomeranchuk, I. Ya. 1958. Zh. Eksp. Teor. Fiz. **35:** 524.
8. Akhiezer, I. A. & E. M. Chudnovsky. 1978. Phys. Lett. **65A:** 433.
9. Chudnovsky, E. M. & A. Vilenkin. 1982. Phys. Rev. **B25:** 4301.
10. Chudnovsky, E. M. 1982. Phys. Lett. **93A:** 97.
11. Wohlfarth, E. P. 1981. Comments Solid State Phys. **10:** 39.
12. Landau, L. D. 1956. Zh. Eksp. Teor. Fiz. **30:** 1058; 1957. Zh. Eksp. Teor. Fiz. **32:** 59; 1958. Zh. Eksp. Teor. Fiz. **35:** 97.
13. Sylin, V. P. 1957. Zh. Eksp. Teor. Fiz. **33:** 495; 1958. Zh. Eksp. Teor. Fiz. **35:** 1243.
14. Pines, D. & P. Nozieres. 1966. The Theory of Quantum Liquids. Benjamin. New York.
15. Akhiezer, I. A. & E. M. Chudnovsky. 1974. Zh. Eksp. Teor. Fiz. **66:** 2303.
16. Akhiezer, I. A., V. G. Seaschenko & E. M. Chudnovsky. 1974. Zh. Eksp. Teor. Fiz. Pis'ma **20:** 739; 1975. Fiz. Tverd. Tela **17:** 2382; Fiz. Nizk. Temp. **1:** 477.
17. Akhiezer, I. A., V. G. Seaschenko & E. M. Chudnovsky. 1974. Fiz. Tverd. Tela **16:** 3684.
18. Seaschenko, V. G. & E. M. Chudnovsky. 1977. Fiz. Tverd. Tela **19:** 1720.
19. Akhiezer, I. A. & E. M. Chudnovsky. 1975. Fiz. Tverd. Tela **17:** 3220.
20. Lifshitz, E. M. & L. P. Pitaevsky. 1978. Statistical Physics, Part 2. Nauka. Moscow.
21. Krive, I. V. & E. M. Chudnovsky. 1976. Zh. Eksp. Teor. Fiz. Pis'ma **23:** 531; Krive, I. V., A. D. Linde & E. M. Chudnovsky. 1976. Zh. Eksp. Teor. Fiz. **71:** 826.
22. Kogan, V. G. 1983. Phys. Rev. **B27:** 3073.
23. Vilenkin, A. Unpublished.
24. Vilenkin, A. In press.
25. Matthias, B. T., A. L. Giorgi, V. O. Struebing & J. L. Smith. 1978. Phys. Lett. **69A:** 221.
26. Monod, P., I. Felener, G. Chouteau & D. Shatiel. 1980. J. Phys. (Paris) Lett. **41:** L-511.
27. Akhiezer, I. A. & E. M. Chudnovsky. 1975. Fiz. Tverd. Tela **17:** 1907.
28. Platzman, P. M. & P. A. Wolff. 1973. Waves and Interactions in Solid State Plasmas. Academic Press. New York.

# Spin Correlations and Magnetization Law in Ferromagnets with Random Anisotropy

EUGENE M. CHUDNOVSKY

*Prospekt Gagarina 199*
*Kharkov 310080, Union of Soviet Socialist Republics*

## INTRODUCTION

The last decade in the physics of magnetism has been marked by the discovery of a number of magnetic systems that can be considered as ferromagnets with random anisotropies. The most known example of such a system is an amorphous ferromagnet. The word "ferromagnet" means that the exchange interaction provides parallel orientation of neighboring atomic spins so that the solid can be characterized by the local magnetization, $\vec{M}(\vec{x})$. The direction of the magnetization at a point $\vec{x}$ depends on the anisotropy field.

In a crystalline ferromagnet, magnetic anisotropy is present due to the spatial anisotropy of a crystal. This anisotropy together with ferromagnetic exchange creates a uniform magnetization, with its direction being defined by the symmetry of the crystal. In reality, however, the ground state of a macroscopic ferromagnetic crystal corresponds to zero net magnetization in zero magnetic field. To decrease the energy of the magnetic field surrounding the solid, its magnetic system splits into domains that differ from each other by the direction of the magnetization.

An amorphous solid is isotropic on large length scales. This isotropy breaks only when we turn to interatomic distances. More rigorously, the anisotropy reveals itself at distances comparable with (or less than) the length scale $R_a$ of a short-range structural order. The direction of local magnetic anisotropy follows the short-range structural order and thus randomly rotates over the solid, with $R_a$ being a typical scale of change.

What would be the behavior of the magnetization in such a solid? On the one hand, anisotropy, which is weak in comparison with exchange, cannot force the magnetization to rotate considerably on the scale $R_a$. One should suppose, therefore, that uniform magnetization spreads over scales that are large compared with $R_a$. On the other hand, for such scales, we have no special direction that would determine the direction of the magnetization in the absence of an external magnetic field. Rigorous analysis of the problem[1–4] shows that an arbitrarily weak random anisotropy destroys ferromagnetic long-range order, even though the characteristic scale $R_f$ of stochastic rotation of the magnetization over the solid is large in comparison with $R_a$. If $\alpha$ is the exchange stiffness and $\beta$ is the anisotropy constant, then $R_f \sim \alpha^2/\beta^2 R_a^3$.[3,4] It should be emphasized that such a behavior of the magnetization has nothing in common with the formation of ferromagnetic domains in a crystalline ferromagnet. First, the physical reason for $\langle \vec{M}(\vec{x}) \rangle = 0$ in an amorphous ferromagnet is random anisotropy, not magnetodipole energy. Secondly, ferromagnetic domains in a crystalline solid have uniform magnetization that rotates only in a thin domain wall separating domains, with the sizes of the domains being dependent on the geometry of the sample. Contrariwise, an amorphous ferromagnet is characterized by smooth stochastic rotations of the magnetization over the solid, with $R_f$ being an intrinsic parameter of the disordered media.

Magnetic properties of random anisotropy ferromagnets have been recently reviewed in reference 5. They include large zero-field susceptibility[6,7] $\chi \sim \beta^{-4}$, high sensitivity to the presence of uniform anisotropy,[6–8] low spin-resonance frequency in the long-wavelength limit,[9] etc. Fairly good agreement with the result of $\chi \sim \beta^{-4}$ has been recently obtained for some amorphous rare earth magnets.[10] The approach to saturation in a moderately large magnetic field characterized by the magnetization law, $\delta M/M_0 \sim H^{-1/2}$, has been predicted in reference 6. It was also predicted[5] that the correlation length for transverse components of the magnetization is proportional to $H^{-1/2}$ in this range of fields. Both these effects have been recently observed in amorphous ferromagnets.[11–13]

This paper contains some improvements of the theory. In the next section, we prove the hypothesis[6] that in zero magnetic field, the correlation function for the magnetization is given by

$$\langle \vec{M}(\vec{x}_1)\vec{M}(\vec{x}_2)\rangle = M_0^2 \exp(-|\vec{x}_1 - \vec{x}_2|/R_f), \tag{1.1}$$

where

$$R_f = \frac{60\pi\alpha^2}{\beta^2\int d^3x\Gamma(x)} \tag{1.2}$$

and $\Gamma(x)$ is the correlation function for random anisotropy axes. The third section is devoted to the behavior of a random anisotropy ferromagnet in the external magnetic field. As is known[6] (see also references 14 and 15), in approaching saturation, $\delta M \sim H^{-1/2}$ for $H \ll \alpha M_0/R_a^2$ and $\delta M \sim H^{-2}$ for $H \gg \alpha M_0/R_a^2$. We show that for $\Gamma(x) = \exp(-|\vec{x}|/R_a)$, the approach to magnetic saturation is defined by the law,

$$\frac{\delta M}{M_0} = \frac{1}{15}\left(\frac{\beta R_a^2}{\alpha}\right)^2 \frac{1}{\sqrt{h}(1+\sqrt{h})^3}, \tag{1.3}$$

which includes $H^{-1/2}$ and $H^{-2}$ laws as limiting cases, $h = HR_a^2/\alpha M_0$.

## BEHAVIOR OF THE MAGNETIZATION AT $H = 0$

Consider the energy functional,[6]

$$E = \int d^3x\{\tfrac{1}{2}\alpha(\nabla\vec{M})^2 - \tfrac{1}{2}\beta(\vec{n}\vec{M})^2 + \lambda(\vec{x})\vec{M}^2\}, \tag{2.1}$$

where $\vec{n}(\vec{x})$ is a unit vector defining the direction of local anisotropy; $\lambda(\vec{x})$ plays the role of the local Lagrange multiplier, which is introduced to take into account that $[\vec{M}(\vec{x})]^2 = M_0^2$. We will assume that

$$\langle n_i(\vec{x}')n_j(\vec{x}')n_k(\vec{x}'')n_l(\vec{x}'')\rangle = \tfrac{1}{15}(\delta_{ij}\delta_{kl} + \delta_{ik}\delta_{jl} + \delta_{il}\delta_{jk})\Gamma(\vec{x}' - \vec{x}''), \tag{2.2}$$

where $\Gamma(\vec{x}' - \vec{x}'')$ rapidly goes to zero for $|\vec{x}' - \vec{x}''| > R_a$ and $\Gamma(0) = 1$. Tensor coefficients in equation 2.2 are found from the condition $\vec{x}' = \vec{x}''$.

Variation of $E$ with respect to $\vec{M}(\vec{x})$ gives

$$\alpha\nabla^2\vec{M} + \beta\vec{n}(\vec{n}\vec{M}) = \lambda\vec{M}. \tag{2.3}$$

Multiplying equation 2.3 by $\vec{M}(\vec{x})$, we obtain the formal expression for $\lambda(\vec{x})$:

$$\lambda = \alpha\frac{\vec{M}\nabla^2\vec{M}}{M_0^2} + \beta\frac{(\vec{n}\vec{M})^2}{M_0^2}. \tag{2.4}$$

Both $\vec{n}(\vec{x})$ and $\vec{M}(\vec{x})$ are random functions of $\vec{x}$. Thus, the most general form of $\lambda(\vec{x})$ is given by

$$\lambda(\vec{x}) = \alpha k^2 + \tilde{\lambda}(\vec{x}), \tag{2.5}$$

where $k$ is a constant and $\tilde{\lambda}(\vec{x})$ is an oscillating function of $\vec{x}$. Then equation 2.3 can be rewritten as

$$(\nabla^2 - k^2)\vec{M} = \alpha^{-1}[\tilde{\lambda}\vec{M} - \beta\vec{n}(\vec{n}\vec{M})]. \tag{2.6}$$

It is convenient to present equation 2.6 in the integral form of

$$\vec{M}(\vec{x}) = \alpha^{-1}\int d^3x' G_k(\vec{x} - \vec{x}')[\tilde{\lambda}'\vec{M}' - \beta\vec{n}'(\vec{n}'\vec{M}')], \tag{2.7}$$

where $f' \equiv f(\vec{x}')$ and where

$$G_k(\vec{x}) = -\frac{e^{-k|\vec{x}|}}{4\pi|\vec{x}|} \tag{2.8}$$

is the Green-function of equation 2.6 satisfying the equation

$$(\nabla^2 - k^2)G_k(\vec{x}) = \delta(\vec{x}). \tag{2.9}$$

With the help of equation 2.7, it is easy to obtain

$$\langle\vec{M}(\vec{x}_1)\vec{M}(\vec{x}_2)\rangle = \alpha^{-2}\int\int d^3x'd^3x'' G_k(\vec{x}_1 - \vec{x}')G_k(\vec{x}_2 - \vec{x}'') \cdot \langle[\tilde{\lambda}'\vec{M}' - \beta\vec{n}'(\vec{n}'\vec{M}')][\tilde{\lambda}''\vec{M}'' - \beta\vec{n}''(\vec{n}''\vec{M}'')]\rangle. \tag{2.10}$$

To perform integration, let us return to the expression for $\lambda(\vec{x})$. The first term in equation 2.4 is of the order of $\alpha/R_f^2$, while the second term is of the order of $\beta$. Qualitative consideration of the previous section shows that the scale $R_f$, which characterizes the spatial variation of $\vec{M}(\vec{x})$, is much greater than the scale $R_a$, which characterizes the variation of $\vec{n}(\vec{x})$. We will assume that $R_f^{-2}$, as well as $k^2$, is of the higher than first order in $\beta$. (Further calculation verifies this assumption.) Then, to the lowest order in $\beta$, one can replace $\tilde{\lambda}(\vec{x})$ in equation 2.10 by $\beta(\vec{n}\vec{M})^2/M_0^2$. This gives

$$\langle\vec{M}(\vec{x}_1)\vec{M}(\vec{x}_2)\rangle = \left(\frac{\beta M_0}{\alpha}\right)^2\int\int d^3x'd^3x'' G_k(\vec{x}_1 - \vec{x}')G_k(\vec{x}_2 - \vec{x}'') \cdot \langle n_i'n_j'n_k''n_l''m_j'm_l''(m_i'm_m' - \delta_{im})(m_k''m_m'' - \delta_{km})\rangle, \tag{2.11}$$

where we have introduced $\vec{m}(\vec{x}) = \vec{M}(\vec{x})/M_0$. Taking into account that $G_k(\vec{x})$ and

$\vec{m}(\vec{x})$ slightly vary on the scale $R_a$, and also that $[\vec{m}(\vec{x})]^2 = 1$, with the help of equation 2.2, we obtain

$$\langle \vec{M}(\vec{x}_1)\vec{M}(\vec{x}_2)\rangle = \frac{2}{15}\left(\frac{\beta M_0}{\alpha}\right)^2 \Omega \int d^3x G_k(\vec{x}_1 - \vec{x})G_k(\vec{x}_2 - \vec{x}), \tag{2.12}$$

where

$$\Omega = \int d^3x \Gamma(x). \tag{2.13}$$

The integral in equation 2.12 can be calculated rigorously,

$$\int d^3x G_k(\vec{x}_1 - \vec{x})G_k(\vec{x}_2 - \vec{x}) = \frac{1}{8\pi k} e^{-k|\vec{x}_1 - \vec{x}_2|}, \tag{2.14}$$

which gives

$$\langle \vec{M}(\vec{x}_1)\vec{M}(\vec{x}_2)\rangle = \frac{\beta^2\Omega}{60\pi\alpha^2 k} M_0^2 e^{-k|\vec{x}_1 - \vec{x}_2|}. \tag{2.15}$$

Because $\langle |\vec{M}(\vec{x})|^2\rangle = M_0^2$, the parameter $k$ can be determined from the condition $\vec{x}_1 = \vec{x}_2$:

$$R_f = k^{-1} = \frac{60\pi}{\Omega}\left(\frac{\alpha}{\beta}\right)^2. \tag{2.16}$$

The correlation function is then given by

$$\langle \vec{M}(\vec{x}_1)\vec{M}(\vec{x}_2)\rangle = M_0^2 e^{-|\vec{x}_1 - \vec{x}_2|/R_f}. \tag{2.17}$$

To define the range of validity of equations 2.16 and 2.17, let us consider $\Gamma(x) = \exp(-|\vec{x}|/R_a)$. In this case, we obtain[4,5]

$$R_f = \frac{15}{2}\frac{\alpha^2}{\beta^2 R_a^3}. \tag{2.18}$$

The condition $R_f \gg R_a$ then gives

$$\Lambda \equiv \frac{\beta R_a^2}{\alpha} < 1. \tag{2.19}$$

This inequality provides quantitative separation of random anisotropy ferromagnets into two groups: ferromagnets with weak local anisotropy ($\Lambda < 1$) and ferromagnets with large anisotropy ($\Lambda > 1$).[6] In the latter case, all the atomic spins look along their local anisotropy fields, that is, $R_f = R_a$.

To conclude this section, we note that the Fourier transform of the correlation function (equation 2.17) is

$$S(Q) = \frac{8\pi M_0^2 k}{(Q^2 + k^2)^2}. \tag{2.20}$$

It should correspond to the Lorentzian squared term in the cross section of neutron scattering, which has been recently observed in many amorphous magnetic systems.[12,16]

## MAGNETIZATION LAW

For small $\Lambda$, the system described above is characterized by the large magnetic susceptibility,[6]

$$\chi \sim \frac{R_a^2}{\alpha}\Lambda^{-4}. \tag{3.1}$$

Thus, for a weak magnetic field,

$$H_s \sim \beta M_0 \Lambda^3; \tag{3.2}$$

the regime of approaching saturation should be achieved. This regime is divided into two subregimes:[6] $\delta M \sim H^{-1/2}$ for $H_s \ll H \ll \alpha M_0/R_a^2$, and $\delta M \sim H^{-2}$ for $H \gg \alpha M_0/R_a^2$. Our purpose is to obtain the magnetization law for the whole range of the field $H > H_s$.

Deviation of the magnetization from $M_0$ is defined by the averaged square of its transverse (with respect to the field) component,

$$\frac{\delta M}{M_0} = \frac{1}{2M_0^2}\langle [\vec{M}_\perp(\vec{x})]^2 \rangle. \tag{3.3}$$

The latter is given by[5]

$$\langle M_\perp^2 \rangle = \frac{2}{15}\left(\frac{\beta M_0^2}{4\pi\alpha}\right)^2 \int\int d^3x d^3y \frac{e^{-(|\vec{x}|+|\vec{y}|)/R_\perp}}{|\vec{x}|\cdot|\vec{y}|}\Gamma\left(\frac{|\vec{x}-\vec{y}|}{R_a}\right), \tag{3.4}$$

where

$$R_\perp = \left(\frac{\alpha M_0}{H}\right)^{1/2}. \tag{3.5}$$

Let us consider the reduced magnetic field,

$$h = \left(\frac{R_a}{R_\perp}\right)^2 = \Lambda \frac{H}{\beta M_0}, \tag{3.6}$$

and rewrite equation 3.4 by using dimensionless coordinates, $\vec{\zeta} = \vec{x}/R_a$ and $\vec{\eta} = \vec{y}/R_a$. Then, equation 3.3 can be rewritten as

$$\frac{\delta M}{M_0} = \frac{1}{15}\left(\frac{\Lambda}{4\pi}\right)^2 \int\int d^3\zeta d^3\eta \frac{e^{-\sqrt{h}(|\vec{\zeta}|+|\vec{\eta}|)}}{|\vec{\zeta}|\cdot|\vec{\eta}|}\Gamma(|\vec{\zeta}|-\vec{\eta}|). \tag{3.7}$$

In regard to $\Gamma(\zeta)$, we know that it rapidly goes to zero for large $\zeta$ and $\Gamma(0) = 1$. These properties of $\Gamma(\zeta)$ completely define the magnetization law in two subregimes: $h \ll 1$ and $h \gg 1$. Indeed, for $h \ll 1$, only $\zeta \sim \eta$ effectively contributes to the integral in equation 3.7. It, therefore, reduces to the form

$$\frac{\delta M}{M_0} = \frac{1}{15}\left(\frac{\Lambda}{4\pi}\right)^2 \int d^3\zeta \Gamma(\zeta) \int d^3\eta \frac{e^{-2\sqrt{h}|\vec{\eta}|}}{|\vec{\eta}|^2}. \tag{3.8}$$

Further calculation gives

$$\frac{\delta M}{M_0} = \frac{R_\perp}{2R_f} = \frac{\sqrt{\alpha}}{2R_f}\left(\frac{M_0}{H}\right)^{1/2}, \tag{3.9}$$

where $R_f$ is defined by equation 2.16.

For $h \gg 1$, only small $\zeta$ and $\eta$ contribute to the integral in equation 3.7; this allows one to put $\Gamma(|\vec{\zeta} - \vec{\eta}|) = 1$. It gives

$$\frac{\delta M}{M_0} = \frac{1}{15}\left(\frac{\Lambda}{4\pi}\right)^2 \left(\int d^3\zeta \frac{e^{-\sqrt{h}|\vec{\zeta}|}}{|\vec{\zeta}|^2}\right). \tag{3.10}$$

Performing the integration, we obtain

$$\frac{\delta M}{M_0} = \frac{1}{15}\left(\frac{\beta M_0}{H}\right)^2. \tag{3.11}$$

In crossover range of the field, $h \sim 1$, the magnetization law depends on the explicit form of the correlation function $\Gamma(|\vec{\zeta} - \vec{\eta}|)$. For $\Gamma = \exp(-|\vec{\zeta} - \vec{\eta}|)$, equation 3.7 can be rewritten as

$$\frac{\delta M}{M_0} = \frac{\Lambda^2}{30}\int_0^\infty d\zeta \zeta e^{-\sqrt{h}\zeta} \int_0^\infty d\eta \eta e^{-\sqrt{h}\eta} \int_{-\pi}^{\pi} d\varphi \sin\varphi\, e^{-\sqrt{\zeta^2+\eta^2-2\zeta\eta\cos\varphi}}. \tag{3.12}$$

After integration over $\varphi$, we obtain

$$\frac{\delta M}{M_0} = \frac{\Lambda^2}{30}\int_0^\infty d\zeta \int_0^\infty d\eta e^{-\sqrt{h}(\zeta+\eta)}[e^{-|\zeta-\eta|}(|\zeta - \eta| + 1) - e^{-(\zeta+\eta)}(\zeta + \eta + 1)]. \tag{3.13}$$

Further integration gives

$$\frac{\delta M}{M_0} = \frac{\Lambda^2}{15}\frac{1}{\sqrt{h}(1 + \sqrt{h})^3}. \tag{3.14}$$

This formula is valid for the whole range of the magnetic field where $\delta M \ll M_0$. In the limiting cases of $h \ll 1$ and $h \gg 1$, it turns into the formulae obtained above.

Our consideration is applied to systems with a large ratio of exchange to anisotropy. For such systems, the crossover magnetic field, $H \sim \alpha M_0/R_a^2$, may be rather large. (For example, values of parameters, $M_0 = 10^3$ Oe, $\alpha M_0^2 = 10^{-7}$ erg/cm, and $R_a = 5$Å, lead to a crossover field $H \sim 40$ kOe.) This range of fields, however, should be of great interest because it could bring information about structural short-range order in an amorphous ferromagnet. Indeed, if the magnetization law in a solid follows equation 3.14, it should indicate that structural disorder in this solid is described by the correlation function $\Gamma = \exp(-|\vec{x}_1 - \vec{x}_2|/R_a)$. Otherwise, a proper form of the correlation function could be found by fitting $\Gamma$ in equation 3.7 to experimental data for $\delta M(H)$.

## REFERENCES

1. IMRY, Y. & S. MA. 1975. Phys. Rev. Lett. **35:** 1399.
2. PELCOVITS, R. E., E. PYTTE & J. RUDNICK. 1978. Phys. Rev. Lett. **40:** 476.
3. ALBEN, R., J. J. BECKER & M. C. CHI. 1978. J. Appl. Phys. **49:** 1653.
4. CHUDNOVSKY, E. M. & R. A. SEROTA. 1982. Phys. Rev. **B26:** 5272.
5. CHUDNOVSKY, E. M., W. M. SASLOW & R. A. SEROTA. 1986. Phys. Rev. **B33:** 251.
6. CHUDNOVSKY, E. M. & R. A. SEROTA. 1983. J. Phys. **C16:** 4181.
7. CHUDNOVSKY, E. M. 1983. J. Magn. Magn. Mater. **40:** 21.
8. CHUDNOVSKY, E. M. & R. A. SEROTA. 1984. IEEE Trans. Magn. **MAG-20:** 1400.
9. CHUDNOVSKY, E. M. & R. A. SEROTA. 1984. J. Magn. Magn. Mater. **43:** 48.
10. BARBARA, B. & B. DIENY. 1985. Physica **130B:** 245.
11. SELLMYER, D. J. & S. NAFIS. 1985. J. Appl. Phys. **59:** 3584.
12. RHYNE, J. J. IEEE Trans. Magn. To be published.
13. PARK, M. J., S. M. BHAGAT, M. A. MANHEIMER & K. MOORJANI. Preprint.
14. FAHNLE, M. & H. KRONMULLER. 1978. J. Magn. Magn. Mater. **8:** 149.
15. CALLEN, E., Y. LIU & M. C. CHI. 1977. Phys. Rev. **B16:** 263.
16. AEPPLI, G., S. M. SHAPIRO, R. J. BIRGENEAU & H. S. CHEN. 1984. Phys. Rev. **B29:** 2589.

# Repulsive Gravitation[a]

Ø. GRØN
*Oslo College of Engineering*
*Oslo, Norway*
*and*
*Institute of Physics*
*University of Oslo*
*0316 Oslo 3, Norway*

A fundamental property of Newtonian gravitational theory is that the force of gravity is always attractive. This attractive property of Newtonian gravitational theory may be seen as follows: Matter with density $\rho_G$ generates a gravitational potential $\phi$ according to Poisson's equation,[b]

$$\nabla^2\phi = 4\pi\rho_G, \tag{1}$$

and the gravitational potential generates acceleration according to Newton's second law,

$$a = -\nabla\phi. \tag{2}$$

These equations always result in gravitational attraction because matter with negative mass density, $\rho_G < 0$, is assumed not to exist.

Consider now the field equations of general relativity:

$$R^\mu_\nu = 8\pi(T^\mu_\nu - \tfrac{1}{2}\delta^\mu_\nu T), \qquad T = T^0_0 + T^1_1 + T^2_2 + T^3_3. \tag{3}$$

The ${}^0_0$-component takes the form

$$R^0_0 = 8\pi(T^0_0 - \tfrac{1}{2}T) = 4\pi(T^0_0 - T^1_1 - T^2_2 - T^3_3). \tag{4}$$

In a static weak gravitational field, the line element may be written as

$$ds^2 = (1 + 2\phi)dt^2 - (1 - 2\phi)(dx^2 + dy^2 + dz^2), \qquad \phi \ll 1, \tag{5}$$

where $\phi$ is the Newtonian gravitational potential. This form of the line element gives

$$R^0_0 = \nabla^2\phi, \tag{6}$$

so the ${}^0_0$-component of the field equations takes the form

$$\nabla^2\phi = 4\pi(T^0_0 - T^1_1 - T^2_2 - T^3_3). \tag{7}$$

Also, in the case of a static weak field, the geodesic equation reduces to the form of

[a] This is an extended version of a report held at a refusnik seminar arranged by J. Alpert.
[b] In this paper, units so that $c = G = 1$ are used, where $c$ is the velocity of light and $G$ is the gravitational constant.

equation 2. This means that the density of active gravitational mass is given by

$$\rho_G = T^0_0 - T^1_1 - T^2_2 - T^3_3. \quad \textbf{(8)}$$

This equation shows that the stresses of a system contribute to its gravitational mass. Even in the case of a weak gravitational field, this effect may not be negligible. The stresses may even be so strong that they outweigh the mass density, $T^0_0 = \rho$, so that $\rho_G < 0$. In this case, there is gravitational repulsion away from the region. Essentially, the same expression as equation 8 is valid also in the full theory.[1,2] However, in the Newtonian limit, the energy equivalent of the stresses vanishes, and there is no gravitational repulsion.

It is not obvious that there may really exist systems with the extreme stresses that are needed to realize the relativistic phenomenon of repulsive gravitation. If one considers a system kept together by molecular forces, the maximal stretching of it before it is torn apart gives rise to a stress corresponding to a mass density of the order $10^{-10}$ g/cm$^3$. In fact, no material that we know of can tolerate the stresses needed to make them sources of repulsive gravitation.

This might seem to be the end of the phenomenon of repulsive gravitation. However, quantum field theory and Lorentz invariance of the vacuum have surprising gravitational consequences. According to quantum field theory, the vacuum has a nonvanishing energy density, which may, for example, be due to vacuum polarization or Higgs fields. Lorentz invariance of the properties of the vacuum implies that it must be described by an energy-momentum tensor of the form[2]

$$T_{\mu\upsilon} = \rho_0 g_{\mu\upsilon}, \quad \textbf{(9)}$$

where $\rho_0$ is a constant mass density. If we compare the vacuum to a fluid and describe it by an energy-momentum tensor of the form

$$T_{\mu\upsilon} = (\rho + p)\delta^0_\mu \delta^0_\upsilon - p g_{\mu\upsilon}, \quad \textbf{(10)}$$

then equation 9 implies

$$p = -\rho_0. \quad \textbf{(11)}$$

According to this equation, the vacuum is in a state of extreme tension. The gravitational mass density as given by equation 8 is

$$\rho_G = -2\rho_0. \quad \textbf{(12)}$$

The gravitational mass-density of the vacuum is negative. Thus, the vacuum will show a tendency to expand under its own repulsive gravitation.

Still, the vacuum we experience has a negligible energy density. The much greater mass density of matter and radiation is dominating. Therefore, it seems that repulsive gravitation is of no physical significance after all.

However, there might exist extreme situations favoring the realization of repulsive gravitation. In fact, the repulsive property of gravitation has recently appeared in two widely different applications of general relativity. Its role in the construction of general relativistic classical electron models and in the relativistic description of the very early universe has shown that the phenomenon of repulsive gravitation may be a general relativistic effect of great importance.

Let us first consider the classical electron models. Tiwari, Rao, and Kanakamedala[3] have found a class of electromagnetic mass models that are static solutions of the Einstein-Maxwell equations. The interior of these solutions represents a charged fluid obeying the equation of state (equation 11). Thus, it is a charged vacuum fluid and a source of repulsive gravitation. If $r_0$ is the radius of the spherical fluid distribution, the gravitational mass of the system (fluid plus electrical field) is negative for $r < (5/4)r_0$.[1] This seems to make the equilibrium problem of the models even worse than that of the purely electromagnetic classical electron models. However, by generalizing the Tolman-Oppenheimer-Volkov equation to the charged case, it was shown that the system is in hydrostatic equilibrium. It is the pressure gradient of the vacuum fluid that keeps equilibrium with the repulsive gravitational and electrostatic force. This is the Poincaré stress, which had earlier been introduced ad hoc to stabilize the electromagnetic classical electron models. If the energy of the vacuum fluid is due to vacuum fluctuations, then, according to the general relativistic models, the existence of the Poincaré stress is a consequence of the quantum mechanical properties of the vacuum.

Some years ago, J. M. Cohen and M. D. Cohen[4] found a solution of the Einstein-Maxwell equations with the peculiar property that the redshift (from a point in the source to infinity) is maximal at the surface rather than at the center. In other words, there is a blueshift from the center to the surface.

This may now be explained as a consequence of repulsive gravitation. The solution of Cohen and Cohen represents a charged spherical shell with vacuum polarization inside the shell. There is Reissner-Nordström space-time outside the shell and de Sitter space-time inside it. The gravitational mass inside the shell is negative. There is then a lower gravitational potential at the shell than at the center. This is the physical reason of the gravitational blueshift noted by Cohen and Cohen.

A source with allowable physical properties of the Kerr-Newman space-time has recently been found by Lopez.[5] This source consists of a charged rotating shell, which is the surface of an oblate ellipsoid of revolution having a minor axis equal to the classical electron radius (in the case that the charge and mass of the source equals the electron's charge and mass). Inside the shell, there is Minkowski space-time, and outside, there is Kerr-Newman geometry.

Lopez calculated the energy-momentum tensor of the source using distribution theory. It turns out that a much simpler construction of his source may be given,[6] using Israel's formulation[7] of the theory of surface layers in general relativity. The result is a classical electron model (including spin, and with the correct gyromagnetic ratio) that may be described as a gas of freely moving charged "bubbles" with negative energy density moving along a vacuum domain wall.[1]

It was shown by Ipser and Sikivie[8] that domain walls are sources of repulsive gravitation, and also that a spherical domain wall will collapse. The "bubbles" keep the wall static. The velocity field of the bubbles represents rigid rotation with subluminal velocity. This source of the Kerr-Newman field is free of singular properties.

The question of whether gravitation at all, and repulsive gravitation in particular, will turn out to be of importance in more realistic electron models, while taking into account quantum field theory, is still entirely open. However, the relevance of repulsive gravitation in connection with the classical electron models may indicate that it should not be neglected in the construction of quantum field theoretic electron models either.

Neglecting gravitation will always represent an approximation to reality; possibly, even a bad one.

We now turn to the description of the very early universe. By combining the grand unified field theories with general relativity, an interesting possibility appears. In 1981, Guth[9] noted that in the context of grand unified theories with spontaneous symmetry breaking, the critical temperature is of the order $T_{\mathrm{GUT}} \geqq 10^{28}$ K, and that it was possible for the universe to supercool by 28 or more orders of magnitude below the critical temperature. When this happened, the negative gravitational mass of the vacuum grew larger than the positive gravitational mass of matter and radiation. The universe became "vacuum dominated," and repulsive gravitation caused an exponentially accelerating expansion. This is the inflationary era, which solves the flatness and horizon problems of the Friedmann universe models; that is, it explains why the universe is homogeneous and how the spatial geometry became nearly flat.

There is an infinity of possible initial conditions for the expansion pattern of our universe. Only one of these represents an isotropic expansion. Thus, there is a priori a vanishing probability that the universe started with an isotropic initial expansion. As stated by Misner, Thorne, and Wheeler,[11] the fundamental cosmological question then is: why should the isotropic Friedmann metrics be a more accurate approximation to the real universe than an anisotropic Kasner metric?

One way to answer this question is to investigate the behavior of vacuum dominated anisotropic cosmological models. In the most simple case of Bianchi type-I spaces (which represent anisotropic world models with flat three-space), the solution of Einstein's field equations for a vacuum dominated space-time can be expressed as[10]

$$\left.\begin{aligned} &ds^2 = dt^3 - [R_i(t)dx^i]^2, \qquad i = 1, 2, 3, \\ &R_i = 2^{2/3}\sinh^{p_i}(\tfrac{3}{2}H_0 t)\cosh^{2/3-p_i}(\tfrac{3}{2}H_0 t), \\ &H_0 = (8\pi\rho_0/3)^{1/2}, \\ &\sum_{i=1}^{3} p_i = \sum_{i=1}^{3} p_i^2 = 1. \end{aligned}\right\} \tag{13}$$

This solution describes an anisotropic generalization of the flat de Sitter universe model. For $t \ll H_0^{-1}$, the expression for the expansion factors reduces to

$$R_i = R_{io}t^{p_i} \quad \text{(no summation)}, \tag{14}$$

which describes a Kasner universe model. For large $t$, $t \gg H_0^{-1}$, the solution of equation 13 approaches

$$R_1 = R_2 = R_3 = e^{H_0 t}. \tag{15}$$

This represents the isotropic de Sitter universe model with flat three-space. Thus, the solution of equation 13 develops from a Kasner universe with arbitrary anisotropy to the isotropic de Sitter universe.

The expansion anisotropy $A$ is defined as

$$A = \frac{1}{3}\sum_{i=1}^{3}(1 - H_i/H)^2, \qquad H = \frac{1}{3}\sum_{i=1}^{3} H_i, \qquad H_i = \dot{R}_1/R_i. \tag{16}$$

During the Guth inflationary era, which lasts from $t_1 = 1.0 \times 10^{-35}$ s to $t_2 = 1.3 \times 10^{-33}$ s, the expansion anisotropy of the solution of equation 13 decays by a factor of $10^{-168}$. Thus, the repulsive gravitation that causes the exponential expansion of the universe during the inflationary era provides a reason for the observed isotropy of our universe.

The evolution of the expansion factors and the anisotropy at the grand unified theory time of an anisotropic de Sitter universe with plane symmetry is illustrated in FIGURES 1 and 2.

Another important problem that must be investigated by means of more general cosmological models than the Friedmann models is the cosmological rotation problem. This has been stated by Ellis and Olive in the following way:[12] If the universe can rotate, why does it rotate so slowly?

The concept of a rotating universe is strange in that one seems to lack a standard that could represent zero rotation against which a rotation of the universe could be measured. However, the concept can be simply explained in the following way: Imagine a laboratory somewhere in the universe that is equipped with engines that can change the rotation of the laboratory. Observe the motion of a Foucault pendulum and adjust the motion of the laboratory so that the pendulum oscillates in a plane. If one then observes that the stars rotate around the laboratory, the universe is said to be rotating.

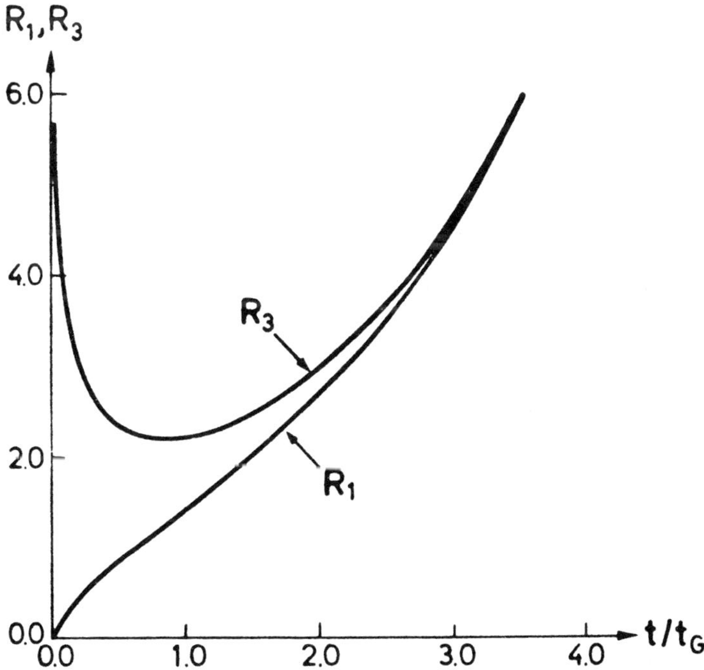


**FIGURE 1.** Expansion factors for a plane-symmetric generalized de Sitter universe. This represents a special case of the solution of equation 13 with $p_1 = p_2 = 2/3$, $p_3 = -1/3$. The radius in cm is less than $10^{-30}$ times the value on the $R$-axis.

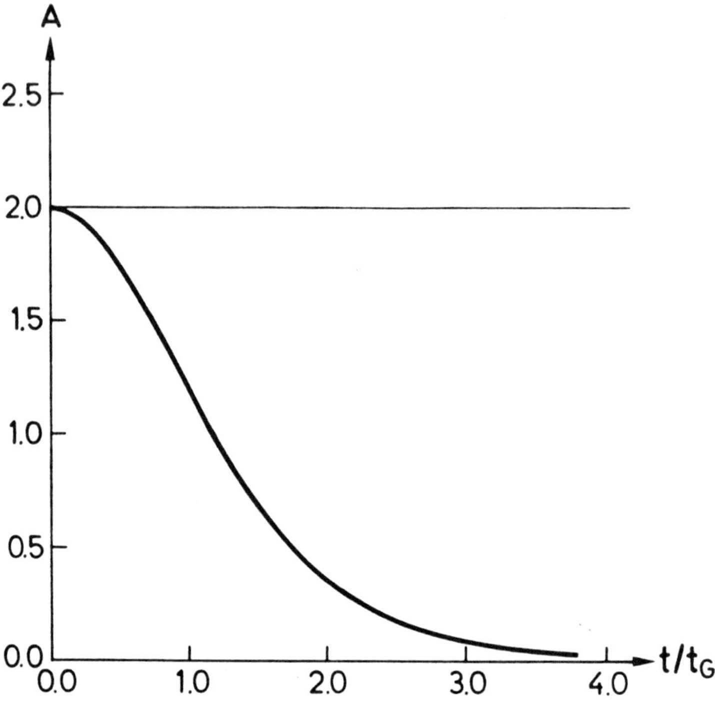


**FIGURE 2.** Average expansion anisotropy for the plane-symmetric generalized de Sitter universe. The Kasner universes have constant anisotropy $A = 2$.

The first rotating universe model was constructed in 1949 by Gödel.[13] This universe model, though, is not able to represent the space-time of our universe because it has no expansion. Several rotating universe models have later been constructed, some of which have expansion.[14–16] However, none of these models develop asymptotically towards a Friedmann model. Thus, neither of them can be realistic models of our universe.

It is remarkable that, here too, repulsive gravitation comes to the rescue. Applying Einstein's field equations to a vacuum dominated Bianchi type-IX cosmological model with rotation and expansion,[17] one finds a solution that can be expressed as

$$\left.\begin{aligned} ds^2 &= (dt + A\omega^1)^2 + B^2[(\omega^1)^2 + (\omega^2)^2 + (\omega^3)^2], \\ \omega^1 &= -\sin x^3 dx^1 + \sin x^1 \cos x^3 dx^2, \\ \omega^2 &= \cos x^3 dx^1 + \sin x^1 \sin x^3 dx^2, \\ \omega^3 &= \cos x^1 dx^2 + dx^3, \\ A &= c_0(c_1^2 e^{2H_0 t} - 1)^{1/2}(c_1^2 e^{2H_0 t} + 1)^{-1} e^{H_0 t/2}, \\ B &= (4H_0 c_1)^{-1}(c_1^2 e^{2H_0 t} + 1)e^{-H_0 t}, \end{aligned}\right\} \quad (17)$$

where $H_0$ is given as in equation 13. The $c_0$ and $c_1$ are integration constants, with $c_1 > 0$. This solution describes rotating generalizations of the de Sitter universe model. The vorticity of these universe models is given by

$$\omega^1 = 8c_0c_1^2H_0^2(c_1^2e^{2H_0t} - 1)^{1/2}(c_1^2e^{2H_0t} + 1)^{-3}e^{5H_0t/2}. \quad \textbf{(18)}$$

The solution found in reference 17 represents the special case $c_1 = 1$ of the general solution given here.

The universe models given by equations 17 and 18 have the notable property that they are singularity free. These models expand from initial values of the expansion factor, $B(0) = (c_1^2 + 1)/4H_0c_1$, that are greater than zero, and from initial values of the vorticity, $\omega^1(0) = 8c_0c_1^2H_0^2(c_1^2 - 1)^{1/2}(c_1^2 + 1)^{-3}$, that are finite.

The evolution of the expansion factor $B$ and the vorticity $\omega^1$ of a rotating de Sitter universe with $c_1 = 3$ at the grand unified theory time is illustrated in FIGURES 3 and 4.

During the Guth inflationary era, the vorticity decays by a factor of $10^{-142}$. Thus, the smallness of the observed upper limit for the rotation of our universe[18] may be explained as a consequence of an early inflationary era with exponential expansion due to repulsive gravitation.

The observed expansion of the universe is what is left of the explosively accelerated expansion caused by the gravitational repulsion during the inflationary era. Recent

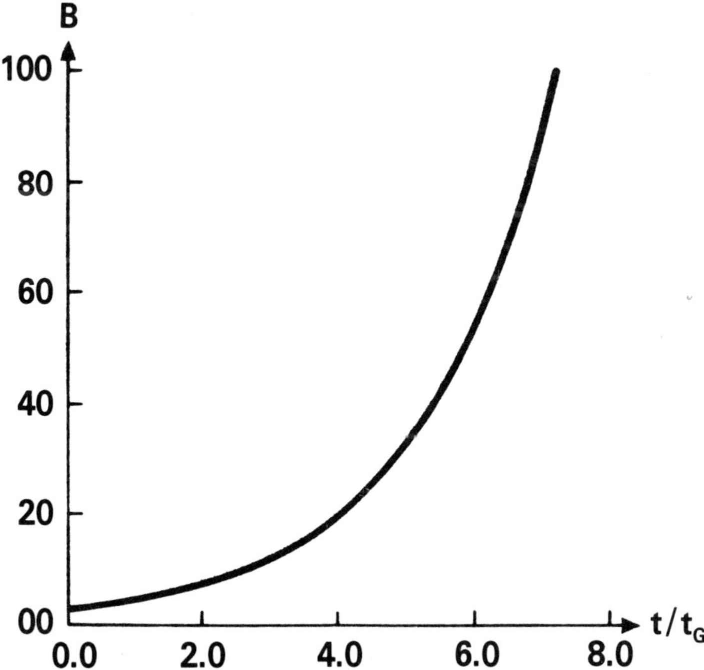


**FIGURE 3.** Expansion factor of a rotating de Sitter universe with $c_1 = 3$. The scale is like that of FIGURE 1.

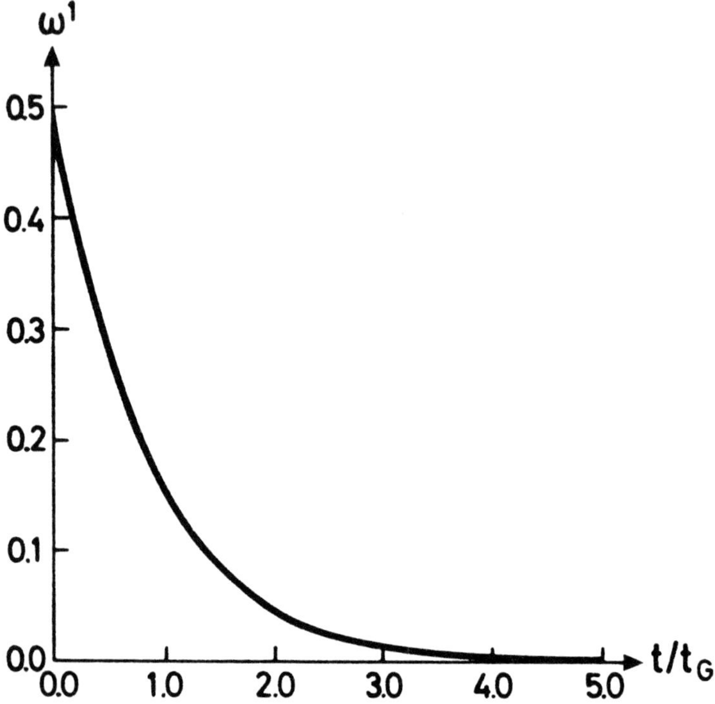


**FIGURE 4.** Cosmic vorticity of the rotating de Sitter universe with $c_1 = 3$. The initial value of the vorticity is arbitrary.

calculations have indicated that quantum gravity effects may lead the universe directly into an inflationary era.[19] If such a scenario is confirmed when more is known about quantum gravity, we would know the answer to the question: Why does the universe expand? The answer is then: once upon a time, the vacuum was stressed, it had negative mass, and gravitation forced the universe to expand.

## ACKNOWLEDGMENT

I would like to thank Jan Frøyland for useful help and stimulating conversations.

## REFERENCES

1. GRØN, Ø. 1985. Phys. Rev. **D31:** 2129.
2. GRØN, Ø. 1986. Am. J. Phys. **54:** 46.
3. TIWARI, R. N., J. R. RAO & R. R. KANAKAMEDALA. 1984. Phys. Rev. **D30:** 489.
4. COHEN, J. M. & M. D. COHEN. 1969. Nuovo Cimento **60:** 241.
5. LOPEZ, C. A. 1984. Phys. Rev. **D30:** 313.
6. GRØN, Ø. 1985. Phys. Rev. **D32:** 1588.

7. ISRAEL, W. 1966. Nuovo Cimento **44B:** 1.
8. IPSER, J. & P. SIKIVIE. 1984. Phys. Rev. **D30:** 712.
9. GUTH, A. H. 1981. Phys. Rev. **D23:** 347.
10. GRØN, Ø. 1985. Phys. Rev. **D32:** 2522.
11. MISNER, C. W., K. S. THORNE & J. A. WHEELER. 1973. Gravitation, chap. 30. Freeman. San Francisco.
12. ELLIS, J. & K. A. OLIVE. 1983. Nature **303:** 679.
13. GÖDEL, K. 1949. Rev. Mod. Phys. **21:** 447.
14. REBOUCAS, M. J. & J. A. S. LIMA. 1982. J. Math. Phys. **23:** 1547.
15. ROSQUIST, K. 1983. Phys. Lett. **97A:** 145.
16. SVIESTINS, E. 1985. Gen. Rel. Grav. **17:** 521.
17. GRØN, Ø. 1986. Phys. Rev. **D33:** 1204.
18. BARROW, J. D., R. JUSZKIEWICZ & D. H. SONODA. 1985. Mon. Not. R. Astron. Soc. **213:** 917.
19. BADMANBHAN, T. 1984. Phys. Lett. **104A:** 196.

# On Optimal Geometry for Generalized Rotational Flow of Elastoviscous Liquids in Thin Gaps

A. I. LEONOV
*Moscow, Union of Soviet Socialist Republics*

## INTRODUCTION

Rotational flows of elastoviscous liquids in sets with simple geometry have been studied in numerous works by various authors. The chief goal of these studies was to find simple situations permitting one to carry out rheometric tests of elastoviscous liquids under stationary and nonstationary conditions. On the basis of these studies, the main kinds of rotational rheometers with different geometries of measuring surfaces were made: cylinder-cylinder, cone-plate, plate-plate, and various modifications of these. A review of the tests of materials on different kinds of rheometers (along with the design descriptions of these) can be found in reference 1.

It is well known that the most familiar feature of elastoviscous liquids in comparison with inelastic non-Newtonian ones is the appearance of normal stresses in shear flows. One can find some applications of this in modern technical plants, such as in the elastic melt extruder employed in polymer processing, as well as in the recently proposed various types of sliding planes and axial bearings.

Rational design of rheometers or other technical plants for elastoviscous liquids has to be based, on the one side, on preliminary (sufficiently profound) knowledge of rheological properties of elastoviscous liquids in shearing flow; on the other side, it has to be based on the choice of optimal geometry of the set. Although the first problem can be treated quite successfully nowadays, the second side of the problem has not (to my knowledge) been discussed in the literature.

Therefore, the problems concerning the choice of the optimal geometry of rotational sets for elastoviscous liquids are the main subject matter of this paper. For this purpose, the formulas of the thrust and torque (dissipation) produced by the flow of elastoviscous liquids in thin gaps on arbitrary surfaces of revolution are derived. Furthermore, the problems of the choice of the set's optimal geometry are solved by use of calculus of variations.

## BASIC RELATIONS

Let us consider a rotational flow of an elastoviscous liquid in a set with a thin gap whose median surface is a surface of revolution. FIGURE 1 demonstrates the geometric scheme of such a set. Here, $z_0(r)$ is the equation of the median surface, $h_0(r)$ is the gap's thickness, and $r$ is the distance of a point $u$ disposed on the median surface from the axis of revolution. It is well known (see, for instance, reference 2) that both main curvature centers, $O_1$ and $O_2$, of a surface of revolution are situated on the same normal to the surface in the point $u$, intersecting the axis of revolution with the angle $\varphi$. Moreover, the second curvature center $O_2$ is disposed on the revolution axis. The principal radii of curvature, $R_1 = |\overrightarrow{O_1M}|$ and $R_2 = |\overrightarrow{O_2M}|$, as well as angle $\varphi$ in FIGURE 1, are defined by familiar formulas:

$$R_1(r) = \frac{(1 + z_0'^2)^{3/2}}{z_0''}, \qquad R_2(r) = \frac{r}{z_0'} \sqrt{1 + z_0'^2},$$

$$\sin\varphi = \frac{z_0'}{\sqrt{1 + z_0'^2}}, \qquad \cos\varphi = \frac{1}{\sqrt{1 + z_0'^2}}. \tag{1}$$

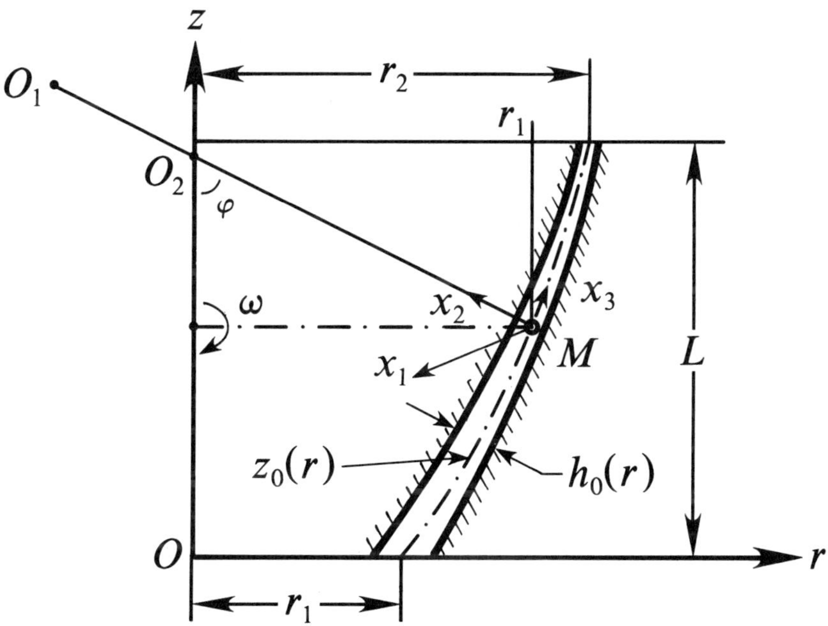


**FIGURE 1.** Geometry of a thin gap on a surface of revolution.

In addition, there is a simple relation that may be easily derived from equation 1:

$$R_1 \cos\varphi \frac{d\varphi}{dr} = 1. \tag{2}$$

For the sake of simplicity, the generator of the median surface will be designated as the meridian, the circle of radius $r$ passing through the median surface will be the parallel, and the gap's thickness will be counted off along the normal to the meridian. In each point of the smooth median surface, it is feasible to introduce a local orthogonal curvilinear coordinate system $x_1$, $x_2$, $x_3$, where $x_1$ is directed along the tangent to the parallel, $x_2$ is directed along the normal, and $x_3$ is directed along the tangent to the meridian.

Now, let us consider the stationary inertialess shearing flow of an elastoviscous liquid produced, for example, by a revolution of internal surface that rotates with constant angular velocity $\omega$ relative to an immovable external surface. In accordance with the main assumption, the flow of the liquid will be treated in the frames of lubrication approximation.[3] This is where the fields of shear rate and stresses are assumed to be locally homogeneous, the variations of $h_0(r)$ and $R_i(r)$ are assumed to be smooth, and the thin shell of revolution formed by the gap is assumed to be "gently sloping". The latter means

$$h_0(r) \cdot \{\min [R_1(r), R_2(r)]\}^{-1} \ll 1. \tag{3}$$

In the local coordinate system, the velocity vector in the gap is of the form,

$$\vec{v} = \{\dot{\gamma} x_2, 0, 0\}, \qquad \dot{\gamma} = \omega r / h_0(r), \tag{4}$$

where $\dot{\gamma}$ is the local shear rate, and the local stress system corresponding to simple shear is given by

$$P_{11}, P_{22}, P_{33}, P_{12}(\dot{\gamma}); P_{ii} = -P + P'_{ii}(\dot{\gamma}), P_1(\dot{\gamma}) = P_{11} - P_{22}, P_2(\dot{\gamma}) = P_{22} - P_{33}. \tag{5}$$

Here, $P$ is the pressure in incompressible liquid, $P'_{ii}(\dot{\gamma})$ are normal stresses due to shear, and $P_1(\dot{\gamma})$ and $P_2(\dot{\gamma})$ are the first and second normal stress differences, respectively. As is well known, the rheological behavior of elastoviscous liquids in stationary shear flows is entirely described by viscometric functions $P_1(\dot{\gamma})$, $P_2(\dot{\gamma})$, and $P_{12}(\dot{\gamma})$, whose form will be discussed in detail below.

We next need to consider the equilibrium conditions for a small liquid element in the gap with provision for the stress system.[5] Along the meridianal direction $x_3$, the equilibrium condition of the element is

$$\frac{d}{d\varphi}(P_{33} h_0 R_2 \sin\varphi) - P_{11} h_0 R_1 \cos\varphi - P_{22} R_2 \sin\varphi \frac{dh_0}{d\varphi} = 0.$$

Using the relation $R_2 \sin\varphi = r$ and equation 2, this equation can be rewritten in the form,

$$\frac{d}{dr}(r h_0 P_{33}) = P_{11} h_0 + P_{22} r h'_0(r);$$

that, in turn, results in

$$\frac{dP_{33}}{dr} = \frac{P_1 + P_2}{r} + \frac{P_2 h'_0}{h_0}. \tag{6}$$

The equilibrium condition of the liquid's element in the direction along the normal to the surface of revolution with provision for the "gently sloping" condition (equation 3) brings us to the conclusion of local homogeneity of the stress field in the gap. This usually takes place in the flows under consideration within the frames of lubrication approximation. After all, the condition of torque equality for a thin ring gives

$$dM = 2\pi r^2 P_{12} ds, \qquad ds = \sqrt{1 + z_0'^2}\, dr, \tag{7}$$

where $dM$ is the elementary torque for a ring with generator's length $ds$ along the meridian.

From equation 6, it is easy to obtain the distribution of the gap's normal stresses along the radius (or meridian). This is no different from that adduced in reference 4:

$$P_{11} = -P_0 + P_1(\dot{\gamma}) + P_2(\dot{\gamma}) - \int_r^{r_2} \left[\frac{P_1 + P_2}{x} + P_2 \frac{h_0'(x)}{h_0(x)}\right] dx,$$

$$P_{22} = -P_0 + P_2(\dot{\gamma}) - \int_r^{r_2} \left[\frac{P_1 + P_2}{x} + P_2 \frac{h_0'(x)}{h_0(x)}\right] dx,$$

$$P_{33} = -P_0 - \int_r^{r_2} \left[\frac{P_1 + P_2}{x} + P_2 \frac{h_0'(x)}{h_0(x)}\right] dx. \quad (8)$$

Here, $P_0$ is a certain external pressure given at $r = r_2$.

The total thrust produced by the liquid flowing through the gap is

$$T = -2\pi \int_{r_1}^{r_2} r P_{22} \cos \varphi \, ds.$$

After substituting here the formula for $P_{22}$ from equation 8 and making allowance for the fact that $\cos \varphi \, ds/dr = 1$ (resulting from equation 1), the latter equation after easy calculations takes the form,

$$T = \pi(r_2^2 - r_1^2)P_0 + \pi \int_{r_1}^{r_2} \left(r - \frac{r_1^2}{r}\right)\left(P_1 - P_2 \frac{\dot{\gamma}'}{\dot{\gamma}} r\right) dr - 2\pi r_1^2 \int_{r_1}^{r_2} \frac{P_2}{r} dr. \quad (9)$$

This relation at $r_1 = 0$ yields the formula cited in reference 4. The formulas for the torque $M$, the gap's volume $V$, and the area of median surface $F$ are of the form:

$$M = 2\pi \int_{r_1}^{r_2} r^2 P_{12}(\dot{\gamma}) \sqrt{1 + z_0'^2} \, dr, \quad (10)$$

$$V = 2\pi \int_{r_1}^{r_2} h_0(r) r \sqrt{1 + z_0'^2} \, dr = 2\pi\omega \int_{r_1}^{r_2} \frac{r^2 \sqrt{1 + z_0'^2} \, dr}{\dot{\gamma}(r)}, \quad (11)$$

$$F = 2\pi \int_{r_1}^{r_2} r \sqrt{1 + z_0'^2} \, dr. \quad (12)$$

For rotational stationary shearing of elastoviscous liquid, the torque $M$ is connected with the total dissipation $D$ by the evident relation

$$D = M\omega. \quad (13)$$

Taking into account the relation $\dot{\gamma} = \omega r/h_0$, equation 13 can be derived immediately:

$$D = 2\pi \int_{r_1}^{r_2} P_{12}(\dot{\gamma}) \cdot \dot{\gamma} h_0(r) r ds = 2\pi\omega \int_{r_1}^{r_2} P_{12}(\dot{\gamma}) r^2 ds = M \cdot \omega.$$

It is natural to treat magnitudes $T$, $M$, $V$, and $F$ as functionals of the functions $h_0(r)$ and $z_0(r)$, where $T$ and $M$ also depend parametrically on the angular velocity $\omega$.

The behavior of functionals is due to the dependence of viscometric functions $P_1$,

$P_2$, and $P_{12}$ on shear rate $\dot{\gamma}$, which we shall now discuss in more detail. First of all, these can be represented in the form,

$$P_1 = G_e\sigma_1(y), P_2 = -G_e\sigma_2(y), P_{12} = G_e\sigma_{12}(y) \qquad (y = \dot{\gamma}\theta), \tag{14}$$

where $G_e$ is a certain "high elasticity" modulus, and $\theta$ is a characteristic relaxation time ($\theta = \eta_0/G_e$, with $\eta_0$ being the largest Newtonian viscosity). The behavior of dimensionless functions $\sigma_1$, $\sigma_2$, and $\sigma_{12}$ has presently been studied rather well both theoretically and experimentally. (For some experimental evidences, see, for instance, reference 5.) Functions $\sigma_1$ and $\sigma_2$ are even, whereas $\sigma_{12}$ is odd. If we confine ourselves to the study of only the case of $y > 0$ ($\dot{\gamma} > 0$), one can consider all three viscometric functions $\sigma_1$, $\sigma_2$,

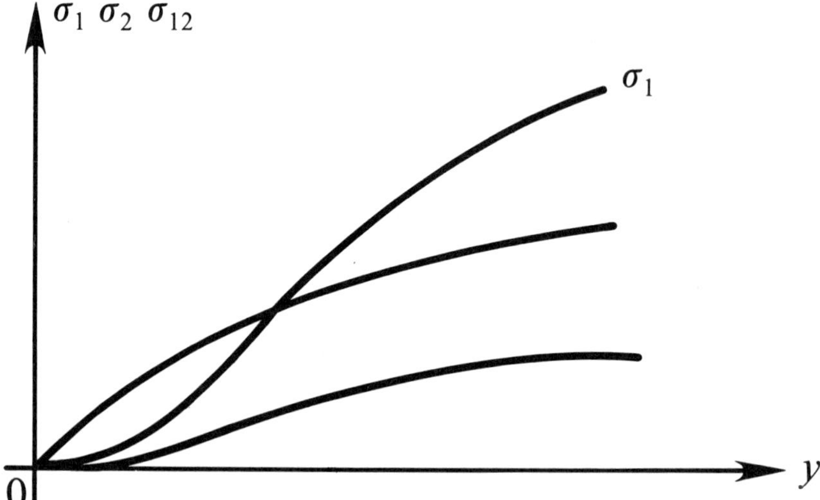


**FIGURE 2.** Qualitative behavior of a viscometric function in simple shear.

and $\sigma_{12}$ as positive. All of them also increase monotonically. Moreover, $\sigma_2/\sigma_1 \ll 1$ and decreases monotonically as $y$ increases (see FIGURE 2).

Let us further suppose that at small and large values of $y$, there are the asymptotic expressions for the viscometric functions:

$$\sigma_i \approx c_i y^2 \qquad (c_2/c_1 \approx 0.3); \qquad \sigma_{12} \approx y \; (y \ll 1); \tag{14a}$$

$$\sigma_1 \approx a y^{\alpha}; \qquad \sigma_2 \approx a_1 y^{\alpha_1}; \qquad \sigma_{12} \approx b y^{\beta} \; (y \gg 1) \tag{14b}$$

(where $0 < \beta \leq \alpha < 1$, $\alpha_1 \ll \alpha$, $a_1 \lesssim a$); let us also suppose that for arbitrary positive $h > 0$ within the whole interval $y > 0$, the following inequalities are suggested to hold true:

$$[y^{1+n}\sigma_{12}'(y)]' > 0; \qquad [y^{1+n}\sigma_i'(y)]' > 0; \qquad \sigma_1' - 2\sigma_2 > 0. \tag{15}$$

These asymptotic formulas (equations 14a and 14b) and inequalities (equation 15) are usually well fulfilled in reality.

Furthermore, for the sake of convenience, the following dimensionless variables, parameters, and functions can be introduced:

$$x = r/r_0 \quad (x_1 = r_1/r_0, x_2 = r_2/r_0), \quad h(x) = h_0/r_0, \quad z(x) = z_0/r_0,$$

$$k = x/h(x), \quad \Omega = \omega\theta, \quad l = L/r_0, \quad \sigma_i = \sigma_i(\Omega k), \quad \sigma_{12} = \sigma_{12}(\Omega k),$$

$$J_T = \frac{T - J_1(r_2^2 - r_1^2)P_0}{J_1 r_0^2 G_e}, \quad J_M = \frac{M}{2\pi r_0^3 G_e}, \quad J_V = \frac{V}{2\pi r_0^3}, \quad \text{and} \quad J_F = \frac{F}{2\pi r_0^2}, \tag{16}$$

where $r_0$ is a certain characteristic radius that, for the sake of generality, we shall initially assume to be different from $r_2$. In dimensionless variables (equation 16), equations 9–12 for the functionals take the forms:

$$J_T = \int_{x_1}^{x_2} \left(x - \frac{x_1^2}{x}\right)\left[\sigma_1(\Omega k) + \sigma_2(\Omega k)\frac{k'}{k}x\right]dx + 2x_1^2 \int_{x_1}^{x_2} \sigma_2(\Omega k)\frac{dx}{x}, \tag{17}$$

$$J_M = \int_{x_1}^{x_2} x^2 \sigma_{12}(\Omega k)\sqrt{1 + z'^2}dx, \tag{18}$$

$$J_V = \int_{x_1}^{x_2} x^2 \sqrt{1 + z'^2}\frac{dx}{k}, \tag{19}$$

$$J_F = \int_{x_1}^{x_2} x\sqrt{1 + z'^2}dx. \tag{20}$$

Some characteristic extremum problems that arise here in the simple proposition are those related with searching dependences $k(x;\Omega)$ [or $h(x;\Omega)$] and $z(x;\Omega)$ delivering an extremum (minimum or maximum) for functionals $J_T$ or $J_M$ at constant gap's volume, that is, at constant $J_V$.

Instead of the function $k(x)$ and the parameter $\Omega$, it is possible to introduce new variables for the functionals of equations 17–19; namely,

$$\hat{k}(x) = k(x)/J_V, \qquad \hat{\Omega} = \Omega/J_V(\hat{J}_V = 1). \tag{21}$$

If $0 < \hat{k} < \hat{k}(x) < \hat{k}_+$ uniformly on the interval $[x_1, x_2]$, then for power asymptotics equations 14a ($\Omega \ll 1$) and 14b ($\Omega \gg 1$), the functionals $J_T$ and $J_M$ take the forms,

$$J_T = \zeta_n^T k_n \left(\frac{\Omega}{J_V}\right)^n \qquad (n = 2, k_1 = c_1; n = \alpha, k_\alpha = a),$$

$$J_M = \zeta_n^M R_n \left(\frac{\Omega}{J_V}\right)^n \qquad (n = 1, R_1 = 1; n = \beta, R_\beta = b), \tag{22}$$

where the functionals $\zeta_n^T$ and $\zeta_n^M$ reveal already no dependences on $\Omega$ and are determined only by the gap's geometry and by the exponents $\alpha$ and $\beta$ in asymptotic rheological laws (equations 14a and 14b):

$$\zeta_2^T = \int_{x_1}^{x_2} \hat{k}^2(x)\left[x - \frac{x_1^2}{x}(1 - 2c_2/c_1) + (c_2/c_1)\hat{k}'/\hat{k}(x^2 - x_1^2)\right]dx;$$

$$\zeta_\alpha^T = \int_{x_1}^{x_2}\left(x - \frac{x_1^2}{x}\right)\hat{k}^\alpha dx; \int_{x_1}^{x_2} x^2\sqrt{1 + z'^2}\,\frac{dx}{\hat{k}(x)} = 1;$$

$$\zeta_1^M = \int_{x_1}^{x_2} x^2\,\hat{k}(x)\sqrt{1 + z'^2}\,dx; \zeta_\beta^M = \int_{x_1}^{x_2} x^2\hat{k}^\beta(x)\sqrt{1 + z'^2}\,dx. \tag{23}$$

The next two sections will be devoted to a study of easy variational problems on the conditional extremum of the functionals of equations 17 and 18 at constant $J_V$ (equation 19). These problems will be seen to be the classical problems of calculus of variations.[6]

## ISOPERIMETRIC PROBLEMS ON EXTREMUM OF THE TORQUE

These problems very often arise in practice (for instance, in the design of rheometers) and have a very easy proposition:

> To seek the extremum of the functional of equation 18 under the condition of constancy for the functional of equation 19.

We shall make some remarks concerning the proposition of the problem. The parameter $\Omega$ is considered to be fixed, whereas both the functions $k(x)$ and $z(x)$, as well as the bounding points $x_1$ and $x_2$, are permitted to vary. Because only the derivative $z'(x)$ is present in all functionals, it could be possible to introduce a new variable, $u(x) = z'(x)$, and to treat further only its variations. However, such an approach will be obtained as a special case within the procedure accepted below. Hereafter, if an extremum for the functionals includes the function $z(x) \equiv 0$, we shall call such a gap's geometry as plane and the flow as torsional. In this event, everywhere below the lower surface will be considered to be a plate that results from the fact that presenting here the equation of the median surface, $z = h(x)/2$, within the accuracy of the theory, can be treated as $z = 0$.

In order to find out the solution of the problem, let us form the functional,

$$G_M = J_M + \lambda J_V = \int_{x_1}^{x_2} F(x, k, z'; \lambda, \Omega)dx,$$

$$F = [\sigma_{12}(k\Omega) + \lambda/k]x^2\sqrt{1 + z'^2}, \tag{24}$$

where Lagrange's multiplier $\lambda$ is subject to determination.

For the functional of equation 24, Euler's equations have the form,

$$F'_k = 0, \qquad F'_{z'} = c = \text{const.},$$

or

$$\Omega\sigma'_{12}(\Omega k) - \frac{\lambda}{k^2} = 0, \tag{25}$$

$$F'_{z'} = x^2[\sigma_{12}(\Omega k) + \lambda/k]\,\frac{z'}{(1 + z'^2)^{1/2}} = c = \text{const.} \tag{26}$$

Equation 25 produces the solution $k$ = const., and then $h = x/k$, that is, the gap's thickness is linearly distributed along the radius. In equation 26, $\lambda = k^2\Omega\sigma'_{12}(\Omega k)$ and the term in the square brackets takes the form,

$$\sigma_{12}(\Omega k_0) + \lambda/k_0 = \left[\frac{d}{dy}\,(y\sigma_{12}(y))\right]_{y=\Omega k_0} = \psi_0(\Omega k_0) > 0.$$

If we further accept $c/\psi_0 = q$ = const. from equation 26, one can obtain

$$z' = \frac{q}{\sqrt{x^4 - q^2}}, \qquad \sqrt{1 + z'^2} = \frac{x^2}{\sqrt{x^4 - q^2}}. \tag{27}$$

Hereafter, without loss of generality, we shall accept $z \geq 0$. Then, depending on the sign of $q$, it results from equation 27 that the two possibilities, $z' > 0$ $(q > 0)$ or $z' < 0$ $(q < 0)$, may take place. The curves of the first kind increase, and for these kinds, the natural condition is $z_1(x_1) = 0$; the curves of the second kind decrease and have as a natural condition $z_2(x_2) = 0$. Then,

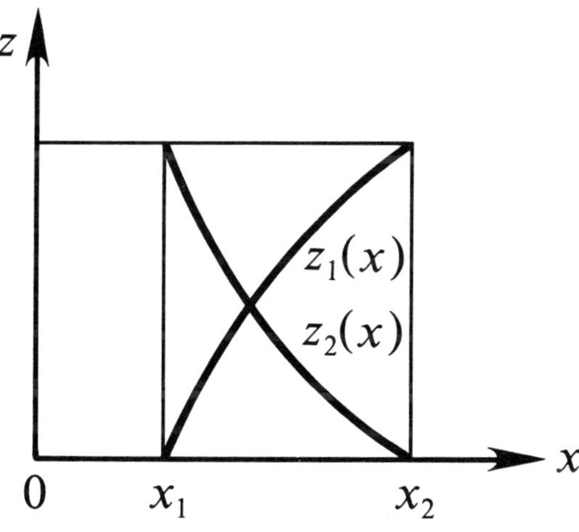


**FIGURE 3.** Qualitative form of dependences $z_i(x)$ according to equations 28.

$$z_1(x) = |q|\int_{x_1}^{x}\frac{d\xi}{\sqrt{\xi^4 - q^2}}, \qquad z_2(x) = |q|\int_{x}^{x_2}\frac{d\xi}{\sqrt{\xi^4 - q^2}}. \tag{28}$$

As it follows from equation 28, $z_1(x)$ is convex and $z_2(x)$ is concave (see FIGURE 3), while $|q| \leq x_1^2$.

Let us elucidate what the restrictions have to impose on the functions that are allowed to vary. Because in the formula (equation 24) for $F$, the value $k'(x)$ is absent, it is becoming evident in the points $x_1$ and $x_2$ that the variations $\delta k$ can be arbitrary. In order to provide the arbitrariness of variations $\delta z$ in points $x_1$, $x_2$, it is necessary to satisfy the well-known conditions of $F'_{z'}|_{x_1} = F'_{z'}|_{x_2} = 0$. However, by virtue of equation 26, it is admissible only if $z' = 0$ [or $z(x) = 0$], that is, for the plane geometry. If, however, $z \not\equiv 0$, the restrictions on variations $\delta z$ in points $x_1, x_2$ are

$$\delta z|_{x_1} = \delta z|_{x_2} = 0.$$

Now, let us make clear if it is possible to vary the bounds $x_1$ and $x_2$. The corresponding conditions are as follows:

$$[F - z'F'_{z'}]_{x_i}\delta x_i = \left[\frac{z'^2}{\sqrt{1 + z'^2}}\right]_{x_i} \psi_0(\Omega k_0)x_i^2\delta x_i = 0 \qquad (i = 1, 2).$$

From this, it follows that $\delta x_1 = \delta x_2 = 0$ because the term in the square brackets in the latter expression does not pass through zero. Evidently, in this case, the magnitudes $x_i$ have to be given a priori; therefore, without loss of generality, one can put $x_2 = 1$, thus accepting $r_0 = r_2$ in equation 16. By the same method, the radial clearance of the set is given. As to the parameter $x_1$, its choice is determined by purely design reasons and there are no restrictions of extremum kind imposed on the value $x_1$.

In the case of $k$ = const. (which is discussed), the functionals of equations 17 and 18 are of the forms,

$$J_T = \left(\frac{1 - x_1^2}{2} + x_1^2 lux_1\right)\sigma_1(\Omega k) - 2x_1^2 lux_1 \cdot \sigma_2(\Omega k);$$

$$J_M = \sigma_{12}(\Omega k)kJ_V; \qquad k = \frac{1}{J_V}\int_{x_1}^{1} \frac{x^4 dx}{\sqrt{x^4 - q^2}}. \tag{29}$$

Now, this question remains to be elucidated: what kind of extremum (maximum or minimum) is secured for the functional $G_M$ by extremal $k$ = const. and $z'$ taken according to equation 27? The well-known answer to that question is given by the sign of the second variation of the functional $G_M$ from equation 24:

$$\delta^2 G_M = \frac{1}{2}\int_{x_1}^{1} [F''_{kk} \cdot (\delta k)^2 + F''_{z'z'}(\delta z')^2]dx. \tag{30}$$

Here, $F''_{kk}$ and $F''_{z'z'}$ are taken along the extrema and are of the forms,

$$[F''_{kk}]_{k_0,\tilde{z}} = k_0^{-2}x^2\sqrt{1 + \tilde{z}'^2}\,[y^2\sigma'_{12}(y)]'_{y-\Omega k_0},$$

$$[F''_{z'z'}]_{k_0,\tilde{z}} = x^2(1 + \tilde{z}'^2)^{-3/2}[y\sigma_{12}(y)]'_{y-\Omega k_0}. \tag{31}$$

Out of both of the equalities of equation 31 (with provision for equation 15), it follows that $F''_{kk} > 0$, $F''_{z'z'} > 0$, which results in $\delta^2 G_M > 0$. The latter is witnessed for the minimum $J_M$ from equation 18 under the condition of $J_V$ = const. As one can see from equation 29, this minimum is accomplished with $q = 0$, that is, $z \equiv 0$ (the set with plane geometry). This minimum, obviously, can be obtained from the Eulerian equation, $F'_{z'} = 0$, if $z'(x)$ is treated there as an independent varying function. Of course, the condition $F'_{z'} = 0$ can be considered as more crude in comparison with equation 26 because no restrictions are imposed here on variations in points $x_1$, $x_2$. Therefore, this condition along with equation 25 briefly leads to the final result: $k = k_0$ = const., $\tilde{z} \equiv 0$.

In this case, the formulas in equation 29 take the forms,

$$J_T = \sigma_1(\Omega k)\left(\frac{1 - x_1^2}{2} + x_1^2 lux_1\right) - \sigma_2(\Omega k)2x_1^2 lux_1,$$

$$J_M = (1/3)\sigma_{12}(\Omega k)(1 - x_1^3); \quad k = (1 - x_1^3)/(3J_v), \tag{32}$$

and the functionals $\zeta_n^T$ and $\zeta_n^M$ are given by

$$\zeta_2^T = \frac{(1 - x_1^3)^2}{18}[1 - x_1^2 + 2(1 - 2c_2/c_l)x_1^2 lux_1],$$

$$\zeta_\alpha^T = \frac{1}{2}\left(\frac{1 - x_1^3}{3}\right)^\alpha (1 - x_1^2 + 2x_1^2 lux_1),$$

$$\zeta_1^M = \frac{1}{9}(1 - x_1^3)^2, \qquad \zeta_\beta^M = \left(\frac{1 - x_1^3}{3}\right)^{1+\beta}. \tag{33}$$

As already mentioned, the parameter $x_1$ $(0 \leq x_1 < 1)$ must be determined from purely design reasons. In particular, such serious reasons as the need for condensations and bearings demand $x_1 = 0$. Then, equations 32 and 33 are essentially simplified to

$$J_T = (1/2)\sigma_1(\Omega/3J_v), \qquad J_M = (1/3)\sigma_{12}(\Omega/3J_v), \qquad [h(x) = 3J_v x], \tag{32a}$$

$$\zeta_2^T = \frac{1}{18}, \qquad \zeta_\alpha^T = \frac{1}{2 \cdot 3^\alpha}, \qquad \zeta_1^M = \frac{1}{9}, \qquad \zeta_\beta^M = 3^{-(1+\beta)}. \tag{33a}$$

According to additional design reasons, there may be a need to construct a three-dimensional geometry of a gap with vertical dimensionless size $l$. Then, the choice of the gap's configurations from the class of functions in equation 28 occurs to be impossible in practice. Actually, out of equation 28, one can reveal

$$l = |q| \int_{x_1}^1 \frac{dx}{\sqrt{x^4 - q^2}} \leq x_1^2 \int_{x_1}^2 \frac{dx}{\sqrt{x^4 - x_1^4}} = f(x_1).$$

It is easy to see that the function $f(x_1)$ goes through a maximum. Numerical calculations showed the value of the maximum $f(x_1^0) \approx 0.047$ at $x_1^0 \approx 0.217$. Evidently, such a maximal value of $l$ is disposed within the same order of ten as the gap's thickness; therefore, within the accuracy of the theory, the geometry of such a set is not different from the plane one.

In addition to the regular solution of the problem, there is also a peculiar one corresponding to $k$ = const., $z' \rightarrow \infty$ in Euler's equations 25 and 26. As it follows from equation 26, such a solution does exist if $x$ = const., and along with the condition $k$ = const., it also results if $h$ = const. This solution predicts the geometry of the gap to be a vertical cylinder of dimensionless length $l$ and constant small thickness $h$; that is, it is an ordinary Couette viscometer. Here, without loss of generality, it is admissible to accept $r_0 = r_2$ in equation 16, and hence $x_1 = x_2 = 1$. In this case, the corresponding formulas for the functionals of equation 17–19 are

$$J_T = 0, \quad J_M = \sigma_{12}(\Omega k)l \qquad (k = 1/h = l/J_v). \tag{34}$$

As one can see, at the same shear rate ($k$ = const.) when the values $J_M$ defined by equations 32a and 34 coincide, the value $l = 1/3$. If $l > 1/3$, the value $J_M$ defined by equation 32a is less than $J_M$ defined by equation 34; at $l < 1/3$, the whole situation is contrary. Evidently, the geometry of Couette's rheometer delivers an absolute minimum for $J_T$, which is $J_T = 0$.

In comparison with the cone-plate, the Couette's rheometer with the same $J_M$ possesses the obvious constructive shortcomings: a large axis dimension and the need for bearings and condensations in the bottom part ($z = 0$) of the set. An example of this is the following: the joining of the cylinder-cylinder viscometer with a bottom, which secures the minimal value of torque (FIGURE 4).

The problem, evidently, has great practical importance because the design scheme treated here is employed very often in practice. The results obtained in this section point out the fact $l_0 = 0$, so the bottom configuration corresponds to the flat set's geometry of cone-plate type. Then, by virtue of equations 32a and 34, the expressions

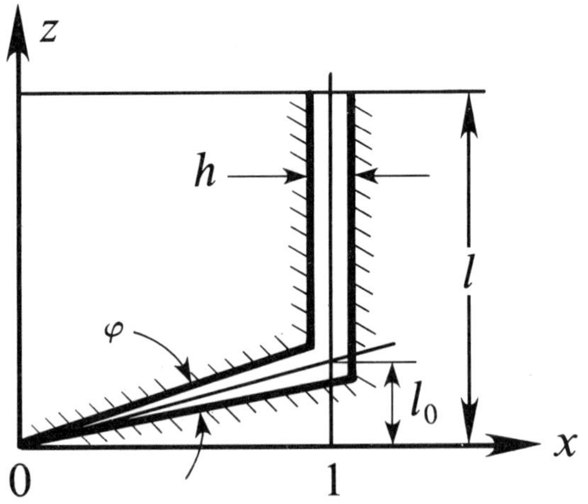

**FIGURE 4.** Sketch of a cylinder-cylinder rheometer with a bottom of cone-plate geometry.

for the functionals $J_T$, $J_M$, and $J_v$ take the forms,

$$J_T = (1/2)\sigma_1(\Omega k_1), \qquad J_M = (1/3)\sigma_{12}(\Omega k_1) + l\sigma_{12}(\Omega k_2),$$

$$J_v = 1/(3k_1) + l/k_2 \qquad (k_1 = 1/\alpha,\ k_2 = 1/h).$$

Here, the parameter $l$ is supposed to be given, whereas $k_1$ and $k_2$ are geometrical characteristics of the gaps for cone-plate and Couette's viscometers that are sought after. This problem leads to the search of $\min_{k_1,k_2} J_M$ under the condition $J_v$ = const. This problem is a standard problem of conditional extremum for a function of two variables and its solution is of the form,

$$k_1 = k_2 = k = \frac{1 + 3l}{3J_v}. \tag{36}$$

If, for instance, the dimensional thickness of the gap $h_0$ in the cylindrical part of the complex set and its radius $r_0$ are given, then the angle $\varphi$ of its cone-plate bottom is given by $\alpha = h_0/r_0$. Namely, such a distribution that is continuous on $h$ in the point of joining guarantees the least dissipation of liquid in the gap under the condition $V$ = const.

## ISOPERIMETRIC PROBLEMS ON EXTREMUM OF THRUST

This problem is also proposed very easily: to seek an extremum of the functional in equation 17 under the condition of constancy of the functional in equation 19.

As mentioned above, the absolute minimum (equal to zero) for the functional is given by the geometry of the vertical Couette's rheometer (equation 34). However, this geometry is not good enough due to the design reasons. Therefore, the problem arises here to find an extremum of the functional in equation 17 with the condition of equation 19. If the extremum responds for the minimum of equation 17, this minimum will be evidently local.

For the solution of the problem, we compose the functional

$$G_T = J_T + \lambda J_v = \int_{x_1}^{x_2} \Phi(x, k, k', z'; x_1, \lambda, \Omega)\, dx,$$

$$\Phi = \left(x - \frac{x_1^2}{x}\right)[\sigma_1(\Omega k) + \sigma_2(\Omega k)k'x/k] + 2x_1^2\sigma_2(\Omega k)/x + \frac{\lambda x^2}{k}\sqrt{1 + z'^2}, \quad \textbf{(37)}$$

where Lagrange's multiplier $\lambda$ is again subject to determination. For the functional in equation 37, Euler's equation are of the form,

$$\Phi'_{k'} - \frac{d}{dx}\Phi'_{k'} = 0, \qquad \Phi'_{z'} = c,$$

which results in the relations,

$$(x^2 - x_1^2)\sigma_1'(\Omega k)\Omega/x + \sigma_2'(\Omega k)2x_1^2\Omega/x - 2x\Omega\sigma_2(\Omega k)/(\Omega k) - \frac{\lambda x^2}{k^2}\sqrt{1 + z'^2} = 0, \quad \textbf{(38)}$$

$$\frac{\lambda x^2}{k}\frac{z'}{\sqrt{1 + z'^2}} = c. \quad \textbf{(39)}$$

Let us consider what the restrictions have to impose on the functions varied in the limit points $x_1$ and $x_2$. In order to treat integrated terms of the first variation of the functional in equation 37 so that they would vanish, it is necessary that

$$[\Phi'_{k'}\delta k]_{x_i} = \left[(x^2 - x_1^2)\frac{\sigma_2(\Omega k)}{k}\delta k\right]_{x_i} = 0,$$

$$[\Phi'_{z'}\delta z]_{x_i} = \left[\lambda\frac{x^2}{k}\cdot\frac{z'}{\sqrt{1 + z'^2}}\cdot\delta z\right]_{x_i} = 0 \qquad (i = 1, 2).$$

The first of these conditions points out that $\delta k$ can be arbitrary at $x = x_1$, whereas at $x = x_2$, the condition $\delta k|_{x_2} = 0$ has to be imposed. With provision of equation 39, the second condition shows that if $z \neq 0$ ($c \neq 0$ in equation 39), then $\delta x|_{x_1} = \delta z|_{x_2} = 0$.

Let us make clear the feasibility of variations of the bound points $x_1$ and $x_2$. The relevant conditions

$$\left[\int_{x_1}^{x_2} dx \frac{\partial \Phi}{\partial x_i} + \Phi - k\Phi'_{k'} - z'\Phi'_{z'}\right]_{x_i} \delta x_i = 0 \qquad (i = 1, 2)$$

(with account of equation 37) take the forms,

$$\left\{2\sigma_2(\Omega k_1) + \frac{\lambda x_1}{k_1\sqrt{1 + z_1'^2}} - 2\int_{x_1}^{x_2}\left[\sigma_1(\Omega k) + \sigma_2(\Omega k)\left(\frac{k'x}{k} - 2\right)\right]\frac{dx}{x}\right\} x_1 \delta x_1 = 0, \quad \textbf{(40)}$$

$$\left\{\frac{x_2^2 - x_1^2}{x_2}\sigma_1(\Omega k_2) + 2\frac{x_2^2}{x_1}\sigma_2(\Omega k_2) + \frac{\lambda x_2^2}{k_2\sqrt{1 + z_2'^2}}\right\}\delta x_2 = 0, \quad \textbf{(41)}$$

where $k_i = k(x_i)$ and $i = 1, 2$.

By virtue of equation 38, it is possible to establish that $\lambda > 0$. Then, equation 41 will be satisfied if only $\delta x_2 = 0$, that is, $x_2 =$ const. In this case, it is admissible to accept $x_2 = i$, thus choosing $r_0 = r_2$ in equation 16. As to equation 40, if $x_1 \neq 0$, this value can be determined by proposing that the terms in the braces vanish. If, however, the terms do not vanish, then $x_1$ cannot be varied and it has to be introduced from other (nonextremal) reasons.

Let us now appeal to equation 39. Determining $z'$ from this, we find

$$z'(x) = \frac{ck}{\sqrt{\lambda^2 x^4 - c^2k^2}}, \qquad z_1(x) = |c|\int_{x_1}^{x}\frac{k(\xi)d\xi}{\sqrt{\lambda^2\xi^4 - c^2k^2(\zeta)}}, \quad \textbf{(42)}$$

$$z_2(x) = \int_{x_1}^{x_2}\frac{|c|k(\xi)d\xi}{\sqrt{\lambda^2\xi^4 - c^2k^2(\xi)}}.$$

From this, as well as from the previous section, it follows that depending on the sign of $c$ here, there can, in general, exist the two plots illustrated by FIGURE 3. If $x_1 = 0$, the easy study of equation 38 with provision for the expression of $z'(x)$ from equation 42 shows that at $c \neq 0$, the function $h \sim 1/x$ as $x \to 0$, which contradicts the presumption of the smallness of $h(x)$. In order to eliminate this inconsistency, one has to put $c = 0$ in equation 39, which results in the condition $z \equiv 0$ for the extremal. Because at $x_1 \neq 0$ no contradictions of such a kind exist, it is feasible to consider extremum solutions of Euler's equations 38 and 39 within the class of extremals obeying the conditions

$$k|_{x-1} = k_1 = \text{const.}; \qquad z_1(x_1) = 0, \quad z_1(1) = l \qquad [z_2(x_1) = l, z_2(1) = 0]. \quad \textbf{(43)}$$

In addition, if $x_1$ is varied, this parameter due to equation 40 is determined from the equation,

$$2\sigma_2(\Omega k_1) + \frac{\lambda x_1}{k_1\sqrt{1 + z_1^2}} = 2\int_{x_1}^{1}\left[\sigma_1(\Omega k) + \sigma_2(\Omega k)\left(\frac{k'x}{k} - 2\right)\right]\frac{dx}{x}. \quad \textbf{(44)}$$

The set of equations 38 and 39, along with equations 19, 43, and 44, is closed in regard to the unknown functions $k(x)$ and $z(x)$ and the parameters $x_1$, $\lambda$, and $c$.

Equation 38 at $x = x_1$ with provision for equation 39 takes the form,

$$2k_1[\Omega k_1 \sigma_2'(\Omega k_1) - \sigma_2(\Omega k_1)] = \frac{\lambda_2 x_1^3}{\sqrt{\lambda^2 x_1^4 - c^2 k_1^2}}.$$

The right-hand side of the equation is positive at $x_1 \neq 0$, whereas its left-hand side is positive only if $k_1 < y_*/\Omega$, where $y_*$ is the root of equation $[\sigma_2(y)/y]' = 0$, which always exists due to asymptotic equations 14a and 14b for $\sigma_2(y)$. The value $y_* \approx (a_1/c_2)^{1/(2-\alpha_1)}$ can be approximately estimated by equating both asymptotic equations in the point $y = y_*$.

Furthermore, for equation 38 (with account of equation 39), the following fundamental fact may be proved: the left-hand side of equation 38 at fixed $x$ increases with the increase of $k$; this results from equations 14a and 14b and the inequalities of equation 15. Using this result, it is also easy to show that a solution $\{\tilde{k}(x), \tilde{z}(x)\}$ of Euler's equations 38 and 39 always exists. Moreover, $\tilde{k}(x)$ increases monotonically, and in the vicinity of point $x_1$, it is determined from quadratic asymptotic equation 14a for $\sigma_i(y)$.

To elucidate the type of the extremal $\{\tilde{k}, \tilde{z}\}$, it is necessary to study the second variation of the functional of equation 37, taking into account the possibility of $x_{11}$ variation:

$$\delta^2 G_T = \int_{x_1}^{1} [Q(\delta k)^2 + M(\delta z')^2]\, dx + P(\delta x_1)^2,$$

$$2Q = \left[\Phi''_{kk} - \frac{d}{dx}\Phi''_{kk'}\right]_{\tilde{k},\tilde{z}}, \quad 2M = \Phi''_{z'z'}|_{\tilde{k},\tilde{z}},$$

$$P = \left\{\frac{\partial}{\partial x_1}\left[\int_{x_1}^{1} \frac{\partial \Phi}{\partial x_1}\, dx + \Phi - k'\Phi'_{k'} - z'\Phi'_{z'}\right]\right\}_{\tilde{k}_1,\tilde{z}_1}.$$

If we utilize here the formula for $\Phi$ from equation 37 and the equilibrium conditions of equations 38 and 39, it is possible to prove that $P > 0$, $M > 0$, $Q > 0$. For instance, the formula for $Q$ can be rewritten in the form,

$$\frac{2Q}{\Omega^2} = \frac{x^2 - x_1^2}{x} \cdot \frac{1}{y}[y^2 \sigma_1'(y)] + \frac{2x_1^2}{x}\frac{1}{y^2}[y^2 \sigma_2'(y)]' + 2xy^2\left[\frac{\sigma_2(y)}{y^3}\right],$$

where $y = \Omega k(a)$. By virtue of the asymptotic equations 14a and 14b and the inequalities of equation 15, there is no difficulty in noticing that the magnitude $Q < 0$.

These facts show that the extremal $\{\tilde{k}(x), \tilde{z}(x)\}$ delivers a local minimum for the functional $J_T$ (equation 17) under the conditions of constancy of the functional in equation 19.

If the magnitude of $l$ in equations 43 is arbitrary, then the minimum of the functional in equation 17 is attained at $l = 0$, that is, $c = 0$ or $z = 0$, because any variations with arbitrary values $z(x)$ in the interval limits are admissible to comparison. The same results follow immediately and from an analysis of equation 38. For the problem concerned, the latter implies that the local minimum is achieved for the plane geometry $\{k(x), \tilde{z}(x) \equiv 0\}$, for which equation 44 takes the easy form:

$$\Omega k_1\sigma_2'(\Omega k_1) = \int_{x_1}^{1}\left[\sigma_1(\Omega k) + \sigma_2(\Omega k)\left(\frac{k'x}{k} - 2\right)\right]\frac{dx}{x}. \qquad \textbf{(44a)}$$

In the two asymptotic cases, one can obtain the solution of the set of equations 38 at $z' = 0$ and equation 44a with the condition of equation 19 of volume constancy.

### *Case 1*

Let $\Omega \ll 1$. Then for functions $\sigma_1$ and $\sigma_2$, the quadratic asymptotics are fulfilled uniformly in $x$. For determinacy, let the relation $c_2/c_1 = 1/4$ be satisfied. Then, the solution of equation 38 with $z' = 0$ is given by

$$k(x) = \frac{qx}{(x^2 - \frac{2}{3}x_1^2)^{1/3}}, \qquad h(x) = (1/q)(x^2 - \tfrac{2}{3}x_1^2)^{1/3}, \qquad q = \left(\frac{2\lambda}{3c_1\Omega^2}\right)^{1/3}. \qquad \textbf{(45)}$$

From equation 19, one can find

$$q = \frac{3}{8J_v}\,[(1 - \tfrac{2}{3}x_1^2)^{4/3} - (x_1^2/3)^{4/3}], \qquad \textbf{(46)}$$

and equation 44a yields the transcendent equation for $x_1$,

$$(3x_1^2)^{2/3} = \frac{9}{4}\,[(1 - \tfrac{2}{3}x_1^2)^{1/3} - (x_1^2/3)^{1/3}] - \frac{1 - x_1^2}{6} - \frac{x_4^2}{9}\ln\left(\frac{3 - 2x_1^2}{x_1^2}\right),$$

whose solution is $x_1 \approx 0.50$.

The substitution of equation 45 into equations 17 and 18 with the account of equation 46 produces the formulas like equation 22,

$$J_T = \zeta_2^T c_1(\Omega/J_v)^2, \qquad J_M = \zeta_1^M(\Omega/J_v),$$

where the functionals $\zeta_2^T$ and $\zeta_1^M$ are determined as follows:

$$\zeta_2^T = \frac{9}{64}S_{4/3}^2\left(\frac{13}{32}S_{4/3} - \frac{5x_1^2}{12}S_{1/3}\right), \quad \zeta_1^M = \frac{3}{8}S_{4/3}\left(\frac{3}{10}S_{5/3} + \frac{x_1^2}{2}S_{2/3}\right),$$

$$S_n(x_1^2) = (1 - \tfrac{2}{3}x_1^2)^n - (x_1^2/3)^n. \qquad \textbf{(47)}$$

In the particular case when $x_1 = 0$, equations 45–47 take the forms,

$$k(x) = \frac{3}{8J_v}x^{1/3}, \qquad h(x) = \frac{8}{3}J_v x^{2/3},$$

$$\zeta_2^T(0) \approx 0.0571, \qquad \zeta_1^M \approx 0.1125. \qquad \textbf{(48a)}$$

In another particular case, namely, $x_1 = 0.5$, equations 45–47 yield

$$k(x) = \frac{0.280x}{J_v(x^2 - 0.167)^{1/3}}, \qquad h(x) = 3.57J_v(x^2 - 0.167)^{1/3},$$

$$\zeta_2^T(0.5) \approx 0.0197, \quad \zeta_1^M \approx 0.0915. \qquad \textbf{(48b)}$$

*Case 2*

Let $\Omega \gg 1$. Then outside a certain vicinity of the point $x_1$, it is admissible to employ the power asymptotic equation 14b and find out an asymptotic solution of equation 38 at $z' = 0$:

$$k \approx q\left(\frac{x^3}{x^2 - x_1^2}\right)^{1/(1+\alpha)}, \qquad h \approx \frac{1}{q}\left(\frac{x^2 - x_1^2}{x_1^{2-\alpha}}\right)^{1/(1+\alpha)}, \qquad q = \left(\frac{\lambda}{a\alpha\Omega^\alpha}\right)^{1/(\alpha+1)} \tag{49}$$

This weak singular solution is incorrect only within the vicinity of the point $x_1$ where, as mentioned above, the quadratic asymptotic equation 14a is fulfilled. However, at large $\Omega$, this vicinity is getting extremely small and it gives practically no contribution at all to the integrals. Therefore, the asymptotic solution in equation 49 is quite acceptable to use it for calculations. From equation 19, one can find

$$q = \frac{1}{J_v} R_1(x_1, \alpha), \qquad R_1 = \int_{x_1}^{1} x^{(2\alpha-1)/(1+\alpha)} (x^2 - x_1^2)^{1/(1+\alpha)}\, dx. \tag{50}$$

For the determination of $x_1$, let us treat equation 44a as follows: in the left-hand side, we shall utilize quadratic asymptotic equation 14a, and in its right-hand side for $\sigma_1$, we shall employ asymptotic equation 14b, neglecting $\sigma_2$ in comparison with $\sigma_1$. As for the result, we obtain

$$\frac{c_1}{2a}\left(\frac{a\alpha x_1}{2}\right)^{2/3} \Omega^{(2-\alpha)/3}\left(\frac{R_1}{J_v}\right)^{(2-\alpha)/3(\alpha+1)} = R_2(x_1, \alpha) = \int_{x_1}^{1} \frac{x^{(2\alpha-1)/(1+\alpha)}\, dx}{(x^2 - x_1^2)^{\alpha/(1+\alpha)}}. \tag{51}$$

Because the integrals $R_1$ and $R_2$ converge, it results (from equation 51) at great $\Omega$ that $x_1 \sim \Omega^{-(1-\alpha/2)}$, that is, $x_1 \longrightarrow 0$ when $\Omega \longrightarrow \infty$.

After substituting equation 49 into equations 17 and 18 and after taking into account equation 50, one can obtain again the formulas like equation 22:

$$J_T = \zeta_\alpha^T a(\Omega/J_v)^\alpha, \qquad J_M = \zeta_\beta^M b(\Omega/J_v)^\beta,$$

where the magnitudes of $\zeta_\alpha^T$ and $\zeta_\beta^M$ are given by

$$\zeta_\alpha^T = R_1^{1+\alpha}, \qquad \zeta_\beta^M = R_1^\beta R_3, \qquad R_3 = \int_{x_1}^{1} x^{(3\beta+2\alpha-2)/(1+\alpha)} (x^2 - x_1^2)^{-\beta/(1+\alpha)}\, dx. \tag{52}$$

In the particular case of $x_1 = 0$, we have

$$R_1(0, \alpha) = \frac{\alpha + 1}{3\alpha + 2}, \qquad R_3(0, \alpha, \beta) = \frac{\alpha + 1}{3\alpha + \beta + 3}, \qquad k = \frac{\alpha + 1}{3\alpha + 2} \cdot \frac{x^{1/\alpha}}{J_v}$$

$$h(x) = \frac{3\alpha + 2}{\alpha + 1} J_v x^{\alpha/1+\alpha}, \qquad \zeta_\alpha^T\big|_{x_1=0} = \left(\frac{\alpha + 1}{3\alpha + 2}\right)^{\alpha+1},$$

$$\zeta_\beta^M\big|_{x_1=0} = \left(\frac{\alpha + 1}{3\alpha + 2}\right)^\beta \frac{\alpha + 1}{3\alpha + \beta + 3}. \tag{53}$$

If $x_1 \neq 0$, there is a need to calculate the integrals $R_i$ numerically, while giving up the values of parameters $\alpha$ and $\beta$. Some examples will be represented in the final part of this paper.

## GAP'S GEOMETRY PRODUCING VERY GREAT THRUST WITH ALMOST MINIMAL TORQUE

An essential feature of the results obtained in the previous section is the fact that the extremals delivered only a local minimum for the functional of equation 17 under the condition of equation 19 of volume constancy. However, in practice for the rotational sets (elastic melt extruder, axial bearings, etc.), the contrary problem occurs very often: to determine a gap's geometry in which the thrust being as great as possible could arise under minimal torque and a given gap's volume. As follows from the results of the previous two sections, such a problem has no exact analytical solution. Nevertheless, it has evident practical interest, especially at great angular velocities of revolution $\Omega$. In this section, we shall further consider a natural situation for applications when $x_1 = 0$ for all the functionals of equations 17–19.

There are two possibilities to elevate the thrust under a given gap's volume: first, by involving in study the three-dimensional geometry of the set, and second, by finding out a particular distribution $k(x)$ for the flat situation. Although the first way leads to the increase of $J_T$ [due to the growth of $k(x)$ at $z \neq 0$] in comparison with the plate geometry, nonetheless, for the most part, it results in the increase of $J_M$ on account of the growth of also the multiplier $\sqrt{1 + z'^2}$ in the integrand of equation 18. Thus, for our purposes, only the second way remains here, that is, the choice of $k(x)$ [or $h(x)$] in the case of plate geometry.

From equations 17 and 18, one can easily perceive that the only method to attain the increase of $J_T$ without essential variation of $J_M$ is to make a distribution $k(x)$ singular at $x = 0$. This distribution is localized in a narrow vicinity of the axis of revolution $(0, x_0)$ and in the rest of the interval's part $(x_0,1)$ to utilize the cone-plate geometry for the gap, which corresponds to the minimum of $J_M$.

Furthermore, to simplify the next few calculations and to keep in mind the fact that the profile satisfying the formulated conditions corresponds to great shear rates in the gap, we shall proceed from asymptotic equations 14 for the viscometric functions. Then, by virtue of equations 22 and 23, it is feasible to introduce the functionals $\zeta_\alpha^T$ and $\zeta_\beta^M$, which at given $\alpha$ and $\beta$ depend only upon the gap's geometry:

$$\zeta^T = \int_0^1 x[\hat{k}(x)]^\alpha \, dx, \tag{54}$$

$$\zeta^M = \int_0^1 x^2[\hat{k}(x)]^\beta \, dx, \tag{55}$$

$$1 = \int_0^1 \frac{x^2 dx}{\hat{k}(x)} \tag{56}$$

Here, $0 < \beta \leq \alpha < 1$; equation 56 results immediately from equation 19 and function $\hat{k}(x)$ is given by

$$\hat{k}(x) = k(x)/J_v = \frac{x}{\hat{h}(x)} \qquad [\hat{h}(x) = J_v h(x)]. \tag{57}$$

In equations 54 and 55, the subscripts $\alpha$ and $\beta$ were omitted to simplify the designations of magnitudes $\zeta_\alpha^T$ and $\zeta_\beta^M$.

In order to study the possibility of employing the singular profile $\hat{k}(x)$ in near-axis zone, let us initially consider the functionals in equations 54 and 55 on the class of curves

$$\hat{k}(x) = \frac{x^n}{3 - n} \qquad [\hat{h}(x) = (3 - n)x^{1-n}], \tag{58}$$

where equation 56 was already used. From equation 58 and the condition $\hat{k} > 0$, the first restriction on $n$ is as follows: $n > 3$. Substituting equation 58 into equations 54 and 55, we obtain

$$\zeta_n^T = \frac{1}{(3 - n)^\alpha (2 + n\alpha)}, \qquad \zeta_n^M = \frac{1}{(3 - n)^\beta (3 + n\beta)}. \tag{59}$$

From equation 59, it results that as $n \to 3$, both $\zeta_n^T$ and $\zeta_n^M$ tend to infinity. However, if $n \to -2/\alpha$, then only $\zeta_n^T \to \infty$, thus making the second restriction on $n$ so that

$$-2/\alpha < n < 3. \tag{60}$$

Moreover,

$$\zeta_n^T \to \infty, \qquad \zeta_n^M \to \left(\frac{\alpha}{3\alpha + 2}\right)^\beta \frac{\alpha}{3\alpha - 2\beta} = \zeta_0^M \qquad (n \to -2/\alpha).$$

Nonetheless, the magnitude of $\zeta_0^M$ may in 1.8 times the number of cases exceed the minimal value of $\zeta_{\min}^M = 3^{-(\beta+1)}$ corresponding to the cone-plate geometry ($n = 0$). Therefore, the problem arises to design such a profile $\hat{k}_*(x)$ [or $\hat{h}_*(x)$] so that

$$\zeta^M \leq (1 + \epsilon)\zeta_{\min}^M, \tag{61}$$

where $\epsilon$ is an arbitrarily small positive number.

Let $\epsilon > 0$ be given. Let us accept

$$\hat{k}_*(x) = \begin{cases} \nu\left(\dfrac{x_0}{x}\right)^{(2-\delta)/\alpha} & (0 < x \leq x_0) \\ \nu \quad (x_0 \leq x \leq 1) \end{cases}; \qquad \hat{h}_*(x) = \begin{cases} \dfrac{x}{\nu}\left(\dfrac{x}{x_0}\right)^{(2-\delta)/\alpha} & (0 < x \leq x_0) \\ x/\nu \quad (x_0 \leq x \leq 1) \end{cases} \tag{62}$$

where the magnitude $\nu$ is determined from equation 56 and is given by

$$\nu = \frac{3\alpha + (2 - \delta)(1 - x_0^3)}{3(3\alpha + 2 - \delta)}. \tag{63}$$

Also, the magnitude $\delta$ ($0 < \delta \ll 1$) is given and the magnitude $x_0$ ($0 < x_0 < 1$) is subject to determination. Equations 62 correspond to the idea of joining an almost singular power profile with the profile of cone-plate geometry. Substituting equations 62 and 63 into equations 54 and 55, we find

$$\zeta_*^T = \left[\frac{3\alpha + (2 - \delta)(1 - x_0^3)}{3(3\alpha + 2 - \delta)}\right]^\alpha \left(\frac{x_0^2}{\delta} + \frac{1 - x_0^2}{2}\right) \equiv A + B/\delta, \tag{64}$$

$$\zeta_*^M = \left[\frac{3\alpha + (2 - \delta)(1 - x_0^3)}{3(3\alpha + 2 - \delta)}\right]^\beta \left[\frac{x_0^3}{3 - (2 - \delta)\beta/\alpha} + \frac{1 - x_0^3}{3}\right] \leq \frac{1 + \epsilon}{3^{1+\beta}}. \tag{65}$$

Equation 65 was derived by employing the first condition contained in equation 61. Because $0 < \delta \ll 1$, it is possible to drop magnitude $\delta$ in equation 65 as well as everywhere in equation 64, except in the parentheses. Furthermore, from equation 65, one can determine $x_0^3$ versus $\epsilon$ because, as is easily seen, there is the existence of such a monotonically increasing dependence $x_0(\epsilon)$ that $x_0(0) = 0$. Thus, with $\epsilon > 0$, one can always choose such a small $\delta$ so as to satisfy both inequalities in equation 61. By the same method, the simple geometry satisfying all the conditions is determined. Certainly, such a geometry is not single.

However, such a geometry contains a small parameter $\delta$ that distinguishes the nonsingular gap's profile from the singular one ($\delta = 0$); likewise, though, this produces many troubles in manufacturing the set. In consequence of this, a question arises about the events that will happen in reality for a crudely made set of geometry (equation 62) with $\delta = 0$. In such a case, we guess that the most likely event will be the pressing out of one of the surfaces from another under action of normal stresses arising in flowing elastic liquid. This pressing out will lead to the formation of an extremely small clearance of dimensionless thickness $\Delta$ on the axis. As a result, the thrust will take a certain finite value according to a balance of axial forces. In a simple case, such a process takes place if one of the surfaces is freely pressed to another. If, however, these are fastened, then the clearance will be formed as a result of elastic deformations in the set's construction (that is, if certainly, there will not be a mechanical fracture of the system). Thus, in the general case, the set with the profile considered ($\delta = 0$) will be adopted to take arbitrary loads due to the formation of the "floating" (self-controlling) clearance.

As an example of the thrust reduction due to the clearance formation, we shall consider the set with the freely pressed surfaces. Let $\hat{h}_*(x)$, $\nu$, and $x_0$ be determined at $\delta = 0$ from equations 62, 63, and 65, respectively. Then, real dimensionless distributions $\hat{h}_\Delta(x)$ and $\hat{k}_\Delta(x)$ are of the forms,

$$\hat{h}_\Delta(x) = \Delta + \hat{h}_*(x), \qquad \hat{k}_\Delta(x) = \frac{x}{\hat{h}_\Delta(x)},$$

where $\Delta \ll 1$. In such events, $\zeta_\Delta^T = J_1 + J_2$, where

$$J_1 = \int_0^{x_0} \frac{x^{1+\alpha}dx}{\left[\Delta + \frac{x}{\nu}\left(\frac{x}{x_0}\right)^{2/\alpha}\right]^\alpha}, \qquad J_2 = \int_{x_0}^1 \frac{x^{1+\alpha}dx}{(\Delta + x/\nu)^\alpha}.$$

Let us consider the asymptotic behavior of integrals $J_1$ and $J_2$ as $\Delta \to 0$, where the magnitude $\lambda = \nu\Delta/x_0 \ll 1$. If we introduce a new variable $z = (x/\lambda x_0)^{-(\alpha+2)/\alpha}$ in integral $J_1$, we represent $J_1$ in the form,

$$J_1 = \frac{\alpha}{\alpha + 2} x_0^2 \nu^\alpha \int_\lambda^\infty \frac{dz}{z(1 + z)^\alpha};$$

then after partial integration, this results in the following form:

$$J_1 = \frac{\alpha}{\alpha + 2} x_0^2 \nu^\alpha \left[ -\frac{\ln \lambda}{(1 + \lambda)^\alpha} + \alpha \int_0^\infty \frac{luz\, dz}{(1 + z)^{1+\alpha}} - \alpha \int_0^\lambda \frac{luz\, dz}{(1 + z)^{1+\alpha}} \right].$$

The second integral in the equality is tabular and the third one is easily estimated by decomposition of the integrand's denominator in the power series of $z$ by taking into account that $0 < z \le \lambda \ll 1$. As a result, one can obtain

$$J_1 = -\frac{\alpha}{\alpha + 2} x_0^2 \nu^\alpha [lu\lambda + c + \Psi(\alpha) + \alpha\lambda + O(\lambda^2 lu\lambda)],$$

where $c \approx 0.577$ is Euler's constant, $\Psi(\alpha) = (d/d\alpha) lu\Gamma$, and $\Gamma(\alpha)$ is the Gamma-function. (One can find tables and plots of $\Psi(\alpha)$, for instance, in reference 8, p. 118 and 119.) Within interval $0 < \alpha < 1$, the function $\Psi(\alpha)$ increases monotonically from $\Psi(0) = -0.577$ to $\Psi(1) = 0.416$.

The asymptotic formula for $J_2$ may be easily obtained by means of the decomposition of the integrand's denominator in the power series of $z$, which, as a result, gives

$$J_2 = \nu^\alpha \left[ 1 - \frac{x_0^2}{2} - \alpha\lambda x_0(1 - x_0) + O(\lambda^2) \right].$$

Thus, with an accuracy of up to small products of high order, the asymptotic formula for $\zeta_\Delta^T$ is of the form,

$$\zeta_\Delta^T = -\frac{\alpha}{\alpha + 2} x_0^2 \nu^\alpha [\ln \lambda + c + \Psi(\alpha)] + \nu^\alpha (1 - x_0^2/2) = -A_\Delta lu\lambda + B_\Delta. \quad \textbf{(64a)}$$

For the magnitude $\zeta_\Delta^M$ that defies the torque, we have the expression

$$\zeta_\Delta^M = \int_0^{x_0} \frac{x^{2+\beta} dx}{\left[ \Delta + \frac{x}{\nu} \left( \frac{x}{x_0} \right)^{2/\alpha} \right]^\beta} + \int_{x_0}^1 \frac{x^{2+\beta} dx}{(\Delta + x/\nu)^\beta}.$$

After proceeding with the calculations of the integrals by the same method within the accuracy of $O(\lambda)$, we obtain [where $\gamma = (3\alpha - 2\beta)/(2 + \alpha)$]:

$$\zeta_\Delta^M \approx \frac{\alpha}{\alpha + 2} \nu^\beta x_0^3 \left[ \frac{1}{\gamma} - \frac{\beta}{\gamma} \cdot \frac{\Gamma(1 - \gamma)\Gamma(\beta + \gamma)}{\Gamma(1 + \beta)} \lambda^\gamma \right] + \nu^\beta \frac{1 - x_0^3}{3} \quad (0 < \gamma < 1). \quad \textbf{(65a)}$$

Equations 64a and 65a show that in the presence of small clearance ($\lambda \ll 1$), the value of torque almost does not vary, whereas the value of the thrust decreases. This is essentially so because of the formation of the clearance with small thickness $\Delta$ that promotes automatic adaptation of the thrust to external conditions.

## EXAMPLES: DISCUSSION OF THE RESULTS

The variational problems on the extremum of torque or thrust considered in the third and fourth sections of this paper do not exhaust all of the varieties of variational problems that are possible to put into and solve for the functionals of equations 17–20. For instance, we can consider a problem of the extremum of the functional of equation 17 at the constancy of the functional of equation 18. It is of interest to notice here that at $x_1 = 0$ and $\Omega \ll 1$, the thrust's minimum is attained with the plate-plate plane geometry of the gap ($h$ = const., $k = a/h$). In addition to the volume constancy, the condition of constancy of median surface area, vertical dimension, etc., can also be taken into consideration.

Because the sets with plate-plate plane geometry are widely used in practice because of their extreme simplicity in design and production, we shall adduce in this event the expressions for the functionals $\zeta_n^T$ and $\zeta_n^M$ ($x_1 \leq x \leq 1$, $c_2/c_1 = 1/4$):

$$\zeta_2^T = \frac{(1 - x_1^2)^2}{4}\left[\frac{5}{16}(1 - x_1^4) - \frac{3}{8}x_1^2(1 - x_1^2)\right];$$

$$\zeta_\alpha^T = \left(\frac{1 - x_1^2}{2}\right)^\alpha\left[\frac{1 - x_1^{2+\alpha}}{2 + \alpha} - x_1^2\frac{1 - x_1^2}{\alpha}\right],$$

$$\zeta_1^M = \frac{(1 - x_1^2)(1 - x_1^4)}{8}, \qquad \zeta_\beta^M = \left(\frac{1 - x_1^2}{2}\right)^\beta\frac{1 - x_1^{3+\beta}}{3 + \beta}.$$

Furthermore, there is a special interest to gather up data that could demonstrate the existence of the asymptotic equation 14b, which holds true within a wide range of shear rates. As an example, FIGURES 5a and 5b display empirical data for $P_1$ and $P_{12}$ dependences on the generalized shear rate $\eta_0\dot{\gamma}$ that are represented in double logarithmic coordinates. These data are borrowed from reference 7 and are represented for polyisobutylene P-20 (in FIGURE 5a) and for butyl rubbers (for FIGURE 5b). As one can see, in both cases, there is an overlapping range of applicability of the power asymptotic equation 14b for $P_1$ and $P_{12}$, which reach almost to two orders of shear rate. For polyisobutylene P-20, according to FIGURE 5a, $\alpha \approx 0.60$ and $\beta \approx 0.30$, and for butyl rubbers, $\alpha \approx 0.69$ and $\beta \approx 0.25$. Unfortunately,[7] data on the second stress difference are absent and there are also no data on the constants $\eta_0$ and $G_e$ that did not allow for the determination of the values of $a$, $b$, $c_1$, and $c_2$. Nevertheless, the values of exponents $\alpha$ and $\beta$ will be employed below in calculations.

It is of interest to consider now a comparative influence of geometrical parameters for the torsional sets (rotational sets with plate geometry) of the three configurations studied: the set of cone-plate geometry minimizing the torque, the set with gap's geometry (equation 45 or 49) minimizing the thrust, and the set of plate-plate ($h$ = const.) geometry, which is easiest to manufacture. The comparison is accomplished at the same gap's volume, with parameters $r_1$ and $r_2$, and, for determinacy, with the assumption of $c_2/c_1 = 0.25$ and $x_1 = 0.5$. In all three cases, the reports of numerical data for coefficients $\zeta_n^T$ and $\zeta_n^M$ calculated according to the formulas derived above and for $\alpha$, $\beta$ values from FIGURE 5 are represented in TABLE 1.

First of all, in comparison, it is of interest to analyze the vertical columns of digits that are related with the same regimes ($\Omega \ll 1$ or $\Omega \ll 1$), rheology ($\alpha$ and $\beta$), and values of $x_1$.

Let us initially pay attention to the regimes $\Omega \ll 1$ corresponding to the values of the functionals $\zeta_2^T$ and $\zeta_1^M$. For pure torsional sets ($x_1 = 0$), it is worth noticing here that the least value of $\zeta_2^T$ is attained not for the distribution in equation 45, but for the cone-plate geometry that results only from the presence of the additional condition of $\delta k|_1 = 0$ for the distribution in equation 45. The greatest value of $\zeta_2^T$ (roughly, 40% exceeding $\zeta_2^T$ for the cone-plate set) corresponds to the plate-plate set. For ringlike rotational sets ($x_1 = 0.5$), the least value of $\zeta_2^T$ is delivered by the geometry of equation 45. This must be due to the sense of the thrust's local minimum, whereas the cone-plate

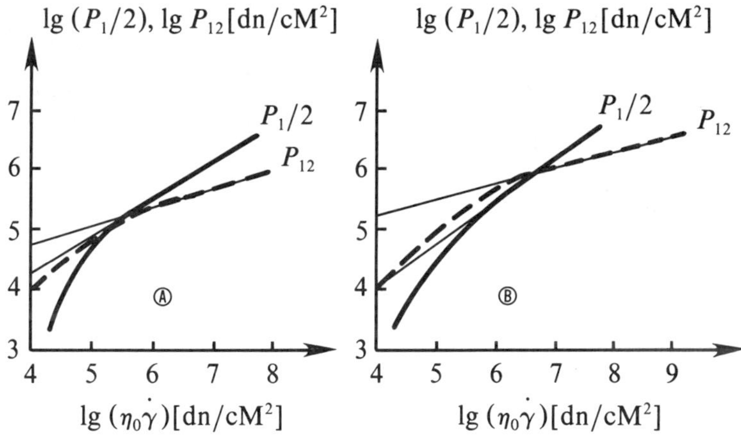


**FIGURE 5.** Plots of the first normal stress difference $P_1/2$ and of the tangential stress $P_{12}$ versus the generalized shear rate $\eta_0\dot{\gamma}$ according to reference 7: (a) polyisobutylene P-20 ($\alpha \approx 0.60$, $\beta \approx 0.30$); (b) butyl rubbers ($\alpha \approx 0.69$, $\beta \approx 0.25$).

thrust exceeds 24% and the plate-plate thrust exceeds 58% of the thrust for the distribution in equation 45. As to the torque coefficient $\zeta_1^M$ in Newtonian range ($\Omega \ll 1$), for pure torsional sets, the minimal torque is attained for the cone-plate geometry and the greatest one is attained for the plate-plate set. The relative difference between these is ≈12.6%. For the ringlike rotational sets ($x_1 = 0.5$), the cone-plate configuration produced the least torque, whereas the largest one was created by the distribution in equation 45.

Within non-Newtonian range ($\Omega \gg 1$), the properties noticed for the coefficient $\zeta_1^M$ are maintained, but for $\zeta_\beta^M$, however, the whole variations of data for the different gap's geometry do not exceed 2%. The same negligible difference is revealed at $\Omega \gg 1$ between the magnitude $\zeta_\alpha^T$ defined by equation 49, which minimizes the thrust, and those for the $k$ = const. and $h$ = const. distributions in both the ringlike and pure torsional sets. Furthermore, as seen from TABLE 1, for the same gap's geometry, the

TABLE 1. Coefficients $\zeta_n^T$ and $\zeta_n^M$ for Different Geometries of Torsional Sets and Rheological Parameters $\alpha$ and $\beta$[a]

| | $\zeta$ | $\zeta_n^T$ | | | $\zeta_n^M$ | | |
|---|---|---|---|---|---|---|---|
| $k(x)$ | $x_1$ | $n = 2$ | $n = \alpha = 0.60$ | $n = \alpha = 0.69$ | $n = 1$ | 0.30 | 0.25 |
| $k$ = const. | $x_1 = 0$ | 0.055 | 0.259 | 0.234 | 0.111 | 0.240 | 0.253 |
| | $x_1 = 0.5$ | 0.0245 | 0.0963 | 0.0862 | 0.085 | 0.201 | 0.214 |
| $k(x)$[b] | $x_1 = 0$ | 0.571 | 0.251 | 0.226 | 0.113 | 0.242 | 0.255 |
| | $x_1 = 0.5$ | 0.0197 | 0.0959 | 0.0816 | 0.0915 | 0.205 | 0.218 |
| $h$ = const. | $x_1 = 0$ | 0.0781 | 0.254 | 0.230 | 0.125 | 0.246 | 0.258 |
| | $x_1 = 0.5$ | 0.0313 | 0.0906 | 0.0896 | 0.0879 | 0.203 | 0.215 |

[a]According to FIGURE 5 ($c_2/c_1 = 0.25$).
[b]According to equations 45 and 49.

**TABLE 2.** Relative Functionals of the Thrust $\chi_i^T$ and Torque $\chi_i^M$ for the Pure Rotational Sets of Various Configurations versus Rheological Parameters $\alpha$ and $\beta$[a]

| $\chi$ | $\chi_1^T$ | $\chi_2^T$ | $\chi_3^T$ | $\chi_1^M$ | $\chi_2^M$ | $\chi_3^M$ |
|---|---|---|---|---|---|---|
| formulas | $\frac{3}{2}\left(\frac{3\alpha + 2}{3\alpha + 3}\right)^{1+\alpha}$ | 1 | $\frac{2}{2 + \alpha}\left(\frac{3\alpha + 2}{2\alpha + 2}\right)^{1+\alpha}$ | 1 | $\left(\frac{3\alpha + 3}{3\alpha + 2}\right)^{\beta} \frac{3\alpha + 3}{3\alpha + 3 + \beta}$ | $\left(\frac{3}{2}\right)^{\beta} \frac{3}{3 + \beta}$ |
| conditions | $0 \le \alpha \le 1$ | | $0 \le \alpha \le 1$ | | $0 \le \beta \le \alpha \le 1$ | $0 \le \beta \le 1$ |
| min $\chi$ | 1 | 1 | 1 | 1 | 1 | 1 |
| conditions | $\alpha = 0$ | | $\alpha = 0$ | | $\beta = 0$ | $\beta = 0$ |
| max $\chi$ | 1.0417 | 1 | 1.0417 | 1 | 1.0286 | 1.125 |
| conditions | $\alpha = 1$ | | $\alpha = 1$ | | $\alpha = \beta = 1$ | $\beta = 1$ |

[a] The subscripts on the $\chi$ are as follows: 1 = cone-plate; 2 = according to equation 53; 3 = plate-plate.

coefficient $\zeta_n^T$ decreases with the increase of $n$ (more rapidly for the pure torsional set). Nonetheless, the data of TABLE 1 give no possibility to make clear the dependences of the relative variations $\zeta_\alpha^T$ and $\zeta_\beta^M$ on $\alpha$ and $\beta$ because at $\Omega \gg 1$, the values $\alpha$ and $\beta$ from FIGURES 5a and 5b are sufficiently close to each another. For the pure torsional sets, such an analysis can be easily carried on theoretically. For this purpose, let us introduce the relative functionals,

$$\chi_i^T = (\zeta_\alpha^T)_i/(\zeta_\alpha^T)_2, \qquad \chi_i^M = (\zeta_\beta^M)_i/(\zeta_\beta^M)_1 \qquad (i = 1, 2, 3),$$

where indexes = 1, 2, 3 denote the configurations of cone-plate geometry (described by equation 53) and plate-plate geometry, respectively. Then, the results of the analysis can be compactly represented in TABLE 2.

From TABLE 2, one can see that in the case of pure rotational sets, the effect of geometry on the thrust at $\Omega \gg 1$ does not exceed 4.2% within the whole admissible range of parameters $\alpha$, $\beta$ ($0 \leq \beta \leq \alpha \leq 1$). With diminishing $\alpha$ and $\beta$, this effect is getting negligible. Similarly, with the increase of degree of non-Newtonian behavior ($\beta \rightarrow 0$), the torque in the pure rotational sets ceases to depend on the gap's geometry.

**TABLE 3.** Numerical Values of Parameters in Equations 62–65 and 64a for Rheological Parameters $\alpha$ and $\beta$[a]

| | | | | $\zeta_*^T = A + B/\delta$ | | $\zeta_\Delta^T = A_\Delta \ln \lambda + B_\Delta$ | | | |
|---|---|---|---|---|---|---|---|---|---|
| Parameters | $\epsilon$ | $x_0$ | $\nu$ | $B$ | $A$ | $A_\Delta$ | $B_\Delta$ | $\zeta_*^M$ | $\zeta_{\min}^M$ |
| $\alpha = 0.60$ $\beta = 0.30$ | 0.05 | 0.536 | 0.306 | 0.175 | 0.141 | 0.0326 | 0.398 | 0.252 | 0.240 |
| $\alpha = 0.69$ $\beta = 0.25$ | 0.05 | 0.657 | 0.287 | 0.120 | 0.182 | 0.0468 | 0.295 | 0.266 | 0.253 |

[a]According to FIGURE 5.

The general conclusion that can be arrived at from the comparison is the fact that the most essential influence of the gap's geometry on both the torque and thrust is revealed at $\Omega \ll 1$. On the other hand, at $\Omega \gg 1$, this effect diminishes so that the variations of the magnitudes for the gaps of the different types are approximately of the order of 1%.

In addition, for the values of $\alpha$ and $\beta$ (according to the data of FIGURE 5), one can carry out the calculations following equations 62–65 and equations 64a and 65a for the gap's geometry. This, of course, produces an extremely great thrust with almost minimal torque. In all equations 62–65 (with the exception of the exponent index in equation 62), it is thus possible to neglect the value of $\delta$ in sum $(2 - \delta)$ by virtue of the fact $\delta \ll 1$. The results of the calculations along with this one corresponding to equation 64a are summarized in TABLE 3.

As a conclusive remark, we shall note that at sufficiently great shear rates, the well-known unstable regimes of flow arise in polymer liquids, which are accompanied by the stick-slip processes. Therefore, the asymptotic laws of equation 14b for viscometric functions will also be restricted from above. In particular, these phenomena may play a very important role in the sets producing large thrust.

## REFERENCES

1. BELKIN, I. M., G. V. VINOGRADOV & A. I. LEONOV. 1968. Rotational sets. Measurements of viscosity and physico-mechanical characteristics of materials. (In Russian.) Mashinostroenie.
2. NORDEN, A. P. 1958. A short course of differential geometry. (In Russian.) Fizmatgiz.
3. SLIOZKIN, N. A. 1955. Dynamics of viscous incompressible fluid. (In Russian.) Gostekhteoretizdat, p. 190–224.
4. HUILGOL, R. R. 1974. On generalized torsional flow. Trans. Soc. Rheol. **18**(no. 2): 191–198.
5. VINOGRADOV, G. V. & A. YA. MALKIN. 1977. Rheology of polymers. (In Russian.) Khimiya. Moscow.
6. GEL'FAND, J. M. & S. V. FOMIN. 1961. Calculus of variations. (In Russian.) Fizmatgiz.
7. YANOVSKII, YU. G. & A. YA. MALKIN. 1970. The correlation of polymer properties at continuous and periodic deformations. *In* Successes in Rheology or Polymers (in Russian), p. 67, 68. Khimiya. Moscow.
8. YANKE, E. & F. EMDE. 1959. Tables of functions with formulas and curves (Russian ed.). Fizmatgiz.

# On Hamiltonian and Lagrangian Formalisms for the KP-Hierarchy of Integrable Equations

L. A. DICKEY

## PART I—GENERALIZATION OF WATANABE'S STRUCTURE

### *Introduction*

The appearance of this paper was stimulated by an article by Watanabe[1] and to some extent by references 2, 3, and 4 in which a Hamiltonian form of the KP-hierarchy was given.

As it was pointed out in reference 5, a great class of integrable equations arises from the fractional powers of a differential operator. Namely, let

$$L_n = \partial^n + v_2\partial^{n-2} + \cdots + v_n, \quad \partial = d/dx.$$

Then, the pseudodifferential operator (PDO),

$$L_n^{m/n} = \partial^m + w_2\partial^{m-2} + \cdots + w_m + w_{m+1}\partial^{-1}{}_+ \cdots,$$

commutes with $L_n$. Also, let $L_n^{m/n}{}_+ = \partial^m + w_2\partial^{m-2} + \cdots + w_m$ and $L_n^{m/n}{}_- = L^{m/n} - L^{m/n}{}_+$. Then we have $[L_n, L^{m/n}{}_+] = -[L_n, L^{m/n}{}_-]$, which implies that $[L_n, L_n^{m/n}{}_+]$ is a differential operator of an order $\leq n-2$. The operators, $L_n$ and $L_n^{m/n}{}_+$, form a Lax pair, so the equation

$$\partial L_n/\partial t = [L_n, L_n^{m/n}{}_+] \quad \textbf{(1.1a)}$$

has sense. This equation is completely integrable. For each $n$, we have a hierarchy of equations with $m = 1, 2, 3, \ldots$. (The KdV equation corresponds to $n = 2$, $m = 3$.)

Sato and other representatives of the Japanese school (see reference 6) noticed that all the hierarchies can be combined into one if the operators $L_n^{1/n}$, for all $n$, are identified. More precisely, if we let

$$L = \partial + u_0\partial^{-1} + u_1\partial^{-2} + \cdots,$$

then we consider the hierarchy of equations

$$\partial_m L = [L^m{}_+, L], \quad m = 1, 2, \ldots. \quad \textbf{(1.2a)}$$

Here, $\partial_m$ denotes the derivative with respect to a variable $x_m$ ($x_1 = x$). The convenience of introducing for every equation its own variable is connected with the fact that all these equations are compatible, which can easily be shown (in other words, the differentiations $\partial_m$ defined by equation 1.2a commute). Therefore, the equations can be considered together. Each possibility of equation 1.2a represents an infinite set of equations for the variables $u_0, u_1, \ldots$.

Equations 1.1a are reductions of equations 1.2a to the manifold of variables $u_i$

defined by

$$(L^n)_- = 0; \tag{1.2b}$$

this creates the possibility of expressing all the $u_i$ starting with $u_{n-1}$ in terms of $u_i$, $i \leqq n - 1$. Equations 1.2b are compatible with equations 1.2a.

From equations 1.2a, the equations

$$\partial_m L^n{}_+ - \partial_n L^m{}_+ = [L^m{}_+, L^n{}_+] \tag{1.3a}$$

follow. Each of these equations is a finite system of equations for $\{u_i\}$. For $n = 2$ and $m = 3$, this is the Kadomtsev-Petviashvili (KP) equation. Therefore, equations 1.3a are called the KP-hierarchy (sometimes, under this name, equations 1.2a are understood). In the first part of this paper, we deal with a Hamiltonian structure for equations 1.2a, and in the second, we deal with that for equations 1.3a. It should be emphasized that these structures are quite different and the connection between them is unclear.

Watanabe[1] has shown that equations 1.2a are Hamiltonians with respect to a symplectic structure that he built by analogy with the structure for equation 1.1a constructed in reference 5; this is explained in the group theory terms by Adler,[7] and Lebedev and Manin.[8]

Using Adler's results, Gelfand and Dickey[9] constructed the second symplectic structure for equation 1.1a. Two symplectic forms give possibility for the construction of the so-called Lennart pair of Hamiltonian operators. Now, we do the same for Watanabe's symplectic form; that is, we construct a one-parametric family of forms that includes the Watanabe form. This makes up the first part of this paper.

The second part deals with equations 1.3a. In the case of $n = 2$, these equations were written in the Hamiltonian form in references 2, 3, and 4. Here, we study the general case. Our approach, though, is quite different. For arbitrary $n$ and $m$, it is senseless to distinguish one of variables $x_m$ and $x_n$ as a time in which the system evolves. It is more natural to use the field (or multitime) Hamiltonian theory. We use it in a form developed by Gelfand and Dickey (for more detail, see reference 10; however, all necessary explanations will be given here).

### *The Hamilton Form of Equations 1.2a—Rings of PDO*

Let $L$ be a little bit more general than above; that is,

$$L = \partial + u_{-1} + u_0\partial^{-1} + u_1\partial^{-2} + \cdots \tag{2.1a}$$

(here, an additional term, $u_{-1}$, is present). Let $R_+$ and $R_-$ be rings of PDO of the form,

$$R_+ = \left\{\sum_{i\geq 0} a_i\partial^i\right\}, \quad R_- = \left\{\sum_{i<0} a_i\partial^i\right\},$$

where in $R_+$, sums are finite, and in $R_-$, they may be infinite. The coefficients $a_i$ belong to a differential algebra $A_u$ generated by $\{u_i\}$; that is, they are differential polynomials in $\{u_i\}$, with the differentiation being supposed with respect to $x$. Let $R = R_+ + R_-$. We introduce also the extended spaces of operators $R'_+$ and $R'_-$:

$$R'_+ = \left\{\sum_{i \geq -1} a_i \partial^i\right\}, \quad R'_- = \left\{\sum_{i<1} a_i \partial^i\right\}.$$

The operators $\Sigma_{i \geq -1} a_i \partial^i$ can also be written as $\Sigma_{i \geq -1} \partial^i \tilde{a}_i$; however, additional terms of the form $\partial^i \tilde{a}_i$, $i < -1$, appearing from the commutational relations must be omitted.

### *The Mapping* H

Now, we define the mapping $H : R'_+ \rightarrow R'_-$,

$$X \in R'_+ \rightarrow H(X) = (\hat{L}X)_- \hat{L} - \hat{L}(X\hat{L})_- = -(\hat{L}X)_+ \hat{L} + \hat{L}(X\hat{L})_+ \in R'_-.$$

Here, $\hat{L} = L - \zeta$ and $\zeta$ is a parameter; the subscripts "+" and "−" denote the projections on $R_+$ and $R_-$ : $(\Sigma a_i \partial^i)_+ = \Sigma_{i \geq 0} a_i \partial^i$ and $(\Sigma a_i \partial^i)_- = \Sigma_{i<0} a_i \partial^i$. Later, we shall also use the subscripts "+′" and "−′", which mean projections on $R'_+$ and $R'_-$:

$$(\Sigma a_i \partial^i)_{+'} = \sum_{i \geq -1} a_i \partial^i \quad \text{and} \quad (\Sigma a_i \partial^i)_{-'} = \sum_{i<1} a_i \partial^i.$$

We also denote Res $\Sigma a_i \partial^i = a_{-1}$. Mappings $H$ of the described form were introduced by Adler.

The mapping $H$ depends on $\zeta$ linearly, $H = H^0 + \zeta H^1$. We have

$$H^0(X) = (LX)_- L - L(XL)_- = -(LX)_+ L + L(XL)_+.$$

Now, we find $H^1(X)$. Let $X = X_+ + a\partial^{-1}$; then

$$\begin{aligned} H^1(X) &= [X_+, L] + [L, X]_+ = [X, L] - [a\partial^{-1}, L] - [X, L]_+ \\ &= [X, L]_- - [a\partial^{-1}, L] = [X - a\partial^{-1}, L]_- - [a\partial^{-1}, L]_+ = [X_+, L]_-. \end{aligned}$$

Thus,

$$H^1(X) = [X_+, L]_-.$$

This mapping can be restricted to $R_+ \rightarrow R_-$; in this form, it was considered by Watanabe.

Also, let

$$R'_+((\zeta)) = \left\{\sum_{i=i_0}^{\infty} X_i \zeta^{-i-1}, X_i \in R'_+\right\}, \qquad R'_-((\zeta)) = \left\{\sum_{i=i_0}^{\infty} X_i \zeta^{-i-1}, X_i \in R'_-\right\}$$

be spaces of a formal series. The mapping $H$ can then be in a natural way extended to $H : R'_+((\zeta)) \rightarrow R'_-((\zeta))$.

### *Resolvent of the Operator* L

Let

$$T(\zeta) = (L - \zeta)^{-1}{}_{+'} = -\left(\sum_0^{\infty} L^i \zeta^{-i-1}\right)_{+'} \in R'_+((\zeta))$$

be a formal series.

*Proposition 1.*

$$H(T(\zeta)) = 0, \quad H(\partial^{-1}) = 0.$$

Proof: Let Res $\hat{L}^{-1} = a$. We have

$$\begin{aligned} H(T(\zeta)) &= (\hat{L}\hat{L}^{-1}{}_{+'})_{-}\hat{L} - \hat{L}(\hat{L}^{-1}{}_{+'}\hat{L})_{-} \\ &= (\hat{L}\hat{L}^{-1}{}_{+})_{-}\hat{L} - \hat{L}(\hat{L}^{-1}{}_{+}\hat{L})_{-} + (\hat{L}a\partial^{-1})_{-}\hat{L} - \hat{L}(a\partial^{-1}\hat{L})_{-} \\ &= -(\hat{L}\hat{L}^{-1}{}_{-})_{-}\hat{L} + \hat{L}(\hat{L}^{-1}{}_{-}\hat{L})_{-} - (\hat{L}a\partial^{-1})_{+}\hat{L} + \hat{L}(a\partial^{-1}\hat{L})_{+} \\ &= (\hat{L}\hat{L}^{-1}{}_{-})_{+}\hat{L} - \hat{L}(\hat{L}^{-1}{}_{-}\hat{L})_{+} - (\hat{L}a\partial^{-1})_{+}\hat{L} + \hat{L}(a\partial^{-1}\hat{L})_{+} \\ &= (\hat{L}(\hat{L}^{-1}{}_{-} - a\partial^{-1}))_{+}\hat{L} - \hat{L}((\hat{L}^{-1}{}_{-} - a\partial^{-1})\hat{L})_{+} = 0. \end{aligned}$$

The second equality is obvious.

*Corollary.* Every linear combination

$$X(\zeta) = c(\zeta)T(\zeta) + d(\zeta)\partial^{-1}, \tag{2.3a}$$

with coefficients $c(\zeta) = \Sigma_{i_1}^{\infty}c_i\zeta^{-i}$, $d(\zeta) = \Sigma_{i_2}^{\infty}d_i\zeta^{-i}$ (where $c_i$, $d_i$ = constants), satisfies the equation $H(X(\zeta)) = 0$.

*Proposition 2.* Ker $H$ consists of linear combinations of equation 2.3a. Proof: Let $X(\zeta) = \Sigma_{i_0}^{\infty}X_i\zeta^{-i-1}$, with $X_{i_0} \neq 0$, and $H(X(\zeta)) = 0$. The latter equation can be written as a recurrency system

$$H^0(X_i) + H^1(X_{i+1}) = 0; \quad i = i_0 - 1, i_0, i_0 + 1, \ldots. \tag{2.3b}$$

We write $X_i = \Sigma_{\alpha=-1}^{\infty}X_{i,\alpha}\partial^{\alpha}$.

*Lemma 1.* If all $X_{i,-1}$ and $X_{i,0}$ have no constants, then $X(\zeta) = 0$ ($X_{i,\alpha} \in A_u$, i.e., they are differential polynomials in $\{u_j\}$; the constant in such a polynomial is uniquely determined). Proof: We make use of the almost obvious lemma 2.

*Lemma 2.* If $Y = \Sigma_0^N y_\alpha\partial^\alpha$, with $y_\alpha \in A_u$, and $[L, Y]_- = 0$, then $Y = y_0 =$ const.

Now equation 2.3b yields $H^1(X_{i_0}) = 0$, that is, $[(X_{i_0})_+, L]_- = 0$. According to lemma 2, $(X_{i_0})_+ =$ const. From the assumption of lemma 1, $(X_{i_0})_+ = 0$; hence, $X_{i_0}$ has the form $X_{i_0} = b\partial^{-1}$. From the next equality, we have $H^0(X_{i_0}) + H^1(X_{i_0+1}) = 0$, that is, $(LX_{i_0})_+L - L(X_{i_0}L)_+ = [(X_{i_0+1})_+, L]_-$. Taking + from both sides, we then obtain $((LX_{i_0})_+L - L(X_{i_0}L)_+)_+ = 0$ and $(-(LX_{i_0})_-L + L(X_{i_0}L)_-)_+ = 0$, that is, Res$[L, X_{i_0}] = 0$. Thus Res $(\partial b\partial^{-1} - b) = 0$ and $b' = 0$, that is, $b =$ const. By assumption, all constants are zero, $b = 0$, which proves lemma 1.

We continue the proof of proposition 2. What are the constants in the terms $X_{i,-1}$ and $X_{i,0}$ of equation 2.3a? The expansion $T(\zeta) = \Sigma_{i=0}^{\infty}\Sigma_{\alpha=-1}^{\infty}T_{i\alpha}\partial^\alpha\zeta^{-i-1}$ contains the only constant in the term $T_{00} = 1$. The expansion $\partial^{-1} = \Sigma\Sigma X_{i\alpha}\,\partial^\alpha\zeta^{-\alpha-1}$ consists of one term, $X_{-1,-1}\partial^{-1} = \partial^{-1}$; the others, $X_{i\alpha} = 0$. Then, it is obvious that by choosing the appropriate coefficients, $c(\zeta)$ and $d(\zeta)$, we can obtain an arbitrary set of constants in the terms $X_{i,-1}$ and $X_{i,0}$. This set uniquely determines the $X(\zeta)$ (lemma 1). Thus, linear combinations (equation 2.3a) exhaust the kernel of the mapping $H$. The operator $T(\zeta)$ is therefore called the resolvent of $L$.

## *Family of Symplectic Forms*

Let $\tilde{A}_u = A_u/\partial A_u$ be the space of "functionals." The natural projection $A_u \to \tilde{A}_u$ will be denoted as an integral, $f \in A \to \tilde{f} = \int f dx \in \tilde{A}_u$ (in analytical terms, this means that we consider such a class of functions $\{u_i(x)\}$ and such a definition of an integral so that $\int f' dx = 0$ for any differential polynomial $f(u)$; for example, the class of functions that tends to zero with all their derivatives when $|x| \to \infty$ or periodic, etc.). It is not difficult to prove that for any two PDO, namely $P$ and $Q$, $\int \operatorname{Res}[P, Q]dx = 0$.

To each $a = \Sigma_{-1}^{\infty} a_j \partial^{-j-1} \in R'_-$, we assign a differentiation $\partial_a$ in $A_u$ acting according to the rule,

$$\partial_a f = \sum_{i,j} a_j^{(i)} \partial f / \partial u_j^{(i)}.$$

We shall call $\partial_a$ a vector field. This vector field commutes with $\partial$ and therefore can be considered also in $\tilde{A}_u$: $\partial_a \tilde{f} = \int \partial_a f dx$. It can act also in $R'_+$ and $R'_-$: $\partial_a \Sigma X_i \partial^i = \Sigma \partial_a X_i \cdot \partial^i$. Evidently, $\partial_a L = a$.

In a usual way, we introduce the variational derivatives:

$$\delta/\delta u_i = \sum_{\alpha} (-\partial)^{\alpha} (\partial/\partial u_i^{(\alpha)}) \quad \text{and} \quad \delta/\delta L = \sum_{-1}^{\infty} \partial^i \cdot \delta f/\delta u_i \in R'_+ .$$

*Lemma 1.*

$$\partial_a \tilde{f} = \int \operatorname{Res} a \frac{\delta f}{\delta L} dx.$$

The proof can be obtained from integration by parts.

*Proposition 1.* The relation

$$[\partial_{H(X)}, \partial_{H(Y)}] = \partial_{H([X,Y]_L + \partial_{H(X)} Y - \partial_{H(Y)} X)},$$

where

$$[X, Y]_L = (X(\hat{L}Y)_+ - (X\hat{L})_- Y)_{+'} - (X \leftrightarrow Y),$$

holds. Proof: Here and throughout the rest of this paper, we repeat proofs of similar theorems in reference 9.

We begin with

$$[\partial_{H(X)}, \partial_{H(Y)}] = \partial_{\partial_{H(X)} H(Y) - \partial_{H(Y)} H(X)}.$$

Furthermore,

$$\begin{aligned} a \equiv \partial_{H(X)} H(Y) - \partial_{H(Y)} H(X) &= (H(X)Y)_- \hat{L} + (\hat{L}Y)_- H(X) - H(X)(Y\hat{L})_- \\ &- \hat{L}(YH(X))_- + H(\partial_{H(X)} Y) - (X \leftrightarrow Y) = (((\hat{L}X)_- \hat{L} - \hat{L}(X\hat{L})_-)Y)_- \hat{L} \\ &+ (\hat{L}Y)_- ((\hat{L}X)_- \hat{L} - \underline{\hat{L}(X\hat{L})_-}) - ((\underline{\hat{L}X)_- \hat{L}} - \hat{L}(X\hat{L})_-)(Y\hat{L})_- \\ &- \hat{L}(Y((\hat{L}X)_- \hat{L} - \hat{L}(X\hat{L})_-))_- + H(\partial_{H(X)} Y) - (X \leftrightarrow Y). \end{aligned}$$

The underlined terms cancel out [if terms hidden under the symbol $(X \leftrightarrow Y)$ are taken into account]. We have

$$a = (((\hat{L}X)_- \hat{L}Y)_- - (\hat{L}(X\hat{L})_- Y)_- + (\hat{L}Y)_-(\hat{L}X)_-)\hat{L}$$
$$+ \hat{L}((X\hat{L})_-(Y\hat{L})_- - (Y(\hat{L}X)_-\hat{L})_- + (Y\hat{L}(X\hat{L})_-)_-)_- + H(\partial_{H(X)}Y) - (X \leftrightarrow Y).$$

We also transform separately

$$((\hat{L}X)_-\hat{L}Y)_- + (\hat{L}Y)_-(\hat{L}X)_- - (X \leftrightarrow Y) = ((\hat{L}X)_-\hat{L}Y - (\hat{L}X)_-(\hat{L}Y)_-)_-$$
$$- (X \leftrightarrow Y) = ((\hat{L}X)_-(\hat{L}Y)_+)_- - (X \leftrightarrow Y) = (\hat{L}X(\hat{L}Y)_+)_- - (X \leftrightarrow Y)$$

and

$$((X\hat{L})_-(Y\hat{L})_- + (Y\hat{L})(X\hat{L})_-)_- - (X \leftrightarrow Y) = -((X\hat{L})_+ Y\hat{L})_- - (X \leftrightarrow Y).$$

Now

$$a = (\hat{L}(X(\hat{L}Y)_+ - (X\hat{L})_-Y))_-\hat{L} + \hat{L}((-(X\hat{L})_+Y + X(\hat{L}Y)_-)\hat{L})_-$$
$$+ H(\partial_{H(X)}Y) - (X \leftrightarrow Y)$$
$$= (\hat{L}(X(\hat{L}Y)_+ - (X\hat{L})_-Y))_-\hat{L} - \hat{L}((X(\hat{L}Y)_+ - (X\hat{L})_-Y)\hat{L})_-$$
$$+ H(\partial_{H(X)}Y) - (X \leftrightarrow Y) = (\hat{L}(X(\hat{L}Y)_+ - (X\hat{L})_-Y))_+\hat{L}$$
$$+ \hat{L}((X(\hat{L}Y)_+ - (X\hat{L})_-Y)\hat{L})_+$$
$$+ H(\partial_{H(X)}Y) - (X \leftrightarrow Y) = H([X, Y]_L + \partial_{H(X)}Y - \partial_{H(Y)}X)$$

as required.

The expression $[X, Y]_L$ is linear with respect to $\zeta$: $[X, Y]_L = [X, Y]_L^0 + \zeta[X, Y]_L^1$,

$$[X, Y]_L^0 = (X(LY)_+ - (XL)_-Y)_{+'} - (X \leftrightarrow Y).$$

Let $X = X_+ + X_{-1}\partial_y^{-1}$ $Y = Y_+ + Y_{-1}\partial^{-1}$; then

$$[X, Y]_L^1 = -(XY_+ - X_-Y)_{+'} - (X \leftrightarrow Y)$$
$$= -(XY - XY_{-1}\partial^{-1} - X_{-1}\partial^{-1}Y)_{+'} - (X \leftrightarrow Y).$$

This means

$$[X, Y]_L^1 = -[X_+, Y_+].$$

Consider now the space of vector fields having the form $\partial_{H(X)}$ with a fixed value of the parameter $\zeta$. In this space, we define a 2-form $\omega$:

$$\omega(\partial_{H(X)}, \partial_{H(Y)}) = \int \operatorname{Res} H(X)Y\,dx = -\int \operatorname{Res} H(Y)X\,dx.$$

Remark: This form can be defined for two vector fields even in the case when only one of $a$ and $b$ has the form $H(x)$.

*Proposition 2.* The form $\omega$ is closed. Proof: We find

$$d\omega(\partial_{H(X)}, \partial_{H(Y)}, \partial_{H(Z)}) = \partial_{H(X)}\,\omega(\partial_{H(Y)}, \partial_{H(Z)}) - \omega([\partial_{H(X)}, \partial_{H(Y)}], \partial_{H(Z)}) + \text{c.p.}$$

The symbol "c.p." denotes the summation over all cyclic permutations of $X$, $Y$, and $Z$. Furthermore,

$$\begin{aligned}
\partial_{H(X)}\omega(\partial_{H(Y)}, \partial_{H(Z)}) &= \partial_{H(X)} \int \operatorname{Res} H(X)Z\,dx \\
&= -\partial_{H(X)} \int \operatorname{Res}\,((\hat{L}Y)_+\hat{L} - \hat{L}(Y\hat{L})_+)Z\,dx \\
&= -\int \operatorname{Res}\,((H(X)Y)_+\hat{L} + (\hat{L}Y)_+H(X) - H(X)(Y\hat{L})_+ - \hat{L}(YH(X))_+)Z\,dx \\
&\quad + \int \operatorname{Res} H(\partial_{H(X)}Y)Z\,dx + \int \operatorname{Res} H(Y)\partial_{H(X)}Z\,dx.
\end{aligned}$$

It is thus easy to transform this into

$$\int \operatorname{Res} H(X)[Y, Z]_L\,dx + \int \operatorname{Res}\,\{-H(Z)\partial_{H(X)}Y + H(Y)\partial_{H(X)}Z\}\,dx.$$

On the other hand,

$$-\omega([\partial_{H(X)}, \partial_{H(Y)}], \partial_{H(Z)}) = \int \operatorname{Res} H(Z)\{[X, Y]_L + \partial_{H(X)}Y - \partial_{H(Y)}X\}\,dx.$$

Now

$$\begin{aligned}
d\omega(\partial_{H(X)}, \partial_{H(Y)}, \partial_{H(Z)}) &= 2\int \operatorname{Res} H(X)[Y, Z]_L\,dx + \text{c.p.} \\
&= 2\int \operatorname{Res} H(X)\{Y(\hat{L}Z)_+ - (Y\hat{L})_-Z\}\,dx + \text{p.} \\
&= 2\int \operatorname{Res}\,((\hat{L}X)_-\hat{L} - \hat{L}(X\hat{L})_-)Y(\hat{L}Z)_+\,dx \\
&\quad + 2\int \operatorname{Res}\,((\hat{L}X)_+\hat{L} - \hat{L}(X\hat{L})_+)(Y\hat{L})_-Z\,dx + \text{p.}
\end{aligned}$$

The symbol "p." denotes the summation over all permutations, with each term being multiplied by sgn of the permutation. Two terms cancel out:

$$\begin{aligned}
&-\int \operatorname{Res} \hat{L}(X\hat{L})_-Y(\hat{L}Z)_+\,dx + \int \operatorname{Res}\,(\hat{L}X)_+\hat{L}(Y\hat{L})_-Z\,dx + \text{p.} \\
&\qquad = -\int \operatorname{Res}\,(\hat{L}Z)_+\hat{L}(X\hat{L})_-Y\,dx + \int \operatorname{Res}\,(\hat{L}X)_+\hat{L}(Y\hat{L})_-Z\,dx + \text{p.} = 0.
\end{aligned}$$

Two terms remain:

$$2\int \operatorname{Res}\,(\hat{L}X)_-\hat{L}Y(\hat{L}Z)_+\,dx - 2\int \operatorname{Res} Z\hat{L}(X\hat{L})_+(Y\hat{L})_-\,dx + \text{p.} \qquad \textbf{(2.4a)}$$

Now we prove another lemma.

*Lemma.* For any three PDO (A, B, and C), we have

$$\int \operatorname{Res} AB_+C_-\,dx + \text{c.p.} = \int \operatorname{Res} ABC\,dx.$$

Before the proof of the lemma, we note that this lemma immediately yields the required assertion; the equation 2.4a vanishes in an obvious way.

Proof of the lemma:

$$\begin{aligned}
\int \operatorname{Res} AB_+C_-\,dx + \text{c.p.} &= \int \operatorname{Res}\,(A_+B_+C_- + A_-B_+C_-)\,dx + \text{c.p.} \\
&= \int \operatorname{Res}\,(A_+B_+C + A_-BC_-)\,dx + \text{c.p.}
\end{aligned}$$

$$= \tfrac{1}{3}\int \mathrm{Res}\,(2AB_+C_- + A_+B_+C + A_-BC_-)\,dx + \text{c.p.}$$
$$= \tfrac{1}{3}\int \mathrm{Res}\,(2AB_+C_- + AB_+C_+ + AB_-C_-)\,dx + \text{c.p.}$$
$$= \tfrac{1}{3}\int \mathrm{Res}\,(AB_+C + ABC_-)\,dx + \text{c.p.}$$
$$= \tfrac{1}{3}\int \mathrm{Res}\,(ABC_+ + ABC_-)\,dx + \text{c.p.}$$
$$= \tfrac{1}{3}\int \mathrm{Res}\,ABC\,dx + \text{c.p.} = \int \mathrm{Res}\,ABC\,dx.$$

Now both the lemma and proposition 2 are proved.

A particular case of $H$ is $H^1$, and the corresponding form is $\omega^1$. As it was already said, $H^1$ can be restricted to $R_+$. Then, we must take $u_1 = 0$. This case was considered by Watanabe.

### *Poisson Bracket*

"Hamiltonians" are arbitrary elements $\tilde{f} \in A_u$. A vector field can be assigned to each Hamiltonian. Namely, let $\tilde{f} = \int f\,dx$ be a Hamiltonian. For any vector field $\partial_{H(X)}$, we have (lemma 1 of the previous section)

$$\partial_{H(X)}\tilde{f} = \int \mathrm{Res}\,H(x)\frac{\delta f}{\delta L}\,dx = \omega(\partial_{H(X)}, \partial_{H(\delta f/\delta L)}).$$

In other words,

$$d\tilde{f} = -i(\partial_{H(\delta f/\delta L)})\omega.$$

This means that the vector field $\partial_{H(\delta f/\delta L)}$ corresponds to the Hamiltonian $\tilde{f}$. We denote $\partial_{\tilde{f}} = \partial_{H(\delta f/\delta L)}$ for simplicity.

The Poisson bracket of two functionals is

$$\{\tilde{f}, \tilde{g}\} = \partial_{\tilde{f}}\tilde{g} = -\partial_{\tilde{g}}\tilde{f} = \omega(\partial_{\tilde{f}}, \partial_{\tilde{g}}).$$

*Proposition 1.* A relation

$$\partial_{\{\tilde{f},\tilde{g}\}} = [\partial_{\tilde{f}}, \partial_{\tilde{g}}]$$

holds. Proof: Let $Z \in R'_+$ be an arbitrary element. We have

$$\begin{aligned}
0 &= d\omega(\partial_{\tilde{f}}, \partial_{\tilde{g}}, \partial_{H(Z)})\\
&= \partial_{\tilde{f}}\omega(\partial_{\tilde{g}}, \partial_{H(Z)}) - \partial_{\tilde{g}}\omega(\partial_{\tilde{f}}, \partial_{H(Z)}) + \partial_{H(Z)}\omega(\partial_{\tilde{f}}, \partial_{\tilde{g}})\\
&\quad -\omega([\partial_{\tilde{f}}, \partial_{\tilde{g}}], \partial_{H(Z)}) + \omega([\partial_{\tilde{f}}, \partial_{H(Z)}], \partial_{\tilde{g}}) - \omega([\partial_{\tilde{g}}, \partial_{H(Z)}], \partial_{\tilde{f}})\\
&= -\partial_{\tilde{f}}\partial_{H(Z)}\tilde{g} + \partial_{\tilde{g}}\partial_{H(Z)}\tilde{f} + \partial_{H(Z)}\{\tilde{f}, \tilde{g}\}\\
&\quad -\omega([\partial_{\tilde{f}}, \partial_{\tilde{g}}], \partial_{H(Z)}) + [\partial_{\tilde{f}}, \partial_{H(Z)}]\tilde{g} - [\partial_{\tilde{g}}, \partial_{H(Z)}]\tilde{f}\\
&= -\partial_{H(Z)}\partial_{\tilde{f}}\tilde{g} + \partial_{H(Z)}\partial_{\tilde{g}}\tilde{f} + \partial_{H(Z)}\{\tilde{f}, \tilde{g}\} - \omega([\partial_{\tilde{f}}, \partial_{\tilde{g}}], \partial_{H(Z)})
\end{aligned}$$

$$= -\partial_{H(Z)}\{\tilde{f}, \tilde{g}\} + \omega(\partial_{H(Z)}, [\partial_{\tilde{f}}, \partial_{\tilde{g}}])$$
$$= \omega(\partial_{H(Z)}, -\partial_{\{\tilde{f},\tilde{g}\}} + [\partial_{\tilde{f}}, \partial_{\tilde{g}}]).$$

The remark that $Z$ is an arbitrary element completes the proof.

*Proposition 2.* The Poisson bracket satisfies the Jacobi identity. Proof: We have

$$0 = d\omega(\partial_{\tilde{f}_1}, \partial_{\tilde{f}_2}, \partial_{\tilde{f}_3}) = \partial_{\tilde{f}_1}\omega(\partial_{\tilde{f}_2}, \partial_{\tilde{f}_3}) - \omega([\partial_{\tilde{f}_1}, \partial_{\tilde{f}_2}], \partial_{\tilde{f}_3}) + \text{c.p.}$$
$$= \partial_{\tilde{f}_1}\{\tilde{f}_2, \tilde{f}_3\} - \omega(\partial_{\{\tilde{f}_1,\tilde{f}_2\}}, \partial_{\tilde{f}_3}) + \text{c.p.},$$
$$\{\tilde{f}_1, \{\tilde{f}_2, \tilde{f}_3\}\} - \{\{\tilde{f}_1, \tilde{f}_2,\}, \tilde{f}_3\} + \text{c.p.} = 2\{\tilde{f}_1, \{\tilde{f}_2, \tilde{f}_3\}\} + \text{c.p.}$$

This is the required identity.

### *Hamiltonians of Equation 2.1a*

We have

$$\int \operatorname{Res} T(\zeta)\, dx = -\sum_0^\infty \zeta^{-k-1} \int \operatorname{Res} L^k\, dx.$$

*Proposition 1.* The relation

$$\frac{\delta}{\delta L}\int \operatorname{Res} L^k\, dx = k(L^{k-1})_+,$$

holds.

This formula can also be written as

$$\frac{\delta}{\delta L}\int \operatorname{Res} T(\zeta)\, dx = -\frac{\partial}{\partial\zeta} T(\zeta).$$

Proof: We have

$$\delta\int \operatorname{Res} L^k\, dx = k\int \operatorname{Res} L^{k-1}\,\delta L\, dx,$$

which gives the required relation.

Let us find the corresponding vector fields. Put $\tilde{J}_k = \int \operatorname{Res} L^k\, dx$, and let $\omega^0$ and $\omega^1$ be the two limiting cases of the symplectic form $\omega$ when $\zeta = 0$ and $\zeta = \infty$ (more exactly, $\omega^1 = \lim_{\zeta\to\infty} \zeta^{-1}\omega$). Then

$$\text{for } \omega^0: \qquad \partial_{\tilde{J}_k} = k\partial_{H^0(L_+^{k-1})} = -k\partial_{[L_+^k, L]};$$
$$\text{for } \omega^1: \qquad \partial_{\tilde{J}_k} = k\partial_{H^1(L_+^{k-1})} = k\partial_{[L_+^{k-1}, L]}.$$

The differential equations that relate to these fields are

$$\dot{L} = -k[L_+^k, L], \qquad \dot{L} = k[L_+^{k-1}, L].$$

These equations differ from equations 1.2a only by insignificant constants. Thus, we have the following proposition.

*Proposition 2.* Equations 1.2a are of Hamiltonian form with respect to both of the symplectic forms, $\omega^0$ and $\omega^1$, where the Hamiltonians are the coefficients of the expansion of $\int$ Res $T\,dx$ in powers of $\zeta^{-1}$.

The commutativity of vector fields $\{\partial_m\}$ implies that the Hamiltonians $\{\tilde{J}_j\}$ are in involution, which also can be proved independently (see below).

*Proposition 3.* The functionals $\tilde{J}_j = \int \text{Res}\, L^j\, dx$ are in involution with respect to both the Poisson brackets.

Proof: First, we take the form $\omega^1$. We find

$$
\begin{aligned}
\{\tilde{J}_i, \tilde{J}_j\} &= \int \text{Res}\, H^1_{J_i} \cdot \delta J_j/\delta L \cdot dx = \text{const.} \cdot \int \text{Res}\, [(L^{i-1})_+, L]_-(L^{j-1})_{+'}, dx \\
&= \text{const.} \cdot \int \text{Res}\, [(L^{i-1})_+, L]_-\, L^{j-1}\, dx \\
&= -\text{const.} \cdot \int \text{Res}\, [(L^{i-1})_+, L]_+\, L^{j-1}\, dx \\
&= -\text{const.} \cdot \int \text{Res}\, [L^{i-1}, L]_+\, L^{j-1}\, dx = 0.
\end{aligned}
$$

For $\omega^0$, the assertion follows from the recurrency relation (equation 2.3b) for the resolvent from the proven involutiveness with respect to the form $\omega^1$.

## PART II—KP-HIERARCHY

### *Introduction to KP-Hierarchy*

Let

$$L = \partial + u_0\partial^{-1} + u_1\partial^{-2} + \cdots$$

be a pseudodifferential operator (PDO). We consider the set of equations

$$\partial_m L = [L^m_+, L], \qquad m = 1, 2, \ldots, \tag{3.1a}$$

where $\partial_m = \partial/\partial x_m$, with $\{x_m\}$ being a set of independent variables, $x_1 = x$ (e.g., see reference 6), and the subscript "+" meaning $(\Sigma\, a i \partial^i)_+ = \Sigma_{i\geq 0}\, a i \partial^i$ (one singles out the pure differential part of PDO). All of the equations 3.1a are compatible (or the differentiations $\partial_m$ commute). From equations 3.1a, the equations

$$\partial_m L^n_+ - \partial_n L^m_+ = [L^m_+, L^n_+] \tag{3.1b}$$

follow for every $m$ and $n$. These equations form the Kadomtsev-Petviashvili (KP) hierarchy. The simplest of them, when $n = 2$ and $m = 3$, is called the KP-equation.

In the first part of this paper, we studied the Hamiltonian structure of equations 3.1a. Now, we will do the same for equations 3.1b. Apparently, they are not connected with each other.

Moreover, our approach here is quite different. Variables $x_n$ and $x_m$ are equal in rights, so it is unreasonable to single out one of them as a time variable. We believe the field formalism to be the most appropriate for this purpose.

Reference 10 describes all the details of this formalism. However, in order to make the exposition in this paper self-contained, we explain this formalism in the next section with the help of an example that is a little bit simpler than our problem.

### *Field (Multitime) Hamiltonian Formalism*

We start with the usual canonical Hamiltonian equations

$$\frac{dq^{(k)}}{dt} = \frac{\partial H}{\partial p^{(k)}}, \qquad \frac{dp^{(k)}}{dt} = -\frac{\partial H}{\partial q^{(k)}}, \tag{3.2a}$$

which are written in canonical variables $\{p^{(k)}, q^{(k)}\}$. If we take a symplectic form $\Omega = \Sigma \delta p^{(k)} \delta q^{(k)}$, then equations 3.2a can be written as

$$\delta H = -i(\tilde{\partial}_t)\,\Omega. \tag{3.2b}$$

(The vector field $\tilde{\partial}_t$ is defined by this rule: for every variation $\delta f$, the relation $i(\tilde{\partial}_t)\delta f = \partial f/\partial t$ holds.) Indeed, the left-hand side of equation 3.2a is $\Sigma(\partial H/\partial q^{(k)} \cdot \delta q^{(k)} + \partial H/\partial p^{(k)} \cdot \delta p^{(k)})$ and the right-hand side is $-\Sigma(\partial p^{(k)}/\partial t \cdot \delta q^{(k)} - \partial q^{(k)}/\partial t \cdot \delta p^{(k)})$; by equating one to another, we get equations 3.2a.

Now, let us have many time variables $t^1, t^2, \ldots, t^n$. The canonical equations are such examples (see de Donder[11] and below). To each coordinate $q^{(k)}$, $k = 1, \ldots, m$, we assign $n$ adjoint momenta $p^{(k)j}$, $j = 1, \ldots, n$. The equations are

$$\frac{\partial q^{(k)}}{\partial t^j} = \frac{\partial H}{\partial p^{(k)j}}, \quad j = 1, \ldots, n; \qquad \sum_{j=1}^{n} \frac{\partial p^{(k)j}}{\partial t^j} = -\frac{\partial H}{\partial q^{(k)}}. \tag{3.2c}$$

Let us write this system of equations in a form similar to equation 3.2b. This form has the advantage of being invariant, independent of the choice of variables. In particular, we do not need to bother with the search of canonical variables.

To this end, besides the variations $\delta q^{(k)}, \delta p^{(k)j}$, we must introduce the differentials of the time variables $dt^1, dt^2, \ldots, dt^n$, with the assumption that all of the differentials, both the variations and the time differentials, are anticommuting. Furthermore, we introduce vector fields $\tilde{\partial}_k$ with the following rule of substitution into a variation—$i(\tilde{\partial}_k)\delta f = \partial f/\partial t^k$—and vector fields $\partial_k$ that can be substituted into the time differentials—$i(\partial_k)dt^j = \delta^j_k$. The other relations are that the operators $i(\partial_k)$ are anticommuting with the variation operator $\delta$, $i(\partial_k)\delta + \delta i(\partial_k) = 0$, while $i(\tilde{\partial}_k)\delta + \delta i(\tilde{\partial}_k) = \partial/\partial t^k$; for any form $A$, we have $dA = \Sigma\, dt^l \partial A/\partial t^l$ and $i(\tilde{\partial}_k)d + di(\tilde{\partial}_k) = 0$; we also have $i(\tilde{\partial}_k)dt^j + dt^j\, i(\tilde{\partial}_k) = 0$.

Put $\Omega = -\Sigma_{k,j}\, i(\partial_j)\delta q^{(k)} \delta p^{(k)j}\, dt^1 \cdots dt^n$, $\mathcal{H} = H dt^1 \cdots dt^n$. Equations 3.2c are equivalent to

$$\delta \mathcal{H} = \sum_l dt^l i(\tilde{\partial}_l)\,\Omega. \tag{3.2d}$$

Indeed, the left-hand side is $\Sigma_k\, (\partial H/\partial q^{(k)} \cdot \delta q^{(k)} + \Sigma_j\, \partial H/\partial p^{(k)j} \cdot \delta p^{(k)j})\, dt^1 \cdots dt^n$ and the right-hand side is $-\Sigma\, dt^l i(\tilde{\partial}_l) i(\partial_j) \delta q^{(k)} \delta p^{(k)j}\, dt^1 \cdots dt^n = \Sigma\, i(\tilde{\partial}_l)\delta q^{(k)} \delta p^{(k)l} \cdot dt^1 \cdots dt^n = \Sigma(\partial q^{(k)}/\partial t^{(l)} \cdot \delta p^{(k)l} - \partial p^{(k)l}/\partial t^l \cdot \delta q^{(k)}) dt^1 \cdots dt^n$. This yields equations 3.2c.

The Hamiltonian equations can be obtained from a Lagrangian by the Legendre transformation. Let us see what this means for the equations 3.2c or 3.2d. Let a Lagrangian depend on the first derivatives of the coordinates (we take it for simplicity because the results of our analysis are applicable to the general case; see reference 10, which will be used further), $L = L(q^{(l)}, q^{(l)}_{,j}, t^1, \ldots t^n)$, where $q^{(l)}_{,j} = \partial q^{(l)}/\partial t^j$. We have

$$\delta L = \sum \frac{\partial L}{\partial q^{(l)}} \delta q^{(l)} + \sum \frac{\partial L}{\partial q^{(l)}_{,j}} \delta q^{(l)}_{,j}$$

$$= \sum_l \left( \frac{\partial L}{\partial q^{(l)}} - \sum_j \frac{\partial}{\partial t^j} \left( \frac{\partial L}{\partial q^{(l)}_{,j}} \right) \right) \delta q^{(l)} + \sum_j \partial_j \left( \sum_l \frac{\partial L}{\partial q^{(l)}_{,j}} \delta q^{(l)} \right).$$

Put $\mathcal{L} = L dt^1 \ldots dt^n$ and $p^{(l)j} = \partial L / \partial q^{(l)}_{,j}$. Then, the last equation takes the form

$$\delta \mathcal{L} = \sum_l \frac{\delta L}{\delta q^{(l)}} \delta q^{(l)} \, dt^1 \ldots dt^n + \sum_j \partial_j \sum_l p^{(l)j} \delta q^{(l)} dt^1 \ldots dt^n$$

$$= \sum_l \frac{\delta L}{\delta q^{(l)}} \delta q^{(l)} dt^1 \ldots dt^n + d \sum_j i(\partial_j) \sum_l p^{(l)j} \delta q^{(l)} \, dt^1 \ldots dt^n.$$

Hence,

$$\delta \mathcal{L} = \sum_l \frac{\delta L}{\delta q^{(l)}} \delta q^{(l)} \, dt^1 \ldots dt^n - d\Omega^{(1)}, \tag{3.2e}$$

where $\Omega^{(1)}$ is such a form that $\Omega$ (see above) is $\delta\Omega^{(1)}$:

$$\Omega = \delta \Omega^{(1)}. \tag{3.2f}$$

Put

$$\mathcal{H} = -\mathcal{L} + \sum_j dt^j i(\tilde{\partial}_j) \, \Omega^{(1)}. \tag{3.2g}$$

Then

$$\delta \mathcal{H} = \sum \frac{\delta L}{\delta q^{(l)}} \delta q^{(l)} \, dt^1 \ldots dt^n - d\Omega^{(1)} + \sum_j dt^j i(\tilde{\partial}_j) \delta \Omega^{(1)} - \sum_j dt^j \frac{\partial}{\partial t^j} \Omega^{(1)}$$

$$= \sum \frac{\delta L}{\delta q^{(l)}} \delta q^{(l)} \, dt^1 \ldots dt^n + \sum_j dt^j i(\tilde{\partial}_j) \delta \Omega^{(1)}.$$

Thus, the set of equations $\{\delta L/\delta q^{(l)} = 0\}$ is equivalent to equations 3.2d.

We summarize the obtained results. By having a Lagrangian, one can construct the whole Hamiltonian formalism. First, $\delta\mathcal{L}$ must be represented in the form of equations 3.2e, which can always be performed by integration by parts. Then, one builds $\Omega$ and $\mathcal{H}$ according to equations 3.2f and 3.2g. The Hamiltonian form for the equation will be equation 3.2d. We once more emphasize that this procedure does not require canonical variables.

Now, we explain the connection between the field formalism and the single-time Hamiltonian formalism. At first, we use the canonical variables and then we do it in an invariant form.

Let

$$T^l_j = -\delta^l_j L + \sum_k q^{(k)}_{,j} p^{(k)l} \tag{3.2h}$$

be the energy-momentum tensor. Next, we choose one of the variables $t^1, \ldots, t^n$ as a single time in which the system evolves; let it be $t^1$. Other variables $t^2, \ldots, t^n$ play the role of indices that numerate the degrees of freedom of the system. The energy-momentum vector can be obtained by integration of the energy-momentum tensor over the plane $t^1$ = const.:

$$P_j = \int_{t^1-\text{const.}} \mathcal{T}_j, \qquad \mathcal{T}_j = T_j^l i(\partial_l)\, dt^1 \cdots dt^n.$$

The component $P_1$ of this vector is the energy of the system, that is, the Hamiltonian

$$h = \int_{t^1-\text{const.}} (-L + \sum_k q^{(k)}_{,1} p^{(k)1})\, dt^2 \cdots dt^n.$$

The symplectic form is obtained by integration of $\Omega$ over the same plane:

$$\omega = \int_{t^1-\text{const.}} \Omega = \int_{t^1-\text{const.}} \sum_k \delta p^{(k)1} \delta q^{(k)}\, dt^2 \cdots dt^n$$

The Hamilton equation has the form

$$\delta h = -i(\tilde{\partial}_1)\omega. \tag{3.2i}$$

It is easy to see that this equation is, indeed, equivalent to $\{\delta L/\delta q^{(k)} = 0\}$.

Now, we do the same in an invariant form. The energy-momentum tensor is a set of forms

$$\mathcal{T}_j = -i(\partial_j)\mathcal{L} + i(\tilde{\partial}_j)\Omega^{(1)};$$

then we put $h = \int_{t^1-\text{const.}} \mathcal{T}_1$ and $\omega = \int_{t^1-\text{const.}} \Omega$, and write equation 3.2i.

The Hamiltonian $\mathcal{H}$ can be expressed in terms of the energy-momentum tensor as

$$\mathcal{H} = \sum_j dt^j \mathcal{T}_j + (n-1)\mathcal{L}.$$

The definitions given here are slightly different from those in reference 3.

### *An Example: The KP-Equation*

This equation is

$$-3u_{yy} + (4u_t - u''' - 6uu')' = 0.$$

Put $u = \varphi'$. The equation takes the form

$$-3\varphi_{yy} + 4\varphi_t' - \varphi^{\mathrm{IV}} - 6\varphi'\varphi'' = 0. \tag{3.3a}$$

To obtain the Lagrangian of this equation, we multiply the left-hand side of it by $\varphi$ and divide every term in it by its degree in $\varphi$ and its derivatives:

$$\mathcal{L} = (-\tfrac{3}{2}\varphi_y^2 + 2\varphi_t\varphi' + \tfrac{1}{2}\varphi''^2 - \varphi'^3)\, dxdydt. \tag{3.3b}$$

The variation of this Lagrangian is

$$\delta\mathcal{L} = (3\varphi_{yy} - 4\varphi_t' + \varphi^{\mathrm{IV}} + 6\varphi'\varphi'')\delta\varphi\, dxdydt - d\Omega^{(1)},$$

where

$$\Omega^{(1)} = 3\varphi_y\delta\varphi_y\, dxdt + 2\varphi'\delta\varphi\, dxdy - (-2\varphi_t\delta\varphi - \varphi''\delta\varphi' + \varphi'''\delta\varphi + 3\varphi'^2\delta\varphi)\, dydt.$$

This gives

$$\frac{\delta L}{\delta\varphi} = 3\varphi_{yy} - 4\varphi_t' + \varphi^{\mathrm{IV}} + 6\varphi'\varphi'',$$

which means that $\mathcal{L}$ is really the Lagrangian of equation 3.3a. Furthermore,

$$\begin{aligned}\Omega = \delta\Omega^{(1)} &= 3\delta\varphi_y\delta\varphi\, dxdt + 2\delta\varphi'\delta\varphi\, dxdy \\ &\quad -(-2\delta\varphi_t\delta\varphi - \delta\varphi''\delta\varphi' + \delta\varphi'''\delta\varphi + 6\varphi'\delta\varphi'\delta\varphi)\, dydt, \\ \mathcal{H} &= -\mathcal{L} + \{dxi(\tilde{\partial}) + dti(\tilde{\partial}_t) + dyi(\tilde{\partial}_y)\}\,\Omega^{(1)} \\ &= (-{}^3\!/\!_2\varphi_y^2 + 2\varphi_t\varphi' + {}^1\!/\!_2\varphi''^2 - 2\varphi'^3 - \varphi'\varphi''')\, dxdydt.\end{aligned}$$

The Hamiltonian form of this equation is

$$\delta\mathcal{H} = \{dxi(\tilde{\partial}) + dyi(\tilde{\partial}_y) + dti(\tilde{\partial}_t)\}\,\Omega.$$

It can be independently verified that this equation is, indeed, equivalent to equation 3.3a.

The energy-momentum tensor is the set of three forms:

$$\begin{aligned}T_x &= -i(\partial)\mathcal{L} + i(\tilde{\partial})\Omega = 3\varphi_y\varphi'\, dxdt + 2\varphi'^2\, dxdy \\ &\quad + ({}^1\!/\!_2\varphi''^2 - 2\varphi'^3 + \varphi'''\varphi' + {}^3\!/\!_2\varphi_y^2)\, dydt, \\ T_y &= -i(\partial_y)\mathcal{L} + i(\tilde{\partial}_y)\Omega = 2\varphi'\varphi_y\, dxdy + ({}^3\!/\!_2\varphi_y^2 + 2\varphi_t\varphi' + {}^1\!/\!_2\varphi''^2 \\ &\quad - \varphi'^3)\, dxdt + (2\varphi_t\varphi_y + \varphi''\varphi_y' - \varphi'''\varphi_y - 3\varphi'^2\varphi_y)\, dydt, \\ T_t &= -i(\partial_t)\mathcal{L} + i(\tilde{\partial}_t)\Omega = ({}^3\!/\!_2\varphi_y^2 - {}^1\!/\!_2\varphi''^2 + \varphi^3)\, dxdy \\ &\quad + 3\varphi_y\varphi_t\, dxdt + (2\varphi_t^2 + \varphi''\varphi_t' - \varphi'''\varphi_t - 3\varphi'^2\varphi_t)\, dydt.\end{aligned}$$

(The Hamiltonian written above is $\mathcal{H} = dxT_x + dyT_y + dtT_t + 2\mathcal{L}$.) The Hamiltonian of the single-time formalism is the coefficient in $dxdy$ in the expression of $T_t$:

$$h = \int_{t=\text{const.}} \left(\frac{3}{2}\varphi_y^2 + \frac{1}{2}\varphi''^2 - \varphi'^3\right) dxdy.$$

This coincides with the Hamiltonian that is usually written for the KP-equation (e.g., see reference 4). The symplectic form of the single-time theory is

$$\omega = \int_{t=\text{const.}} \Omega = 2\int_{t=\text{const.}} \delta\varphi'\delta\varphi\, dxdy.$$

## *The General Case*

We return now to equation 3.1b. As it became clear in the previous section, instead of variables $\{u_i\}$, one must use a new set of variables $\{\varphi_i\}$ generalizing the substitution

$u = \varphi'$ of that section. This will be

$$L = \varphi\partial\varphi^{-1},$$

where $\varphi = \Sigma_0^\infty \varphi_i\partial^{-i-1}$, $\varphi_0 = 1$, is a formal series of new variables; in fact, only a finite set of them is used, namely, $i \le \max(n, m) - 1$.

As we have seen earlier in the procedure of constructing a Lagrangian, we use an important operation: dividing all the terms of a differential polynomial in $\{\varphi_i\}$ by its degrees in $\{\varphi_i\}$ and their derivatives. This operation can be performed in the following way. Take a real parameter, $p$, and substitute $p\varphi_i$ for $\varphi_i$ wherever they stand in a differential polynomial $f(\varphi)$. Then we write the integral: $\int_0^1 p^{-1}f(p\varphi)\,dp$. The result of this integration will be just as required. Therefore, we introduce now $\varphi_p = 1 + p\,\Sigma_1^\infty \varphi_i\partial^{-i}$.

*Proposition 1.* Equation 3.1b can be obtained from the Lagrangian,

$$\mathcal{L} = \mathrm{Res}\left\{-\int_0^1 p^{-1}[(\varphi_p\partial^m\varphi_p^{-1})_+, (\varphi_p\partial^n\varphi_p^{-1})_+]\varphi_p^{-1}\,dp \right.$$

$$\left. + \partial^n\varphi^{-1}\partial\varphi/\partial x_m - \partial^m\varphi^{-1}\partial\varphi/\partial x_n\right\} dx\,dx_m dx_n \qquad \left(\mathrm{Res}\sum a_i\partial^i = a_{-1}\right).$$

Proof: In essence, the proof repeats the proof of a similar statement in reference 3. First, we prove that

$$\delta\,\mathrm{Res}\int_0^p p^{-1}[(\varphi_p\partial^m\varphi_p^{-1})_+, (\varphi_p\partial^n\varphi_p^{-1})_+]\varphi_p^{-1}\,dp$$

$$= -\mathrm{Res}\,[(\varphi_p\partial^m\varphi_p^{-1})_+, (\varphi_p\partial^n\varphi_p^{-1})_+]\delta\varphi_p\cdot\varphi_p^{-1} + \partial(\quad), \qquad \textbf{(3.4a)}$$

where $\partial(\quad)$ symbolizes the derivative of a form $\Sigma\,a_i\delta\varphi_i$ with respect to $x$. The expression $\partial(\quad)$ appears usually as a result of the fact that Res $[A, B] = \partial(\quad)$ for any pseudodifferential operator A and B.

In order to prove equation 3.4a, it is sufficient to verify the equation obtained by differentiation with respect to $p$:

$$\delta\,\mathrm{Res}\,[(\varphi_p\partial^m\varphi_p^{-1})_+, (\varphi_p\partial^n\varphi_p^{-1})_+]\varphi_p^{-1}$$

$$= -\mathrm{Res}\,[(\varphi_p\partial^m\varphi_p^{-1})_+, (\varphi_p\partial^n\varphi_p^{-1})_+]\delta\varphi_p\cdot\varphi_p^{-2}$$

$$-\mathrm{Res}\left(p\frac{\partial}{\partial p}[(\varphi_p\partial^m\varphi_p^{-1})_+, (\varphi_p\partial^n\varphi_p^{-1})_+]\right)\cdot\delta\varphi_p\cdot\varphi_p^{-1} + \partial(\quad)$$

(it was taken into account that $p\partial\varphi_p/\partial p = \varphi_p - 1$). We transform

$$p\frac{\partial}{\partial p}(\varphi_p\partial^m\varphi_p^{-1})_+ = ((\varphi_p - 1)\partial^m\varphi_p^{-1} - \varphi_p\partial^m\varphi_p^{-1}(\varphi_p - 1)\varphi_p^{-1})_+$$

$$= -[\varphi_p^{-1}, (\varphi_p\partial^m\varphi_p^{-1})_+]_+,$$

$$p\frac{\partial}{\partial p}(\varphi_p\partial^n\varphi_p^{-1})_+ = -[\varphi_p^{-1}, (\varphi_p\partial^n\varphi_p^{-1})_+]_+,$$

$$\delta(\varphi_p\partial^m\varphi_p^{-1})_+ = [\delta\varphi_p \cdot \varphi_p^{-1}, (\varphi_p\partial^m\varphi_p^{-1})_+]_+,$$

$$\delta(\varphi_p\partial^n\varphi_p^{-1})_+ = [\delta\varphi_p \cdot \varphi_p^{-1}, (\varphi_p\partial^n\varphi_p^{-1})_+]_+.$$

We also put $S = \varphi_p^{-1}$, $T = \delta\varphi_p \cdot \varphi_p^{-1}$, $U = (\varphi_p\partial^m\varphi_p^{-1})_+$, and $V = (\varphi_p\partial^n\varphi_p^{-1})_+$. Now, we must prove that

$$\operatorname{Res}\{([[T, U]_+, V] + [U, [T, V]_+])S + [U, V]TS$$
$$-[U, V]ST - [[S, U]_+, V]T - [U, [S, V]_+]T\} = \partial(\quad).$$

Two of the terms are

$$\operatorname{Res}\{[U, [T, V]_+]S - [[S, U]_+, V]T = \operatorname{Res}\{[S, U] \cdot [T, V]_+$$
$$+ [T, V][S, U]_+\} + \partial(\quad) = \operatorname{Res}\{[T, V] \cdot [S, U]\} + \partial(\quad).$$

Similarly,

$$\operatorname{Res}\{[[T, U]_+, V]S - [U, [S, V]_+]T\} = \operatorname{Res}\{[T, U] \cdot [V, S]\} + \partial(\quad).$$

Our expression takes the form

$$\operatorname{Res}\{-[T, U] \cdot [S, V] + [T, V] \cdot [S, U] + [U, V] \cdot [T, S]\} + \partial(\quad)$$
$$= \operatorname{Res} T\{[U, [V, S]] + [V, [S, U]] + [S, [U, V]]\} + \partial(\quad) = \partial(\quad),$$

which proves equation 3.4a. Put $p = 1$:

$$\delta \operatorname{Res} \int_0^1 p^{-1}[(\varphi_p\partial^m\varphi_p^{-1})_+, (\varphi_p\partial^n\varphi_p^{-1})_+]\varphi_p^{-1}\,dp$$
$$= -\operatorname{Res} \varphi^{-1}[(\varphi\partial^m\varphi^{-1})_+, (\varphi\partial^n\varphi^{-1})_+]\delta\varphi - \partial\omega_1,$$

where $\omega_1$ is a form. Then, we calculate the rest of the variation:

$$\delta \operatorname{Res}(\partial^n\varphi^{-1}\delta\varphi/\partial x_m - \partial^m\varphi^{-1}\delta\varphi/\partial x_n)$$
$$= \frac{\partial}{\partial x_m}\operatorname{Res}(\partial^n\varphi^{-1}\delta\varphi) - \frac{\partial}{\partial x_n}\operatorname{Res}(\partial^m\varphi^{-1}\delta\varphi)$$
$$+ \operatorname{Res}\left(\partial^n\varphi^{-1} \cdot \frac{\partial\varphi}{\partial x_m}\varphi^{-1}\delta\varphi - \partial^m\varphi^{-1}\frac{\partial\varphi}{\partial x_n}\varphi^{-1}\delta\varphi\right.$$
$$\left.-\varphi^{-1}\frac{\partial\varphi}{\partial x_m}\partial^n\varphi^{-1}\delta\varphi + \varphi^{-1}\frac{\partial\varphi}{\partial x_n}\partial^m\varphi^{-1}\delta\varphi\right) + \partial\omega_2$$
$$= \frac{\partial}{\partial x_m}\operatorname{Res}(\partial^n\varphi^{-1}\delta\varphi) - \frac{\partial}{\partial x_n}\operatorname{Res}(\partial^n\varphi^{-1}\delta\varphi)$$
$$+ \operatorname{Res}\varphi^{-1}\left(\frac{\partial L_+^m}{\partial x_n} - \frac{\partial L_+^n}{\partial x_m}\right)\delta\varphi + \partial\omega_2,$$

where $\omega_2$ is another form. Thus,

$$\delta\mathcal{L} = \text{Res}\left\{\varphi^{-1}\left(-\frac{\partial L_+^n}{\partial x_m} + \frac{\partial L_+^m}{\partial x_n} + [L_+^m, L_+^n]\right)\delta\varphi\right\} dx dx_m dx_n$$
$$+ d\{-\omega dx_m dx_n + \text{Res}(\partial^n\varphi^{-1}\delta\varphi\, dx dx_n + \partial^m\varphi^{-1}\delta\varphi\, dx dx_m)\}, \quad \textbf{(3.4b)}$$

where $\omega = \omega_1 + \omega_2$. This implies the expression for the variational derivative:

$$\frac{\delta L}{\delta\varphi} = \left\{\varphi^{-1}\left(-\frac{\partial L_+^n}{\partial x_m} + \frac{\partial L_+^m}{\partial x_n} + [L_+^m, L_+^n]\right)\right\}_+ .$$

By equating this to zero, we get an equation that is equivalent to equation 3.1b. This completes the proof.

*Proposition 2.* The 1-form corresponding to the Lagrangian is

$$\Omega^{(1)} = -\omega\, dx_m dx_n + \text{Res}(\partial^n\varphi^{-1}\delta\varphi\, dx dx_n + \partial^m\varphi^{-1}\delta\varphi\, dx dx_m)$$

and the 2-form is

$$\Omega^{(2)} = \delta\Omega^{(1)} = -\delta\omega\, dx_m dx_n - \text{Res}(\partial^n\varphi^{-1}\delta\varphi\varphi^{-1}\delta\varphi\, dx dx_n + \partial^m\varphi^{-1}\delta\varphi\varphi^{-1}\delta\varphi\, dx dx_m).$$

The Hamiltonian is

$$\mathcal{H} = \mathcal{L} + (dx\, i(\tilde{\partial}) + dx_m i(\tilde{\partial}_m) + dx_n i(\tilde{\partial}_n))\Omega^{(1)}$$
$$= (-i(\tilde{\partial})\omega - \text{Res}\int_0^1 p^{-1}[(\varphi_p\partial^m\varphi_p^{-1})_+, (\varphi_p\partial^n\varphi_p^{-1})_+]\varphi_p^{-1}\, dp)\, dx dx_m dx_n$$

and the Hamiltonian form of the equation is

$$\delta\mathcal{H} = (dx\, i(\tilde{\partial}) + dx_m i(\tilde{\partial}_m) + dx_n i(\tilde{\partial}_n))\Omega.$$

All this immediately follows from equation 3.4b.

## REFERENCES


1. WATANABE, Y. 1983. Hamiltonian structure of Sato's hierarchy of KP equations and a coadjoint orbit of a certain formal Lie group. Lett. Math. Phys. **7**(no. 2): 99–106.
2. ZAKHAROV, V. E. & E. SHULMAN. 1980. Degenerative dispersion laws, motion invariants, and kinetic equations. Physica **D1**(no. 2): 191–202.
3. REYMAN, A. & M. SEMENOV-TJAN-SHANSKY. 1984. Hamilton structure of the Kadomtsev-Petviashvili type equations. Zap. Nauchni Sem. LOMI **133**.
4. CASE, K. M. 1985. Symmetries of the higher-order KP equations. J. Math. Phys. **26**(no. 6): 1158–1159.
5. GELFAND, I. M. & L. A. DICKEY. 1976. Fractional powers of operators and Hamiltonian systems. Funkz. Anal. Jego Prilozh **10**(no. 4): 13–29.
6. DATE, E., M. JIMBO, M. KASHIVARA & T. MIWA. 1983. Transformation groups for soliton equations. *In* Non-linear Integrable Systems—Classical Theory and Quantum Theory. Proc. RIMS Symposium, Singapore. M. Jimbo & T. Miwa, Eds.
7. ADLER, M. 1979. On a trace functional for formal pseudodifferential operators and the symplectic structure of the Korteweg–de Vries equation. Invent. Math. **50:** 219–248.

8. LEBEDEV, D. R. & JU. I. MANIN. 1979. Hamiltonian operator of Gelfand-Dickey and coadjoint representation of the Volterra group. Funkz. Anal. Jego Prilozh **13**(no. 4): 40–46.
9. GELFAND, I. M. & L. A. DICKEY. 1978. Family of Hamiltonian structures connected with integrable nonlinear differential equations. Preprint IPM no. 136.
10. DICKEY, L. A. Multi-time Lagrangian and Hamiltonian formalism and integrable systems. To be published.
11. DE DONDER. No source listed.

# Tunneling Soliton in the Equations of Reaction-Diffusion Type

M. I. FREIDLIN

Consider the Cauchy problem:

$$\frac{\partial \tilde{u}^\epsilon(t, x)}{\partial t} = \frac{\epsilon^2}{2} \Delta \tilde{u}^\epsilon + c\left(x, \int_{R^r} \tilde{u}^\epsilon(t, y)\, dy\right)\tilde{u}^\epsilon(t, x),$$

$$t > 0, \quad x \in R^r, \quad \tilde{u}^\epsilon(0, x) = g(x). \quad \mathbf{(1)}$$

Here, $g(x)$ is a continuous nonnegative function with a bounded support $G_0$, $\epsilon > 0$. The function $c(x, v)$, with $x \in R^r$, $v \in R^1$, is Lipschitz continuous and bounded. Moreover, a bounded Lipschitz-continuous function $a(x) > 0$, $x \in R^r$, is assumed to exist such that $c[x, a(x)] = 0$, $c(x, v) > 0$ for $v < a(x)$, and $c(x, v) < 0$ for $v > a(x)$. For brevity, we put

$$\int_{R^r} g(x)\, dx \le \max_{x \in G_0 \cup \partial G_0} a(x).$$

Problems of equation 1 arise, for example, in connection with certain evolution models (see reference 1), where the mutual relation between specimens with different genotypes is reduced to a competition for one and the same substrate that they feed on. This causes the fitness coefficient, $c(x, v_t^\epsilon)$, to become negative when the general number of specimens, $v_t^\epsilon = \int_{R^r} \tilde{u}^\epsilon(t, x)\, dx$, exceeds a critical value $a(x)$, with $a(x)$ being different for different genotypes $x$. In such an interpretation, $\tilde{u}^\epsilon(t, x)$ is the density of the number of specimens with the genotype $x$, and $\epsilon$ is the mutation intensity.

It is not hard to check that for small $\epsilon$, significant changes of $\tilde{u}^\epsilon(t, x)$ occur on time intervals of order $\epsilon^{-1}$. Hence, when examining equation 1, a new time scale should be used. We put $u^\epsilon(t, x) = \tilde{u}(\epsilon^{-1}t, x)$. From equation 1, it follows that $u^\epsilon(t, x)$ is the solution of the problem:

$$\frac{\partial u^\epsilon(t, x)}{\partial t} = \frac{\epsilon}{2} \Delta u^\epsilon + \frac{1}{\epsilon} c(x, v_t^\epsilon) u^\epsilon(t, x),$$

$$v_t^\epsilon = \int_{R^r} u^\epsilon(t, y)\, dy, \quad t > 0, \quad x \in R^r, \quad u^\epsilon(0, x) = g(x). \quad \mathbf{(2)}$$

If the Markov process in $R$ is denoted by $(X_t^\epsilon, P_x)$, corresponding to the operator $\epsilon/2\Delta$ (up to the factor $\sqrt{\epsilon}$ it is the Wiener process), then by the Feynman-Kac formula (see, e.g., reference 2), we have the following equation for the function $u^\epsilon(t, x)$:

$$u^\epsilon(t, x) = E_x g(X_t^\epsilon) \exp\left\{\epsilon^{-1} \int_0^t c(X_s^\epsilon, v_{t-s}^\epsilon)\, ds\right\},$$

$$v_t^\epsilon = \int_{R^r} u^\epsilon(t, y)\, dy. \quad \mathbf{(3)}$$

From equation 3, one can readily deduce the existence and uniqueness of the solution of the problem of equation 2 for Lipschitz-continuous $c(x, v)$. This is proved with successive approximations. We use equation 3 for examining the behavior of $u^\epsilon(t, x)$ as $\epsilon \downarrow 0$.

For small $\epsilon$, $u^\epsilon(t, x)$ has a solitonlike shape: $u^\epsilon(t, x)$ tends to $a(\hat{\varphi}_t)\delta(x - \hat{\varphi}_t)$ as $\epsilon \downarrow 0$. Here, $\delta(\ )$ is the Dirac $\delta$-function and $\hat{\varphi}$ is a piecewise continuous function such that $a(\hat{\varphi}_t)$ does not decrease as $t$ grows. We will evaluate $\hat{\varphi}$ later on. Therefore, the soliton has a volume that is nondecreasing with time. It moves with a finite velocity for all $t$, with the possible exception of a countable number of times. At these times, the soliton has jumps. For large $t$, the soliton is concentrated near the point at which

$$\sup_{x \in R^r} a(x)$$

is attained, provided such a point exists. Unless this supremum is achieved, the soliton goes to infinity.

For a strictly nondecreasing right-continuous function $h(t)$: $[0, T] \rightarrow R^1$, $T > 0$, we will define the function $V_h(t, x)$, $t \in (0, T]$, $x \in R^r$:

$$V_h(t, x) = \sup\left\{\int_0^t \left(c[\psi_s, h(s)] - \frac{|\dot{\psi}_s|^2}{2}\right) ds: \quad \psi \in C_{0T}(R^r), \right.$$

$$\left. \psi_0 \in G_0, \psi_t = x, \text{ and } \psi_s \text{ is absolutely continuous}\right\}.$$

The function $V_h(t, x)$ is continuous for $t \in (0, T]$ and $x \in R^r$. As $t \downarrow 0$, $V_h(t, x)$ tends to 0 provided $x \in G_0 \cup \partial G_0$, and $V_h(t, x) \rightarrow -\infty$ for $x \notin G_0 \cup \partial G_0$.

By $\Xi_T$, $T > 0$, we denote the set of measurable functions $\varphi$: $(0, T] \rightarrow R^r$, such that $a(\varphi_t)$ does not decrease, $a(\varphi)_t$ is right continuous for $t > 0$, and $\lim_{t\downarrow 0} a(\varphi_t)$ exists.

A function $\hat{\varphi} \in \Xi_T$ is called a maximal solution of the equation

$$V_{a(\varphi)}(t, \varphi_t) = 0, \qquad t \in (0, T], \tag{4}$$

whenever equation 4 is satisfied for $\varphi_t = \hat{\varphi}_t$ and

$$V_{a(\hat{\varphi})}(t, x) \leq 0 \quad \text{for} \quad t \in (0, T], x \in R^r.$$

Theorem 1: Suppose that equation 4 has a unique maximal solution $\hat{\varphi} \in \Xi_T$. Then, $a(\hat{\varphi}_t)$ is continuous everywhere on $[0, T]$, with the possible exception of (at most) a countable set $\Lambda \subset [0, T]$, and for $t \in [0, T]\backslash\Lambda$, the solution $u^\epsilon(t, x)$ of the problem of equation 2 weakly converges to $a(\hat{\varphi}_t)\delta(x - \hat{\varphi}_t)$ as $\epsilon \downarrow 0$.

Proof: We provide it for the case $x \in R^1$. Then, we will hint at the changes to be done in the proof for $x \in R^r$, $r > 1$.

First of all, by taking into account the hypotheses on $g(x)$, we note that the maximum principle implies

$$0 < u^\epsilon(t, x), \quad 0 < v_t^\epsilon = \int_{R^r} u^\epsilon(t, y)\, dy \leq a_0 = \sup_{x \in R^r} a(x) < \infty. \tag{5}$$

For any $0 < t_1 < t_2$, one can find $\delta_\epsilon = \delta_\epsilon(t_1, t_2)$ such that

$$\lim_{\epsilon \downarrow 0} \delta_\epsilon = 0, \qquad v_{t_1}^\epsilon < v_{t_2}^\epsilon + \delta_\epsilon. \tag{6}$$

Indeed, if such is not the case, then there is a sequence of $\epsilon \downarrow 0$ for which $\lim_{\epsilon\downarrow 0} v_{t_1}^\epsilon = v_{t_1}$, $\lim v_{t_2}^\epsilon = v_{t_2}$, and $v_{t_1} > v_{t_2}$. Suppose that $b \in (v_{t_2}, v_{t_1})$. By the definition of $v_{t_1}$,

$$\lim_{\epsilon\downarrow 0} \left( \int_{\{a(x)<b\}} u^\epsilon(t_1, y)\, dy + \int_{\{a(x)\geq b\}} u^\epsilon(t_1, y)\, dy \right) = v_{t_1}.$$

If

$$\overline{\lim_{\epsilon\downarrow 0}} \int_{\{a(x)\geq b\}} u^\epsilon(t, y)\, dy = \delta_t > 0$$

for some $t \in (0, t_2)$, then

$$v_{t_2} \geq \overline{\lim_{\epsilon\downarrow 0}} \int_{\{a(x)\geq b\}} u^\epsilon(t_2, y)\, dy \geq b,$$

contradicting the choice of $b \in (v_{t_2}, v_{t_1})$. On the other hand, if $\delta_t = 0$ for $t \in (0, t_2)$ and

$$\int_{\{a(x)<b\}} g(x)\, dx < b,$$

then for any $\lambda > 0$ and $t \in (0, t_2)$,

$$w_b^\epsilon(t) = \int_{\{a(x)<b\}} u^\epsilon(t, y)\, dy < b + \lambda,$$

provided $\epsilon$ is small enough. This bound can easily be deduced from equation 3 using the fact that $a(x)$ is Lipschitz continuous. Because $w_b^\epsilon(t_1) < b + \lambda$ and $\delta_{t_1} = 0$, we obtain $v_{t_1} = \lim_{\epsilon\downarrow 0} w_b^\epsilon(t_1) + \delta_{t_1} < b + \lambda$, which contradicts the choice of $b \in (v_{t_2}, v_{t_1})$ for appropriately small $\lambda$. Therefore, equation 6 holds true.

A sequence of functions $v^\epsilon(t)$, $t \in [0, T]$, is said to converge $c$-weakly to $v(t)$ if $\lim_{\epsilon\downarrow 0} v^\epsilon(t) = v(t)$ at every continuity point of the limit function $v(t)$. Of course, for a limit in this sense to be unique, it is necessary that $v(t)$ have quite a lot of continuity points.

From equations 5 and 6, it follows that a subsequent $v_t^{\epsilon'}$ may be singled out of the family $v_t^\epsilon$ that $c$-weakly converges to a function $v_t$. (Henceforth, the prime in $\epsilon'$ will be dropped.) By equation 6, the function $v_t$ does not decrease, and thus, at most, it has a countable number of discontinuity points. For the limit $\lim_{\epsilon\downarrow 0} v^\epsilon$ to be unique, we will consider $v_t$ to be right continuous.

In the sequel, some bounds of $u^\epsilon(t, x)$ in the uniform norm and some bounds of the continuity module of $u^\epsilon(t, x)$ will be useful. Here, to a great degree, we use the fact that $x$ is one-dimensional. For $r = 1$ from equations 2 and 5, we obtain

$$\begin{aligned} 0 \leq u^\epsilon(t, x) &= \frac{1}{\sqrt{2\pi\epsilon t}} \int_{-\infty}^{\infty} g(y) \exp\left\{ -\frac{(y-x)^2}{2\epsilon t} \right\} dy \\ &\quad + \frac{1}{\epsilon} \int_0^t \int_{-\infty}^{\infty} \frac{c(y, v_s^\epsilon) u^\epsilon(s, y)}{\sqrt{2\pi\epsilon s}} \exp\left\{ \frac{-(y-x)^2}{2\epsilon s} \right\} ds dy \\ &\leq \max_{y\in R^1} g(y) + \frac{\sup_{y,v\in R^1} |c(y, v)|}{\sqrt{2\pi}\, \epsilon^{3/2}} \int_0^t \frac{1}{\sqrt{s}} v_s^\epsilon\, ds \leq c_1 + c_2 \epsilon^{-3/2}, \end{aligned} \tag{7}$$

where $c_1$ and $c_2$ are some constants independent of $\epsilon$, with $t \in [0, T]$ and $x \in R^1$.

From equation 2, while relying on equation 7, we have

$$\|u_t^\epsilon\|_{L^2_{[0,T]\times R^1}} + \|u_{xx}^\epsilon\|_{L^2_{[0,T]\times R^1}} \le c_3\epsilon^{-3/2}.$$

Because the norms of $u_t'$ and $u_{xx}''$ in $L^2_{[0,T]\times R^1}$ are bounded, we derive that $u^\epsilon(t, x)$ is Hölder continuous of exponent (at least) ¼ (see reference 5):

$$|u^\epsilon(t + h, x + \delta) - u^\epsilon(t, x)| \le \frac{c_4}{\epsilon^{3/2}}(|h|^{1/4} + |\delta|^{1/4}). \tag{8}$$

We notice that the continuity module can be bounded more exactly; however, for us, equation 8 is sufficient.

Therefore, suppose that as $\epsilon \downarrow 0$, $v_t^\epsilon$ converges $c$-weakly to a bounded right-continuous and strictly nondecreasing function $v_t$, $t \in [0, T]$. We put $\mathscr{E}_t = \{x \in R^1: a(x) = v_t\}$. We show that if $t_0$ is a continuity point of $v_t$ and $x \in \mathscr{E}_{t_0}$, then

$$\overline{\lim_{\epsilon\downarrow 0}}\, \epsilon \ln u^\epsilon(t_0, x) < 0. \tag{9}$$

Indeed, if $a(x) < v_{t_0}$, then one can find a small $h, \delta > 0$, such that $a(y) < v_s$ for $s \in [t_0 - h, t_0]$, $|x - y| < \delta$. Therefore, $c(y, v) \le -c_0 < 0$ for $(s, y) \in \Pi = \{(s, y): t_0 - h \le s \le t_0, |x - y| < \delta/2\}$. Consider the Markov time $\tau^\epsilon = \tau^\epsilon_{h,\delta} = h \Lambda \min\{s: |X_s^\epsilon - x| \ge \delta/2\}$. Using the strong Markov property of the process $(X_t^\epsilon, P_x)$, we derive (from equation 3) the bound

$$0 < u^\epsilon(t_0, x) < P_x\{\tau^\epsilon > h\} \exp\left\{-\frac{c_0 h}{\epsilon}\right\} + P_x\{\tau^\epsilon \le h\} \cdot \max_{(t,y)\in\partial\Pi} |u^\epsilon(t, y)|, \tag{10}$$

where $\partial\Pi$ is the surface of the cylinder $\Pi$. By the properties of the Wiener process,

$$\lim_{\epsilon\downarrow 0} \epsilon \ln P_x\{\tau^\epsilon < h\} = -\alpha < 0,$$

which together with equation 10 implies equation 9. With similar arguments and the bound from equation 8, one can prove equation 9 for the points $x$ such that $a(x) > v_{t_0}$.

As is known,[2,3] the action functional for the family of processes $(X_t^\epsilon, P_x)$ in the space $C_{0T}$ has the form $\epsilon^{-1}S_{0T}(\psi)$ as $\epsilon \downarrow 0$ with

$$S_{0T}(\psi) = \begin{cases} \dfrac{1}{2}\displaystyle\int_0^T |\dot{\psi}_s|^2\, ds, \ \psi \in C_{0T}, \text{ where } \psi \text{ is absolutely continuous,} \\ +\infty, \ \psi \in C_{0T}, \text{ where } \psi \text{ is not absolutely continuous.} \end{cases}$$

If $v_t^\epsilon$ $c$-weakly converges to $v_t$, then by relying on the results of §3.3 in reference 1 and by employing equation 3, we obtain the following expression for the logarithmic asymptotics of $u^\epsilon(t, x)$:

$$\lim \epsilon \ln u^\epsilon(t, x) = V_v(t, x) = \sup\left\{\int_0^t \left[c(\psi_s, v_s) - \frac{1}{2}|\dot{\psi}_s|^2\right] ds: \right.$$

$$\left. \psi_0 \in G_0, \psi_t = x, \text{ where } \psi \text{ is absolutely continuous}\right\}. \tag{11}$$

From equation 9, it follows that $V_v(t, x) < 0$ outside the level set $\mathscr{E}_t = \{x \in R^1: a(x) = v_t\}$. The function $V_v(t, x)$ cannot be negative everywhere in $R^1$; otherwise, it would contradict the condition

$$\lim_{\epsilon\downarrow 0} \int_{R_1} u^\epsilon(t, x)\, dx = v_t > 0.$$

Therefore, one can find a point $\hat{\varphi}_t$ in $\mathscr{E}_t$ such that $V_v(t, \hat{\varphi}_t) \geq 0$. However, $V_v(t, \hat{\varphi}_t)$ cannot be positive because this together with equation 8 would lead to $\lim_{\epsilon\downarrow 0} v_t^\epsilon = \infty$. Hence, there is a point $\hat{\varphi}_t$ such that $V_v(t, \hat{\varphi}_t) = 0$ and $a(\hat{\varphi}_t) = v_t$. From equation 11, we derive the equation for $\hat{\varphi}_t$:

$$0 = V_{a(\hat{\varphi})}(t, \hat{\varphi}_t) = \sup\left\{\int_0^t \left(c[\psi_s, a(\hat{\varphi}_s)] - \frac{|\dot{\psi}_s|^2}{2}\right) ds:\right.$$

$$\left.\psi \in C_{0t}, \psi_0 \in G_0, \psi_t = \hat{\varphi}_t, \text{ where } \psi \text{ is absolutely continuous}\right\}. \quad \textbf{(12)}$$

This equality (equation 12) also holds at the discontinuity points of $a(\hat{\varphi}_t)$, provided that at these points $\hat{\varphi}_t$ is defined by:

$$\hat{\varphi}_t = \overline{\lim_{s\downarrow t}}\, \hat{\varphi}_s.$$

The above reasoning yields $V_{a(\hat{\varphi})}(t, x) \leq 0$, and thus $\hat{\varphi}$ is the maximal solution of the equation $V_{a(\hat{\varphi})}(t, \hat{\varphi}_t) = 0$.

By the condition of the theorem, such a solution is unique; this implies the existence of a unique limit point $v_t = a(\hat{\varphi}_t)$ for the family $v_t^\epsilon$ in the sense of $c$-weak convergence. From this, we conclude that $v_t^\epsilon$ converges to $v_t$ for all ways in which $\epsilon$ tends to 0. The uniqueness of the maximal solution also implies that there is only one point $\hat{\varphi}_t$ in $\mathscr{E}_t$ at which $V_{a(\hat{\varphi})}(t, \hat{\varphi}_t) = 0$. For $x \neq \hat{\varphi}_t$, we have $V_{a(\hat{\varphi})}(t, x) < 0$. From this,

$$\lim_{\epsilon\downarrow 0} u^\epsilon(t, x) = 0 \quad \text{for} \quad x \neq \hat{\varphi}_t,$$

$$\lim_{\epsilon\downarrow 0} \int_{\{|y-\hat{\varphi}_t|<\delta\}} u^\epsilon(t, y)\, dy = v_t,$$

$$\lim_{\epsilon\downarrow 0} \int_{\{|y-\hat{\varphi}_t|\geq\delta\}} u^\epsilon(t, y)\, dy = 0$$

for any $\delta > 0$. This completes the proof of the theorem.

Remark: From the above proof, for $r = 1$, it follows that the solution $u^\epsilon(t, x)$ of equation 2 not only weakly converges to $a(\hat{\varphi}_t)\delta(x - \hat{\varphi}_t)$ as $\epsilon \downarrow 0$, but also that $u^\epsilon(t, x)$ tends to zero exponentially quickly, provided $x \neq \hat{\varphi}_t$. To prove such a pointwise convergence, we need bounds of the type in equation 7 and 8. For $x \in R^r$, $r > 1$, such bounds do not seem to exist. However, for $r > 1$, one can prove that for every $\delta > 0$,

$$\lim_{\epsilon\downarrow 0} \epsilon \ln \int_{D_\delta} u^\epsilon(t, x)\, dx < 0,$$

where $D_\delta = \{x \in R^r: \rho(x, \mathscr{E}_t) > \delta\}$, with $\rho(\ ,\ )$ being the Euclidean distance in $R^r$, $\mathscr{E}_t = \{x \in R^r: a(x) = v_t\}$. From this, one can also deduce the claim of the theorem in the multidimensional case as well.

Now, suppose that $c(x, v) = a(x) - v$ and let $a(x)$ satisfy the above-listed hypotheses. Then, equation 4 takes the form,

$$A(t, \hat{\varphi}_t) = \int_0^t a(\hat{\varphi}_s)\, ds, \qquad t \in (0, T], \tag{13}$$

where the function $A(t, x)$, $t > 0$, $x \in R^1$, is defined by

$$A(t, x) = \sup\left\{\int_0^t \left[a(\psi_s) - \frac{|\dot{\psi}_s|^2}{2}\right] ds : \psi \in C_{0t}(R^r), \psi_0 \in G_0, \psi_t = x\right\}. \tag{14}$$

For $A(t, x)$, one can write down the Hamilton-Jacobi equation, which, along with equation 13 and the maximality condition, may be employed for calculating $\hat{\varphi}_t$.

Next, we consider, in detail, the case where $x \in R^1$, $c(x, v) = a(x) - v$, and $a(x)$ is a piecewise linear function. First, we assume that $a(x) = 1$ for $x < 0$ and $a(x) = x + 1$ for $x > 0$. For the boundedness condition to be fulfilled, we will suppose that for appropriately large $x$, the function $a(x)$ is cut: $a(x) = a(N)$ for $x \geq N$, $N \gg 1$. Such a cutting does not at all affect the soliton movement until it reaches the point $N$, provided that the support $G_0$ of the initial function lies on the left of the point $N$. Let $G_0 = (-h, 0)$, $h > 0$. For $a(x) = x + 1$, the supremum in equation 14 can easily be calculated and we have

$$A(t, x) = \frac{t^3}{24} + t + \frac{tx}{2} - \frac{x^2}{2t}.$$

Equation 13 takes the form

$$\frac{t^4}{12} + t^2\hat{\varphi}_t - \hat{\varphi}_t^2 = 2t\int_0^t \hat{\varphi}_s ds.$$

We put $\phi_t = \int_0^t \hat{\varphi}_s\, ds$. For $\phi_t$, we obtain the equation

$$\frac{t^4}{12} + t^2\phi_t' - \phi_t'^2 = 2t\phi_t, \qquad \phi_0 = 0. \tag{15}$$

From equation 15, we derive

$$\phi_t' = \frac{t^2}{2} + \sqrt{\frac{t^4}{3} - 2t\phi_t}, \qquad \phi_0 = 0. \tag{16}$$

The minus sign before the root gives a solution that does not satisfy the maximality condition. By straightforward substitution, one can make sure that $\phi_t = t^3/6$ is a solution of the problem of equation 16. From equation 16, we derive that $\phi_t' \geq t^2/2$ and thus $\phi_t \geq t^3/6$. On the other hand, $\phi_t$ can exceed $t^3/6$ for no $t > 0$ because in this case, the expression under the root sign in equation 16 would be negative. Therefore, $\phi_t = t^3/6$ is a unique solution of equation 16, the hypotheses of theorem 1 hold, and the soliton movement is controlled by the formula $\hat{\varphi}_t = \phi_t' = t^2/2$. By theorem 1, $u^\epsilon(t, x)$ weakly converges as $\epsilon \downarrow 0$ to $(1 + t^2/2)\delta(x - t^2/2)$ for every $t > 0$.

Now let $a(x)$ be the piecewise linear function shown in FIGURE 1. We will suppose that $0 < \alpha < 1$, $a(-h) = 1$, $a(\mathrm{A}) < a(\mathrm{B})$, and $G_0 = (-h, 0)$. Generally speaking, the soliton formed under the initial condition can move both to the right and to the left of $[-h, 0]$. If $\alpha < 1$ and $a(\mathrm{A}) < a(\mathrm{B})$, then (from the above) one can deduce that the solution $\hat{\varphi}_t$ of equation 4 that takes negative values is not maximal. It is not hard to see that for small $t > 0$ for $a(x)$ (shown in FIGURE 1), the soliton will move to the right according to the same law as in the case of $a(x) = x + 1$ for $x > 0$, that is, $\hat{\varphi}_t = t^2/2$. If (for fixed B and C) $a(\mathrm{C}) - a(\mathrm{B})$ is positive, but not too large, then at time $t = \sqrt{2\mathrm{B}}$, the soliton will reach the point $x = \mathrm{B}$ and will stay there for some given time. Afterwards, the soliton will tunnel into the left neighborhood of the point C. If $a(\mathrm{C}) - a(\mathrm{B})$ is large enough, then it will tunnel into a neighborhood of C without hitting B. To calculate the time of the jump and the soliton position after the jump, we denote $\lambda(t) = t^2/2$ for $t \leq \sqrt{2\mathrm{B}}$ and $\lambda(t) = \mathrm{B}$ for $t > \sqrt{2\mathrm{B}}$. Consider the function

$$A(t, x) = \sup\left\{\int_0^t \left[a(\psi_s) - \frac{|\dot{\psi}_s|^2}{2}\right] ds: \psi \in C_{0t}, \psi_0 = 0, \psi_t = x\right\},$$

where $a(x)$ is the piecewise linear function shown in FIGURE 1. If one writes down the

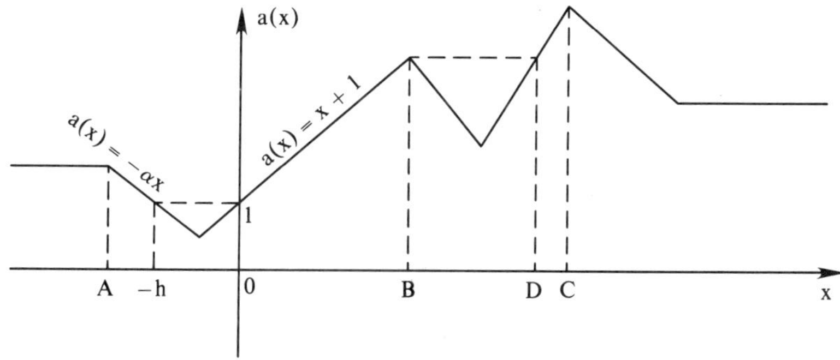


**FIGURE 1.** The piecewise linear function $a(x)$. (See text for details.)

equations of the lines whose segments make up the broken line $a(x)$, then $A(t, x)$ can be evaluated in an explicit way. We find $t^*$ and $x^*$ from the condition

$$t^* = \min\left\{t: A(t, x) = \int_0^t a[\lambda(s)]\, ds \text{ for some } x > \mathrm{B}\right\},$$

$$A(t^*, x^*) = \int_0^{t^*} a[\lambda(s)]\, ds. \qquad (17)$$

For $a(\mathrm{C}) > a(\mathrm{B})$, such a $t^* < \infty$ exists. In the case of general position, the point $x^*$ is defined by equation 17 in a unique way. It is not difficult to check that $x^* \in (\mathrm{D}, \mathrm{C}]$. By theorem 1, $\hat{\varphi}_t = t^2/2$ for $t \in (0, t^*)$. At time $t^*$, tunneling occurs from the point $\lambda(t^*) \leq \mathrm{B}$ into $x^*$. If, as is shown in FIGURE 1, $a(\mathrm{C}) = \max a(x)$, then from the point $x^*$, the soliton arrives at the point C in a finite time and remains there forever. If

$a(\mathrm{C}) < \max a(x)$, then sooner or later there will be tunneling into a neighborhood of a maximum of the function $a(x)$ that is larger than $a(\mathrm{C})$.

We emphasize that the soliton may also have jumps in the case of strictly increasing $a(x)$ whenever $a(x)$ has regions of both slow and sufficiently quick growth.[4,5]

## REFERENCES

1. FREIDLIN, M. I. 1986. Some general properties of evolution processes. Quasi-deterministic approximation. Proceedings of VIIIth International Congress in Math. Physics, Marseille.
2. FREIDLIN, M. I. 1985. Functional Integration and Partial Differential Equations. Princeton Univ. Press. Princeton, New Jersey.
3. FREIDLIN, M. I. & A. D. WENTZELL. 1983. Random Perturbations of Dynamical Systems. Springer-Verlag. New York. (Translated from Russian. 1979. Nauka. Moscow.)
4. FREIDLIN, M. I. 1985. Limit theorems for large deviations and reaction-diffusion equations. Ann. Probability **13**(no. 3): 639–675.
5. VOLEVICH, L. R. & B. P. PANEJAH. 1965. Some spaces of generalized functions and imbedding theorems. Usp. Mat. Nauk (Russian) **20**(no. 1): 3–74.

# Phase Transitions in Driven Diffusive Systems

HERBERT SPOHN

*Universität München*
*Theoretische Physik*
*D-8000 München 2, Federal Republic of Germany*

## INTRODUCTION

Some aspects of phase transitions in a nonequilibrium situation have been advertised widely.[1] Let us recall the Rayleigh-Bénard instability as one prototypical example. A fluid is between two horizontal plates—the upper at temperature $T_1$ and the lower at temperature $T_2$, with $T_2 > T_1$. We are interested in the structure of the steady state depending on the driving force. In our example, this is the temperature difference that is proportional to the Rayleigh number $R$. For small $R$, there is diffusive heat transport. The velocity field is zero. As $R$ increases, buoyancy competes more successfully with friction and gravity. At some critical value $R_c$, it becomes more favorable for the fluid to move; thus, typical roll patterns develop. The correlations in the fluid near the transition have many features in common with a second-order phase transition for a system in thermal equilibrium. However, quite contrary to a usual bulk transition, the correlations in the fluid are on a scale set by the geometry of the experiment. A little fluid element, whether moving or at rest, is in thermal equilibrium without any sign of a phase transition.

My interest here is in a completely different aspect of a nonequilibrium phase transition. We imagine that we have a system that undergoes a second-order phase transition in thermal equilibrium. Now, we can drive the system by either external forces, temperature gradients, or the like. How is then the nature of the equilibrium phase transition changed? Is the transition suppressed or reenforced? Do the critical exponents change continuously with the driving strength? In what sense are exponents universal?

A famous example is the segregation (demixing) transition of a binary fluid under shear. This system was analyzed theoretically by Kawasaki and Onuki[2] and studied experimentally by Beysens and Gbadamassi.[3] They chose a binary mixture of cyclohexane and aniline close to the critical composition. The mixture flows through a pipe from a cylindrical reservoir. The parabolic velocity profile in the pipe enforces a shearing of the binary mixture. As the hydrostatic pressure exerted by the reservoir changes slowly, a whole range of shear rates is scanned. The demixing transition under shear is observed by means of light-scattering techniques. The surprising theoretical prediction of Kawasaki and Onuki is a lowering of the upper critical dimension from 4 to 12/5 because of the shear. Therefore, in three dimensions, the critical exponents of the transition should be of the mean field type. This has been confirmed experimentally. For small shear rates, one even observes a crossover from the Ising to the mean field exponents.

S. Katz, J. L. Lebowitz, and myself investigated a system of comparable nature.[4] Our analysis relies heavily on an exploratory numerical simulation that has been continued with greater precision by J. Marro and co-workers.[5,6] I want to sketch here the present status of the segregation transition in a driven diffusive system, along with including the work in progress of various groups.

The basic structure of our model is perhaps most transparent and familiar in a continuum language. We start from an order parameter field $\phi(x)$. Its (free) energy is given by the Ginzburg-Landau-Wilson functional

$$H(\{\phi\}) = \beta \int dx [|\text{grad}\, \phi(x)|^2 + \alpha\phi(x)^2 + g\,\phi(x)^4], \qquad \textbf{(1.1)}$$

where $\beta$, $g > 0$. A configuration $\phi$ is weighted statistically by the Boltzmann factor, $e^{-H(\{\phi\})}$. As is well known, the system undergoes a segregation transition; in our units, this is as a function of $\alpha$. For $\alpha > \alpha_c$, typically $\phi \simeq 0$; on the other hand, for $\alpha < \alpha_c$, the system segregates into a phase with typical $\phi > 0$ and one with $\phi < 0$. The transition is in the same universality class as the Ising model.

We assume that the dynamics of the model is of such a type as to locally conserve the order parameter field, but to otherwise be purely dissipative. Therefore, $\phi$ is governed by the Cahn-Hilliard equation (Model B of critical dynamics):

$$\frac{\partial}{\partial t}\phi + \text{div}\,\vec{j} = 0, \qquad \textbf{(1.2)}$$

$$\vec{j} = -L\,\text{grad}\,\frac{\delta H}{\delta \phi} + L^{1/2}\vec{j}_{\text{ran}}. \qquad \textbf{(1.3)}$$

In our applications, $\phi$ is (up to a constant) proportional to the density (or the density difference in the case of a binary mixture). Therefore, we refer to $\vec{j}$ as a mass current. Equation 1.2 expresses the local conservation law. Equation 1.3 describes two physical mechanisms to set up a mass current $\vec{j}(x, t)$: By Fick's law, there is a systematic current proportional to the gradient of the chemical potential. In addition, fast microscopic processes are summarized as a random current $\vec{j}_{\text{ran}}(x, t)$. It is Gaussian and $\delta$-correlated in space-time. The $L$ is the Onsager coefficient, which in principle may be $\phi$-dependent, $L > 0$. Equations 1.2 and 1.3 have the stationary solutions,

$$P(\{\phi\}) = \frac{1}{Z} e^{-H(\{\phi\})} e^{h\int dx\phi(x)}, \qquad \textbf{(1.4)}$$

with the free parameter $h$ originating from the conservation law.

We now want to drive the system with a uniform force. If the particles are charged, the driving force could be a constant external electric field. In a closed box, thermal equilibrium would establish itself where the driving force is balanced by the gradient in the chemical potential and where the net current is zero. However, this is not the case we are interested in. Rather, we impose periodic boundary conditions, at least in the direction of the driving force. Then, in the steady state, the force maintains a uniform current. The typical $\phi$-field will be more or less constant provided that no segregation sets in.

Dynamically, we add the term $\sigma\vec{E}$ to equation 1.3, where $\vec{E}$ is the strength of the driving force and $\sigma$ is, in general, the $\phi$- and $\vec{E}$-dependent conductivity. Then, equation 1.3 becomes

$$\vec{j} = -L \operatorname{grad} \frac{\delta H}{\delta \phi} + \sigma\vec{E} + L^{1/2}\vec{j}_{\mathrm{ran}}. \tag{1.5}$$

The problem is to understand the properties of the steady state solution to equations 1.2 and 1.5—in particular, the nature of the phase transition. An immediate difficulty is that, in general, the steady state is no longer of the form of equation 1.4.

Let us now point out two general features of equations 1.2 and 1.5 that also hint at the physical mechanism we are trying to understand. For the moment, let us assume $L$ = constant. Then, if $\sigma(\phi) = c_1 + c\phi$, the Galilei-transformed field $\tilde{\phi}(x, t) = \phi(x - c\vec{E}t, t)$ satisfies equations 1.2 and 1.3. In the moving frame of reference, the steady state is again given by equation 1.4 and, of course, the transition is not at all modified by the driving force. Only if $\sigma$ depends nonlinearly on $\phi$ do we expect some new physics. For example, in a lattice gas with particle-hole symmetry, the critical density is ½. The $\phi$ denotes then the deviation from ½. Also, the jump rate can be chosen such that the conductivity is symmetric around and with a maximum at $\rho$ = ½. The appropriate choice for $\sigma$ is thus $\sigma(\phi) = -c\phi^2$ and, indeed, the signature of the phase transition is modified.

The $L$ and $\sigma$ are two phenomenological coefficients. Can they be chosen independently? The argument given in reference 4 translates in the following way to the model of equations 1.2 and 1.5: Locally, the driving force is (minus) the gradient of a linear potential. Therefore, in a closed box, we expect the equilibrium measure

$$\frac{1}{Z} e^{-H(\{\phi\})} e^{\int dx (h + \vec{E}\cdot x)\phi(x)} \tag{1.6}$$

to be a stationary solution of our dynamics. We also expect the dynamics in this state to satisfy the condition of detailed balance. If we impose both as a condition on equations 1.2 and 1.5, then

$$\sigma = L. \tag{1.7}$$

Repeating the argument from above shows that new physics can be expected only if the Onsager coefficient depends nonlinearly on $\phi$.

As an aside, I mention that the model defined by equations 1.2 and 1.5 has an interesting dynamical behavior even at high temperatures; this corresponds to the case of $L$ = const., $H(\{\phi\}) = \beta \int dx \phi(x)^2$, and $\sigma(\phi) = c\phi^2$ as the crucial nonlinearity. In one space dimension, one then obtains the fluctuating (= randomly stirred) Burgers equation. The spectrum of its current-current correlations diverges for low frequencies as $\omega^{-1/3}$. This implies a nondiffuse (namely, $t^{4/3}$ rather than $t$) spreading of density fluctuations; cf. references 7 and 8 for details and a discussion of dimensions $d > 1$. The randomly stirred Burgers equation is also investigated in reference 9, while the relationship to the statistics of an interface in a random medium is pointed out in reference 10.

## AN EXPLORATORY NUMERICAL STUDY

It is fairly straightforward to implement a dynamics of the type in equations 1.2 and 1.5 as a Monte Carlo simulation. Computations have been carried through in two and three dimensions.[4-6] For simplicity of the exposition, let us restrict ourselves to two dimensions. We choose then a square lattice, $\Lambda$, with periodic boundary conditions. Typically, $\Lambda = 30 \times 30$ to $50 \times 50$. At each lattice site $x \in \Lambda$, there is an occupation variable $\eta_x$ with $\eta_x = 0, 1$ for $x$ empty, respectively occupied. The energy of a configuration $\{\eta_x | x \in \Lambda\} \equiv \eta$ is given by the usual, isotropic, nearest-neighbor Ising Hamiltonian:

$$\beta H(\eta) = -4\beta J \, \Sigma_{x,y,|x-y|=1} \, \eta_x \eta_y. \tag{2.1}$$

Here, $J > 0$ is the coupling constant and $\beta$ is the inverse temperature. The driving force $\vec{E} = (0, E)$ is oriented along the positive two-direction. As exchange rates, we adopt the usual Metropolis rates: Let $\eta^{xy}$ denote the configuration $\eta$ with occupations at sites $x$ and $y$ interchanged and let $c_E(x, y, \eta)$ be the exchange rate for the occupations at sites $x$ and $y$ when the configuration is $\eta$. Then to the value of the change of energy in a jump, we add the work done by the driving force,

$$\Delta U = \beta[H(\eta^{xy}) - H(\eta)] + E(x_2 - y_2)(\eta_x - \eta_y), \tag{2.2}$$

and prescribe the exchange rates as

$$c_E(x, y, \eta) = \begin{cases} 1, \text{ if } \Delta U < 0 \\ e^{-\Delta U}, \text{ if } \Delta U > 0 \end{cases} \tag{2.3}$$

for $|x - y| = 1$ and $c_E(x, y, \eta) = 0$ otherwise. The only difference to the standard equilibrium procedure is the additional term in equation 2.2.

The steady state distribution, $p(\eta)$, of interest is defined by specifying the number of particles, $N = \Sigma_x \eta_x$, and by

$$\Sigma_{x,y,|x-y|=1} \{c_E(x, y, \eta^{xy}) p(\eta^{xy}) - c_E(x, y, \eta) p(\eta)\} = 0. \tag{2.4}$$

Because of the periodic boundary conditions, detailed balance does not hold. Numerically, after some transient time, configurations typical for $p(\eta)$ are generated. By symmetry, the critical density is ½ and it is therefore adopted in all simulations. The density dependence has not yet been explored systematically.

The qualitative nature of the transition observed is best summarized through three typical configurations taken from the Monte Carlo simulation of J. Marro and J. L. Vallés[6] (see FIGURE 1). Here, $T_c(E)$ is the critical temperature that is dependent on the driving force $E$. At high temperatures, the configuration is random with essentially no sign of anisotropy, whereas for $T < T_c(E)$, the system segregates into striplike configurations that are highly anisotropic in the field direction. In fact, the displayed configuration is metastable. For sufficiently long times, surface tension will be further reduced by a merging into a single strip. To save computer time, the above snapshots are for strong $E$. For small $E$, typical configurations look alike provided that the temperature scale is adjusted.

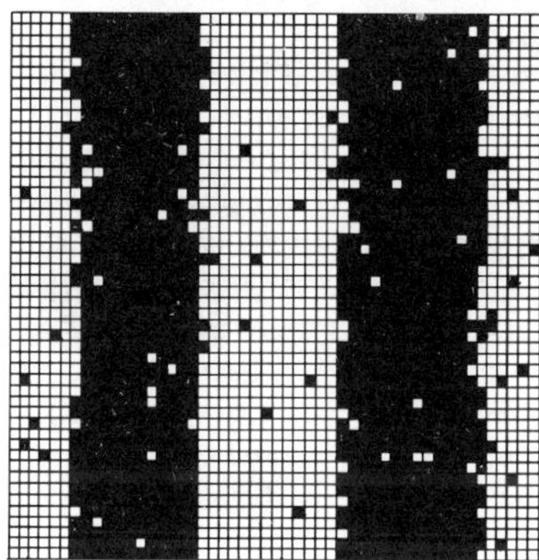

1.19 $T_c(E)$

0.89 $T_c(E)$

0.59 $T_c(E)$

**FIGURE 1.** Typical configurations at three different temperatures in units of $T_c(E)$. The driving field $\vec{E}$ points downwards.

On a more quantitative level, one observes the following features: The critical temperature is shifted to higher values. For large $E$, $T_c(E)/T_c(O) = 1.35$, where $T_c(O)$ is the Onsager value for the critical temperature of the two-dimensional Ising model. This indicates that the driving force suppresses fluctuations. The most accurate values for $T_c(E)$ are obtained by means of a Monte Carlo renormalization.[6,11] The static structure factor is essentially isotropic for large $T$ and develops a strong anisotropy as one passes into the critical region. If one plots the steady state current as a function of the temperature, then it has a kink roughly at $T_c(E)$. Striplike configurations carry a considerably reduced current as compared to random configurations and, therefore, the current senses the change in structure. Also, the specific heat (defined as $\partial\langle H\rangle(T)/\partial T$) has a pronounced maximum near $T_c(E)$. In passing, let me mention that the antiferromagnetic orderung ($J < 0$) is completely suppressed by the driving force. Presumably, this is because a strictly antiferromagnetic particle configuration has no blocking and a maximal current.

In the critical region, the exponent $\beta$ is the most accessible quantity. It expresses the difference in density of the two phases below, but close to $T_c(E)$:

$$\Delta\rho \simeq [T_c(E) - T]^{\beta}. \tag{2.5}$$

Marro and Vallés "measured" $\beta$ for strong $E$ (i.e., jumps opposite to the field essentially do not occur). To test universality, they sampled nearest-neighbor pairs in the field direction $\gamma$ times as often as in the direction orthogonal to the field. We display their results[6] (see FIGURE 2) for $\gamma = 1$ ($*$), 5 (O), 20 ($\times$), and 80 ($\Delta$). Rather convincingly, this yields a critical exponent $\beta \simeq 0.226$ that differs significantly from the Onsager value of 0.125 and the classical (mean field) value of 0.5.

## THEORETICAL ATTEMPTS

As mentioned already, because of the lack of detailed balance, we have no way to obtain the steady solution of equation 2.4. Only for very small systems, as half-filled $2 \times 3$, has the explicit solution been computed.[12,13] In view of this, we have to resort to other methods. Two approaches have been carried through with some success: (i) The mean field type of approximation by van Beijeren and Schulman.[14] Their basic idea is to postulate that jumps in the field direction are more frequent than those orthogonal to it. As in FIGURE 2, the ratio of time scales is denoted by $\gamma$. As $\gamma \to \infty$, the dimension of our problem reduces by one. Between any jump in the one-direction, the system equilibrates, and instead of a master equation for particle configurations in $\Lambda$, we only have to consider the reduced master equation for the column occupation variables $\{n_j(t)\}$. The $n_j(t)$ denotes the number of particles in the $j$-th column at time $t$.

The reduced master equation, though, still does not satisfy detailed balance (with the exception of degenerate cases). Therefore, the original difficulty still remains. However, the reduced master equation contains a small parameter that suppresses fluctuations, namely, $1/N$, where $N$ is the number of sites in a column. Thus, the master equation for the $n_j(t)$ can be studied by asymptotic methods, such as the $\Omega$-expansion of van Kampen[15] and the Freidlin-Wentzell theory of large deviations.[16] In the one-direction, note that the spatial structure is retained. A difficulty is that the jump rates of the reduced master equation are given only implicitly. Fortunately, as

$E \to \infty$, the fast dynamics randomizes each column by itself (no exchanges between different columns are allowed). In this case, the reduced rates are readily computed. The van Beijeren–Schulman model can therefore be applied in the limit of strong driving field and fast jump rates in the direction of the field.

I would like to give the reader a feeling for the structure of the equations so obtained. For $N \to \infty$, $n_j(t)/N$ typically has $N$ changes of magnitude $1/N$ per unit of time. Therefore, $n_j(t)/N$ tends to the deterministic limit $\rho_j(t)$, $0 \leq \rho_j(t) \leq 1$. The column densities are governed by the equation of conservation type:

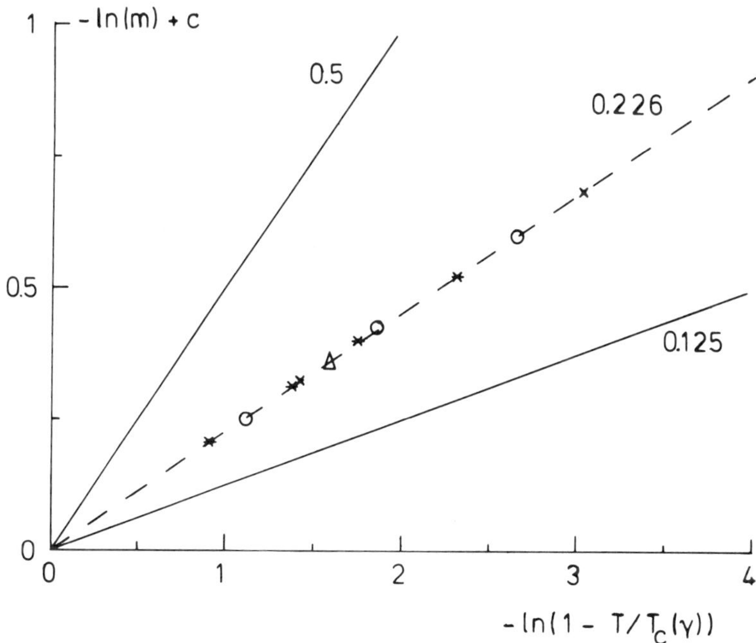


**FIGURE 2.** A log-log plot of the order parameter versus temperature for different sampling ratios $\gamma$.

$$\frac{d}{dt}\rho_j(t) + J_j[\vec{\rho}(t)] - J_{j-1}[\vec{\rho}(t)] = 0. \tag{3.1}$$

The current, $J_j(\vec{\rho})$, between columns $j$ and $j + 1$ is given as the difference of the jumps from $j$ to $j + 1$ minus those from $j + 1$ to $j$:

$$J_j(\vec{\rho}) = R(\rho_{j-1}, \rho_j, \rho_{j+1}, \rho_{j+2}) - R(\rho_{j+2}, \rho_{j+1}, \rho_j, \rho_{j-1}). \tag{3.2}$$

Here,

$$R(\rho_{j-1}, \rho_j, \rho_{j+1}, \rho_{j+2}) = \langle \Phi(\beta[H(\eta^{xy}) - H(\eta)]) \rangle(\vec{\rho}), \tag{3.3}$$

with $x_2 = y_2, x_1 = j, y_1 = j + 1, \eta_x = 1$, and $\eta_y = 0$. The average $\langle \cdot \rangle (\vec{\rho})$ denotes that the $\eta_x$'s are independent, and that in the $j$-th column, $\eta_x = 1$ with probability $\rho_j$. The $\Phi$ is the rate function satisfying $\Phi(x) = \Phi(-x)e^{-x}$ [$\Phi(x) = 1$ for $x \leqq 0$ and $\Phi(x) = e^{-x}$ for $x \geq 0$ in the numerical simulations described in the previous section]. The fluctuations around $\rho_j(t)$ are of the order $N^{-1/2}$ and are Gaussian. At low temperatures, equation 3.1 has many dynamically stable stationary solutions. One then has to go back to the reduced master equation in order to obtain some information about the steady state.

The properties of the van Beijeren–Schulman model have been investigated in considerable detail.[12–14,17] First of all, this model exhibits the segregation transition. For antiferromagnetic coupling, there is no transition. As expected, the critical exponents are mean field, with the exception of the specific heat, where there is some freedom of definition because the temperature does not enter canonically into the steady state distribution. The steady state current as a function of temperature has a kink at $T_c$. A great advantage of the model is the possibility of investigating the rate dependence (i.e., the dependence on $\Phi$) of the results; for example, the critical temperature depends strongly on the rates, whereas the critical exponents are independent of the rates. The model thus reflects qualitatively many features observed in the simulations.

To investigate the low temperature phase diagram, one looks for stationary solutions to equation 3.1 and one has to determine their stability character. One imposes $J_j(\vec{\rho}) = 0$. However, equation 3.2 may then be regarded as a discrete dynamical system with a three-dimensional phase-space. Any trajectory of this discrete dynamical system defines a stationary solution to equation 3.1 with zero current provided that the $\rho_j$'s stay within the physically meaningful domain.

The low temperature phase diagram is complicated and is only partially elucidated.[13] In particular, one finds metastable phases, of which an example can be seen in FIGURE 1. The phase coexistence curve is computed from the Maxwell lever rule and, as a necessary check of consistency, the so-determined densities of the pure phases coincide with the asymptotic densities of the dynamically stable kink solution to equation 3.1. The spinodal is the line of limit of stability for the homogeneous solution to equation 3.1. Between the phase coexistence and the spinodal, in the $\rho$–$T$ plane, one finds dynamically stable, periodic (in space) density profiles. Inside the spinodal and outside the phase coexistence curve, no such stable solutions to equation 3.1 exist. There is no stable period 2 solution. The stable period 3 solution exists below a certain characteristic. Above the period 3 curve, there is a period 4 curve, and below that there exists a stable period 4 solution, etc. The whole structure accumulates at $\rho_c = \frac{1}{2}$ and $T_c$.

The other approach alluded to before is: (ii) The field-theoretic approach.[18–20] References 19 and 20 give a very detailed study of the model in equations 1.2 and 1.5 [with $L$ = const. and $\sigma(\phi) = \phi^2$ near $T_c$] using the methods of critical dynamics. The most striking predictions are: The critical exponent $\beta$, which governs the onset of the spontaneous magnetization, equals $\frac{1}{2}$ for $d > 2$. Above $T_c$, the system is spatially anisotropic with

$$\xi_\parallel = \xi_\perp^{2+(5-d)/3}, \qquad d \leqq 5,$$

and

$$\xi_\parallel = \xi_\perp^2, \qquad d \geqq 5,$$

where $\xi_{\parallel}(\xi_{\perp})$ is the longitudinal (transverse) correlation length. The transverse critical behavior is of mean field type. This is in correspondance to the findings of Kawasaki and Onuki. The longitudinal exponents are determined explicitly. Their upper critical dimension is five.

## CONCLUSIONS

The segregation transition in driven diffusive systems has some interesting features. Most desirable is an experimental realization of such a system. I would like to remind the reader that the model was proposed mostly on the basis of simplicity rather than on the basis of physical realism. Our hope is that the model stimulates further investigations on phase transitions in driven systems.

## ACKNOWLEDGMENT

It is a pleasure to thank J. Krug for highly useful discussions.

### REFERENCES

1. HAKEN. 1977. Synergetics. Springer-Verlag. Berlin/New York.
2. ONUKI, A. & K. KAWASAKI. 1979. Ann. Phys. (N.Y.) **121:** 456; ONUKI, A., K. YAMAZAKI & K. KAWASAKI. 1981. Ann. Phys. (N.Y.) **131:** 217.
3. BEYSENS, D. & M. GBADAMASSI. 1980. Phys. Rev. **A22:** 2250.
4. KATZ, S., J. L. LEBOWITZ & H. SPOHN. 1984. J. Stat. Phys. **34:** 497.
5. MARRO, J., J. L. LEBOWITZ, H. SPOHN & M. H. KALOS. 1985. J. Stat. Phys. **38:** 725.
6. VALLÉS, J. L. & J. MARRO. 1986. J. Stat. Phys. **43:** 441.
7. VAN BEIJEREN, H., R. KUTNER & H. SPOHN. 1985. Phys. Rev. Lett. **54:** 2026.
8. JANSSEN, H. K. & B. SCHMITTMANN. Z. Phys. B. (Condensed Matter).
9. FORSTER, D., D. R. NELSON & M. J. STEPHEN. 1977. Phys. Rev. **A16:** 732.
10. HUSE, D. A., C. L. HENLEY & D. S. FISHER. 1985. Phys. Rev. Lett. **55:** 2924.
11. KATZ, S. 1984 (March). Unpublished.
12. ZHANG, Q. W. Ph.D. thesis. Physics Department, Rutgers University. To be completed.
13. KRUG, J. 1985 (June). Diplomarbeit. Fachbereich Physik, Universität München.
14. VAN BEIJEREN, H & L. S. SCHULMAN. 1984. Phys. Rev. Lett. **53:** 806.
15. VAN KAMPEN, N. G. 1981. Stochastic Processes in Physics and Chemistry. North-Holland. Amsterdam.
16. FREIDLIN, M. I. & A. D. WENTZELL. 1984. Random Perturbations of Dynamical Systems. Springer-Verlag. Berlin/New York.
17. KRUG, J., J. L. LEBOWITZ, H. SPOHN & Q. W. ZHANG. 1986. J. Stat. Phys. **44:** 567.
18. GAWEDZKI, K. & A. KUPIAINEN. 1986. Nucl. Phys. **B269:** 45.
19. LEUNG, K. T. & J. L. CARDY. 1986. J. Stat. Phys. **44:** 567.
20. JANSSEN, H. K. & B. SCHMITTMANN. 1986. Z. Phys. B. (Condensed Matter) **64:** 503.

# Phase Transitions in Biological Membranes[a]

OLE G. MOURITSEN

*Department of Structural Properties of Materials*
*The Technical University of Denmark*
*DK-2800 Lyngby, Denmark*

The most common cellular structure in living organisms is the cell membrane. This membrane not only acts as a passive container, it is also associated with a substantial fraction of biological activity—for example, transport of matter, growth, energy transduction, immunological response, nerve processes, intercellular recognition, and biosynthesis. Despite the fundamental importance of biomembranes to many life processes, the quantitative physical knowledge about membranes has been rather sparse compared to the information that is presently available about other cellular components. However, in recent years, an increasing number of chemists and physicists have taken a growing interest in the study of biological membranes and related systems such as lipid bilayers, soaps, and liquid crystals. Using mainly experimental methods that are well known from condensed-matter physics, these researchers have provided important new information about the structural and dynamical properties of biomembranes. This new approach may be considered as being complementary to the more conventional, phenomenological biochemical and biophysical study of the functioning of membranes. Both approaches, however, endeavor to describe the relationship between the physiological function and the physical structure of the biological membrane.

From a physical-chemistry point of view, a biological membrane is an extremely complicated system consisting of an aqueous biomolecular planar aggregate of predominantly lipids and proteins. A natural first step would therefore be to study simpler and more well-defined model membranes. In fact, in recent years, the main occupation of the physical membranologists[1–8] has been to provide an experimental description of certain synthetic lipid bilayers, single membrane-bound proteins, and reconstituted cell membranes. As a result of the experimental efforts, these model membranes now present themselves as extremely well-characterized systems with properties that are believed to be common to a great variety of biological membranes. At the same time, the experimental work has highlighted an urgent need for a parallel theoretical activity. The theoretical work should involve construction of mathematical models that allow a quantitative interpretation of the experimental measurements and that may furthermore suggest and guide a more detailed experimental program geared towards an assessment of the relevant parameters describing the structure and function of biomembranes.[9–12]

[a]This paper is an abstract of a talk presented at the Moscow Refusnik Seminar on June 28, 1986. The work was supported by the Danish Natural Science Research Council under Grant No. 5.21.99.72.

Lipid bilayers are model membranes that display phase transitions. The phase transition, which is believed to be of biological importance, is the so-called main transition that takes the bilayer from a low-temperature solid phase (gel) to a high-temperature fluid (liquid-crystalline) phase. The properties of the phase transition depend on the lipid molecular properties, in particular, the acyl chain length and the degree of saturation, as well as the type of polar head. Although many-component systems are composed of a large number of different lipid species and proteins, biological membranes often display clear lipid phase transitions. Striking examples are the phase transition of the plasma membrane of the mycoplasma *Acholeplasma laidlawii*[5] and the phase transition of the hypothalamic phospholipid membranes of the garden lizard *Calotes versicolor*.[13] There exists no clear-cut experimental evidence for the phase transition per se as being important for biological processes. However, a large number of strong indications have accumulated, and it is now generally accepted that the thermodynamic state of the lipid membrane is an important regulator of the physiological functions of the membrane, specifically transport and enzymatic activity.[14] Biological activity almost universally requires that the membranes be in their fluid state; in many cases, just above the transition point.

The lipid bilayer phase transition is a strongly first-order endothermic process signaled by strong anomalies in the thermodynamic and mechanical properties, for example: (i) a specific-heat anomaly containing a large transition entropy ($\sim 15k_B$); (ii) a large area expansion ($\lesssim 20\%$) [and a correspondingly large decrease in bilayer thickness because the volume expansion is very small]; and (iii) an increase in the lateral mobility of the membrane components. The source of the large transition entropy is the rotational isomerism of the lipid chains that become conformationally disordered during the transition (chain melting). A proper theory of the lipid bilayer transition must therefore account for the rotational isomerism of the chains, the anisotropic van der Waals forces between the lipid chains, the polar forces between the polar head groups, the excluded volume interactions, and the interactions with water. Obviously, a theoretical description of lipid bilayer melting is thus going to be difficult, depending on the degree of realism of the theoretical model employed.[10] The current rather advanced stage of the experimental characterization of lipid membranes has required theoretical modeling of a considerable complexity. In order to measure up with the details of the experimental data and (maybe more important) in order to give meaningful guidelines for future experimental programs, the theorists have been forced to treat detailed models characterized by a high degree of physical realism. Calculation of the statistical mechanics of these models is often possible only by use of computer simulation techniques.[11] This is particularly true of the multistate microscopic interaction models of lipid bilayers advanced by Pink and co-workers.[9] Computer simulation of these models has given considerable insight regarding the mechanism of the chain melting transition and its static and kinetic behavior.[11,15] In particular, computer simulation—by its capacity of accessing directly the microconfiguration of the system—has provided information on the inhomogeneous lateral structure of lipid bilayers in the transition region.[16,17]

The next step towards a description of biomembranes is to consider lipid bilayers with biologically important "impurities," such as cholesterol and proteins. Basic questions to be answered are related to the possible existence of a structure-function relationship; that is, how do the lipids modulate the protein function and, conversely, how do the proteins affect the lipid behavior? Formulation of reliable theoretical models for membranes containing intrinsic proteins is hampered by the lack of three-dimensional structural information on membrane-bound proteins. So far, only a couple of proteins have had their three-dimensional structure worked out.[6,18] An important step towards a theory proceeds via construction of models for lipid membranes with intrinsic amphiphilic polypeptides, ion-channels, and antibiotics such as gramicidin-A, filipin, and amphotericin-B. These model systems simulate important physical aspects of lipid-protein interactions. Physical realizations of such model systems are, at present, subject to intensive experimental investigation.[19] A general model, the so-called "mattress model," to describe lipid-protein interactions in membranes has been proposed.[20] This model introduces the concept of matching between the hydrophobic cores of proteins and lipid bilayers. It is suggested that hydrophobic matching is required for optimal functioning of membranes and that this matching manifests itself in the lack of perturbation of the lipid order parameters as is invariably observed in lipid-protein recombinants in the fluid phase.[8,21]

Cholesterol has dramatic effects on lipid membrane properties in the fluid as well as the gel phases. It basically acts as an antifreeze: it disorders the gel phase and orders the fluid phase, thus eliminating the phase transition at high concentrations. The phase diagram of lipid-cholesterol bilayers is still controversial and, at the moment, no really successful theories of lipid-cholesterol interactions have been proposed. From a biological standpoint, it appears that cholesterol—the last step in a long evolutionary chain[22]—is a remarkable molecule[23] shaped by nature over more than $10^9$ years. Its unique biological advantage seems to be related to its capacity of increasing the mechanical stability and thickness of fluid membranes without affecting the microviscosity of the bilayer.

Among the important problems in membranology for physicists to consider in the near future are membrane-membrane interactions in relation to cell fusion[24] and intercellular communication. Membrane fusion involves an intricate interplay between long-ranged electrostatic and dispersion forces and extremely short-ranged hydration forces and the modulation of these forces by the presence of specific cations.

## REFERENCES


1. QUINN, P. J. & D. CHAPMAN. 1980. CRC Crit. Rev. Biochem. **8:** 1.
2. ISRAELACHVILI, J. N., S. MARČELJA & R. G. HORN. 1981. Q. Rev. Biophys. **13:** 121.
3. SEELIG, J. 1981. *In* Membranes and Intercellular Communication. R. Balian, M. Cabre & P. F. Devaux, Eds.: 15. North-Holland. Amsterdam.
4. DEVAUX, P. F. 1983. *In* Biological Magnetic Resonance, vol. 5. L. J. Berliner & J. Reuben, Eds.: 183. Plenum. New York.
5. MCELHANEY, R. N. 1984. Biochim. Biophys. Acta **779:** 1.
6. UNWIN, N. & R. HENDERSON. 1984. Sci. Am. **250**(no. 2): 56.
7. SACKMANN, E. 1984. *In* Biological Membranes, vol. 5, p. 105. Academic Press. New York/London.
8. BLOOM, M. & I. C. P. SMITH. 1984. *In* Progress in Protein-Lipid Interactions. A. Watts & J. J. H. H. M. De Pont, Eds.: 61. Elsevier. Amsterdam.
9. CAILLE, A., D. PINK, F. DE VERTEUIL & M. J. ZUCKERMANN. 1980. Can. J. Phys. **58:** 581.
10. NAGLE, J. F. 1980. Ann. Rev. Phys. Chem. **31:** 157.
11. MOURITSEN, O. G. 1984. *In* Computer Studies of Phase Transitions and Critical Phenomena, chap. 5. Springer-Verlag. Heidelberg.

12. ABNEY, J. R. & J. C. OWICKI. 1985. *In* Progress in Protein-Lipid Interactions. A. Watts and J. J. H. H. M. De Pont, Eds.: 1. Elsevier. Amsterdam.
13. DURAIRAJ, G. & I. VIJAYAKUMAR. 1984. Biochim. Biophys. Acta **770:** 7.
14. MCELHANEY, R. N. 1982. *In* Current Topics in Membranes and Transport, vol. 17, p. 317. Academic Press. New York.
15. MOURITSEN, O. G., D. BOOTHROYD, R. HARRIS, N. JAN, T. LOOKMANN, L. MACDONALD, D. A. PINK & M. J. ZUCKERMANN. 1983. J. Chem. Phys. **79:** 2027.
16. MOURITSEN, O. G. 1983. Biochim. Biophys. Acta **731:** 217.
17. MOURITSEN, O. G. & M. J. ZUCKERMANN. 1985. Eur. Biophys. J. **12:** 75.
18. DEISENHOFER, J., O. EPP, K. MIKI, R. HUBER & H. MICHEL. 1985. Nature **318:** 618.
19. HUSCHILT, J. C., R. S. HODGES & J. H. DAVIS. 1985. Biochemistry **24:** 1377; MORROW, M. R., J. C. HUSCHILT & J. H. DAVIS. 1985. Biochemistry **24:** 5396.
20. MOURITSEN, O .G. & M. BLOOM. 1984. Biophys. J. **46:** 141.
21. RIEGLER, J. & H. MÖHWALD. 1986. Biophys. J. **49:** 1111.
22. BLOCH, K. 1983. CRC Crit. Rev. Biochem. **14:** 47.
23. YEAGLE, P. L. 1985. Biochim. Biophys. Acta **822:** 267.
24. PARSEGIAN, V. A., R. P. RAND & D. GINGELL. 1984. *In* Cell Fusion, p. 1. Pitman. London.

# On the Ergodic Decomposition of Gibbs Random Fields for Ferromagnetic Abelian Lattice Models

CHARLES-ED. PFISTER

*Départements de Mathématiques et de Physique*
*Ecole Polytechnique Fédérale*
*CH-1015 Lausanne, Switzerland*

## INTRODUCTION

The notion of Gibbs random field (GRF) was introduced in statistical mechanics by Dobrushin[1] in order to characterize the probability measures that describe equilibrium states. An equivalent characterization was given by Lanford and Ruelle.[2] A basic problem of statistical mechanics is to determine all GRF for a given model. This problem can be solved only in few cases. For example, for the classical Heisenberg model, it is not known whether all ergodic GRF form a single orbit under the action of the symmetry group $SO(3)$ (see reference 3). However, the structure of the set of the translation-invariant GRF is well understood for a class of ferromagnetic lattice models, including such well-known cases as the Ising model, the XY model, and the Potts model.[4] In this discussion, we extend the results of reference 4. We can consider models with arbitrary Abelian compact metrizable symmetry groups.

Our main result can be described as follows: Let $S$ be the (internal) symmetry group of the model and let $S(\beta_0)$ be the largest subgroup of $S$ that leaves all GRF invariant at inverse temperature $\beta_0$. There is a one-to-one correspondence between the ergodic probability measures on $S/S(\beta_0)$ and the extremal translation-invariant GRF (ergodic GRF) if and only if the pressure is differentiable in $\beta$ at $\beta_0$. If the last condition is fulfilled, we say that $\beta_0$ is regular. Because the pressure is a convex function of $\beta$, the set of nonregular $\beta$'s is at most countable. Moreover, using perturbation expansions, it is usually possible to show that large $\beta$ or small $\beta$ are regular.

In the next section, we define the lattice models and consider one example, namely, the XY model without magnetic field. We also define the notion of GRF and recall some standard results that we will use later on. Because our models are defined on a compact Abelian group, it is natural to do a Fourier analysis of the GRF. This analysis is exposed in the third section. It contains the key results for proving our main theorem, which is formulated in the fourth section. The fifth section contains the proof of the results of the third section.

## LATTICE MODELS AND GIBBS RANDOM FIELDS

### *Lattice Models*

A lattice model describes a large system composed of identical small subsystems. These small subsystems are indexed by the elements of a lattice. We always choose the lattice $Z^d$ for the sake of simplicity. The state space of a subsystem is an Abelian compact metrizable group $G$. Thus, for each $x \in Z^d$, $G_x$ is a copy of $G$. Here, an element of $G_x$ is denoted by $\varphi(x)$, and $\underline{\underline{F}}_x$ is the Borel $\sigma$-algebra of $G_x$. The configuration space of the large system is

$$\Omega = \prod_{x \in Z^d} G_x.$$

Its elements are functions $\varphi$, defined on $Z^d$, with values $\varphi(x) \in G$. The $\Omega$ is a compact Abelian group with the product topology; $\underline{\underline{F}}$ is its Borel $\sigma$-algebra. The restriction of $\varphi$ to a subset $\Lambda \subset Z^d$ is denoted by $\varphi(\Lambda)$. It is an element of

$$\Omega(\Lambda) = \prod_{x \in \Lambda} G_x.$$

For any $\Lambda \subset Z^d$, there is a natural decomposition of $\Omega$ as $\Omega(\Lambda) \times \Omega(\overline{\Lambda})$, with $\overline{\Lambda} = Z^d \setminus \Lambda$. Thus, we identify $\varphi \in \Omega$ and $(\varphi(\Lambda), \varphi(\overline{\Lambda})) \in \Omega(\Lambda) \times \Omega(\overline{\Lambda})$. The action of $Z^d$ on $\Omega$ is defined by

$$(\tau_x \varphi)(y) = \varphi(y - x). \tag{2.1}$$

There is also a natural action of $\Omega$ on $\Omega$:

$$(\tau_\psi \varphi)(y) = \varphi(y) + \psi(y). \tag{2.2}$$

An interaction between the subsystems is given by a potential. We use the Fourier decomposition of the potential in order to specify the interaction. Let $\hat{G}_x$ be the dual group of $G_x$ and let $\hat{\Omega}$ be the dual group of $\Omega$. An element $\gamma \in \hat{\Omega}$ is of the form,

$$\gamma(\varphi) = \prod_x \gamma_x(\varphi(x)), \tag{2.3}$$

with $\gamma_x \in \hat{G}_x$, and the product in equation 2.3 contains only a finite number of factors which are different from 1. Let

$$[\gamma] = \{x \in Z^d: \gamma_x \not\equiv 1\}. \tag{2.4}$$

A potential $J$ is a function on $\hat{\Omega}$. We always suppose that $J$ has the following three properties:

(A1) $J(\gamma \circ \tau_x) = J(\gamma), \forall\, x \in Z^d$ (translation invariance);
(A2) $\Sigma_{o \in [\gamma]} |J(\gamma)| < \infty$ (summability);
(A3) $J(\gamma) = J(\gamma^{-1}) \geq 0$ (positivity).

Of course, the main hypothesis is property (A3). This positivity condition corresponds to a ferromagnetic condition.

The Hamiltonian of the model is the formal sum

$$H = - \sum_{\gamma \in \hat{\Omega}} J(\gamma) \cdot \gamma. \tag{2.5}$$

By definition, the internal symmetry group $S$ is the subgroup of $\Omega$ that leaves invariant all $\gamma \in \hat{\Omega}$ with $J(\gamma) > 0$. Thus,

$$S = \{\varphi \in \Omega; \gamma(\varphi) \equiv 1, \forall\, \gamma \text{ with } J(\gamma) > 0\}. \tag{2.6}$$

The $S$ is a closed subgroup of $\Omega$, and its annihilator in $\Omega$ is the subgroup $\Gamma$ of $\hat{\Omega}$ that is generated by the elements $\gamma$ with $J(\gamma) > 0$. Property (A.2) implies that $S$ is globally $Z^d$-invariant.

Let us consider one example (see references 4 and 5 for further examples). Let $G = R/2\pi\, Z$. The $XY$ model without magnetic field is given by the Hamiltonian

$$-k \sum_{\langle xy \rangle} \cos\,[\varphi(x) - \varphi(y)], \qquad k > 0. \tag{2.7}$$

In equation 2.7, we sum over all two-point subsets $\{x, y\}$ with $|x - y| = 1$ (Euclidean norm). The dual of $\Omega$ is the set of functions $\chi_m(\varphi) = \exp im \cdot \varphi$, with $m \cdot \varphi \equiv \Sigma_x\, m(x) \cdot \varphi(x)$ and $m(x) \in Z$, and the support of the function $m$ is finite. Because

$$2 \cos\,[\varphi(x) - \varphi(y)] = \exp\,(i[\varphi(x) - \varphi(y)]) + \exp\,(-i[\varphi(x) - \varphi(y)]), \tag{2.8}$$

it is clear that the Hamiltonian (equation 2.7) is in the class described above. The symmetry group $S$ is given by the elements of $\Omega$ that are constant, that is, $\varphi(x) = \varphi(y)\, \forall\, x, y$. The action of $Z^d$ on $S$ is trivial, with each element of $S$ being a fixed point. Thus, all ergodic probability measures on $S$ are the Dirac measures at $\varphi$, $\varphi \in S$, and $S \simeq R/2\pi\, Z$.

### *Gibbs Random Fields*

Let $\Lambda$ be a finite subset of $Z^d$ and let $\theta \in \Omega$ be given. The energy of the configuration $\varphi(\Lambda)$ in $\Lambda$, with boundary condition $\theta$, is

$$H_\Lambda[\varphi(\Lambda) | \theta(\overline{\Lambda})] = - \sum_{\substack{\gamma \in \Omega \\ [\gamma] \cap \Lambda \neq \phi}} J(\gamma) \cdot \gamma(\varphi), \tag{2.9}$$

where in equation 2.9, $\varphi \equiv (\varphi(\Lambda), \theta(\overline{\Lambda}))$. We can define a finite-volume Gibbs measure on $\Omega(\Lambda)$ by the usual formula,

$$\frac{\exp\,(-\beta H_\Lambda[\varphi(\Lambda) | \theta(\overline{\Lambda})])}{Z_\Lambda(\theta)}\, dm(\varphi(\Lambda)), \tag{2.10}$$

where in equation 2.10, $m(\varphi(\Lambda))$ is the normalized Haar measure on $\Omega(\Lambda)$ and $Z_\Lambda(\theta)$ is the partition function,

$$Z_\Lambda(\theta) = \int_{\Omega(\Lambda)} dm(\varphi(\Lambda)) \exp\,(-\beta H_\Lambda[\varphi(\Lambda) | \theta(\overline{\Lambda})]). \tag{2.11}$$

Of course, if $\theta_1$ and $\theta_2$ coincide on $\overline{\Lambda}$, they give the same Gibbs measure on $\Omega(\Lambda)$. The measure (equation 2.10) uniquely defines a measure on $(\Omega, \underline{\underline{F}})$ that is concentrated on the subset $\{\varphi | \varphi(\overline{\Lambda}) = \theta(\overline{\Lambda})\}$. This measure is denoted by $\mu_\Lambda^\theta$.

*Definition*: Let $I_B$ be the indicator function of $B \in \underline{\underline{F}}$. Let $\underline{\underline{F}}(\overline{\Lambda})$ be the $\sigma$-algebra generated by the cylinder sets with bases in $\overline{\Lambda}$. A probability measure $\mu$ on $(\Omega, \underline{\underline{F}})$ is a

Gibbs random field if and only if for any finite $\Lambda \subset Z^d$ and any $B \in \underline{\underline{F}}$,

$$\mathscr{E}_\mu[I_B|\underline{\underline{F}}(\overline{\Lambda})](\theta) = \mu_\Lambda^\theta(B) \qquad \mu \text{ a.-s..}$$

The set $\Delta(J, \beta)$ of all GRF at inverse temperature $\beta$ is a nonempty metrizable compact set in the weak topology. If $\mu \in \Delta(J, \beta)$, then $\tau_x\mu \in \Delta(J, \beta)$ for all $x \in Z^d$ because the potential is translation-invariant. The measure $\tau_x\mu$ is defined by

$$\int f(\varphi)\, d(\tau_x\mu)(\varphi) := \int f(\tau_x(\varphi))\, d\mu(\varphi). \tag{2.12}$$

Similarly, if $\psi \in S$, $\tau_\psi\mu$ is also in $\Delta(J, \beta)$ because the Hamiltonian is $S$-invariant. In particular, when $\Delta(J, \beta)$ consists of only one measure, this measure is necessarily $Z^d$-invariant and $S$-invariant. By contrast, if there exists a measure $\mu$ in $\Delta(J, \beta)$ that is not $Z^d$-invariant or $S$-invariant, then the symmetry of the model is broken. The next theorem is a standard result in statistical mechanics.[6,7]

*Theorem 1*: The set $\Delta(J, \beta)$ is a Choquet simplex. The set $\Delta_i(J, \beta)$ of all $Z^d$-invariant GRF is also a Choquet simplex. The extremal elements of $\Delta_i$ are ergodic measures and we call them ergodic GRF.

Finally, we define the pressure of the model. It is the function $f(\beta, J)$,

$$f(\beta, J) = \lim_{L\to\infty} \frac{1}{L^d} \ln Z_{\Lambda_L}(\theta), \tag{2.13}$$

where in equation 2.13, $\Lambda_L$ is the subset of $Z^d$:

$$\Lambda_L = \{x = (x^1, \ldots, x^d) \in Z^d; \qquad |x^i| \le L/2\ \forall i\}. \tag{2.14}$$

The limit in equation 2.13 is independent of the choice of $\theta$. The function $f(\beta, J)$ is convex in $\beta$, and we call $\beta_0$ regular if $f(\beta, J)$ is differentiable in $\beta$ at $\beta_0$. The next result is proved in reference 8. The proof is based on the characterization of the $Z^d$-invariant GRF as tangent functionals to the pressure.

*Theorem 2*: The set of nonregular $\beta$'s is at most countable. The $\beta$ is regular if and only if for all $\gamma$, with $J(\gamma) > 0$, and for any $\mu_1$ and $\mu_2$ in $\Delta_i(J, \beta)$, $\mathscr{E}_{\mu_1}(\gamma) = \mathscr{E}_{\mu_2}(\gamma)$.

## FOURIER ANALYSIS OF GIBBS RANDOM FIELDS

We begin by recalling some basic results on the Fourier-Stieltjes transform. (A standard reference is reference 9.) The $\Omega$ is a compact Abelian group and the $\hat{\Omega}$ is a discrete Abelian group. We write the group operation of $\hat{\Omega}$ multiplicatively. The Fourier-Stieltjes transform of a measure $\mu$ on $\Omega$ is the function $\hat{\mu}$ on $\hat{\Omega}$, which is defined by

$$\hat{\mu}(\gamma) = \int_\Omega \overline{\gamma}(\varphi)\, d\mu(\varphi). \tag{3.1}$$

The Fourier-Stieltjes transform uniquely characterizes the measure $\mu$ : $\hat{\mu}_1 = \hat{\mu}_2$ if and only if $\mu_1 = \mu_2$. The weak convergence of the sequence $\mu_n$, $\mu_n \to \mu$ is equivalent to the pointwise convergence of $\hat{\mu}_n$ to $\hat{\mu}$. Moreover, if $\lim_n \hat{\mu}_n(\gamma)$ exists for every $\gamma$, then there exists a measure $\mu$ on $\Omega$ such that $\hat{\mu}_n(\gamma) \to \hat{\mu}(\gamma)$.

Let $S$ be a closed subgroup of $\Omega$ and let $\Gamma$ be its annihilator, that is, $\Gamma = \{\gamma \in \hat{\Omega} : \gamma(\varphi) \equiv 1 \text{ for } \varphi \in S\}$. A measure $\mu$ is $S$-invariant if and only if $\hat{\mu}(\gamma) = 0$ for all

$\gamma \notin \Gamma$. The measure $\mu$ is concentrated on $S$, that is, $\mu(B) = \mu(B \cap S)$ for all $B \in \underline{\underline{F}}$, if and only if $\hat{\mu}$ is constant on the cosets of $\Gamma$, that is, $\hat{\mu}(\gamma \cdot \gamma_0) = \hat{\mu}(\gamma)$ for all $\gamma \in \hat{\Omega}$ and all $\gamma_0 \in \Gamma$.

Let $S \supset S(\beta)$. Then, the annihilator of $S(\beta)$, specifically, $\Gamma(\beta)$, contains $\Gamma$ as a subgroup. The dual of $S$ is isomorphic to $\hat{\Omega}/\Gamma$. For $\gamma \in \hat{\Omega}$, let $\dot{\gamma}$ be the canonical image of $\gamma$ in $\hat{\Omega}/\Gamma$. The annihilator of $S(\beta)$ as a subgroup of $S$ is isomorphic to $\Gamma(\beta)/\Gamma$. Consequently, the dual of $S/S(\beta)$ is isomorphic to $\Gamma(\beta)/\Gamma$. The (normalized) Haar measure on $S$ is written as $m_S$, and if $\varphi \in S$, then $\dot{\varphi}$ is the canonical image of $\varphi$ in $S/S(\beta)$. The (normalized) Haar measure on $S/S(\beta)$ is $m_{S/S(\beta)}$. Let $\nu$ be a measure on $S/S(\beta)$. Then, there is a unique measure $\nu^*$ on $\Omega$ such that

$$\hat{\nu}^*(\gamma) = 0, \qquad \gamma \notin \Gamma(\beta), \tag{3.2}$$

and

$$\hat{\nu}^*(\gamma) = \hat{\nu}^*(\gamma \cdot \gamma_0) = \hat{\nu}(\dot{\gamma}), \qquad \gamma \in \Gamma(\beta), \qquad \gamma_0 \in \Gamma. \tag{3.3}$$

For any continuous function $f$ on $\Omega$, we have

$$\int_\Omega f(\varphi)\, d\nu^*(\varphi) = \int_{S/S(\beta)} d\nu(\dot{\varphi}) \int_{S(\beta)} f(\varphi + \psi)\, dm_{S(\beta)}(\psi). \tag{3.4}$$

The measure $\nu^*$ is $S(\beta)$-invariant and is concentrated on $S$.

Let $J$ be a potential satisfying properties (A1), (A2), and (A3). Let $\beta$ be a fixed inverse temperature. For any $(J, \beta)$, we construct a GRF $\mu^0$ as follows. The family $\{\mu_\Lambda^\theta$, $\Lambda$ finite subset, $\theta(x) \equiv 0\}$ is a net if the subsets $\Lambda$ are ordered by inclusion. This net is always convergent in the weak topology and we set

$$\mu^0 = \lim_\Lambda \mu_\Lambda^0, \qquad \theta \equiv 0. \tag{3.5}$$

The GRF $\mu^0$ has remarkable properties, which are summarized in theorem 3. Let $S(\beta)$ be the subgroup of $S$ that leaves $\mu^0$ invariant.

*Theorem 3*: Let $J$ be a potential satisfying properties (A1), (A2), and (A3). Then,

(a) $\mu^0$ is an extremal GRF in $\Delta(J, \beta)$;
(b) $\mu^0$ is an ergodic GRF;
(c) $\mu^0$ has minimal symmetry in the sense that all GRF are at least $S(\beta)$-invariant;
(d) if $\beta_1 \leq \beta_2$, then $S(\beta_2)$ is a subgroup of $S(\beta_1)$;
(e) the Fourier-Stieltjes transform $\hat{\mu}^0$ of $\mu^0$ is positive, and its support is exactly the annihilator $\Gamma(\beta)$ of $S(\beta)$;
(f) for any $\mu \in \Delta(J, \beta)$, we have $|\hat{\mu}(\gamma)| \leq \hat{\mu}^0(\gamma)\ \forall \gamma \in \hat{\Omega}$;
(g) if $\mu \in \Delta(J, \beta)$ and $\hat{\mu}(\gamma_i) = \hat{\mu}^0(\gamma_i) > 0$, with $i = 1, \ldots,$ then $\hat{\mu}(\gamma) = \hat{\mu}^0(\gamma)$ for all $\gamma$ in the subgroup generated by the $\gamma_i$, $i = 1, \ldots$.

## THE ERGODIC DECOMPOSITION OF GIBBS RANDOM FIELDS

Let $S(\beta) \neq S$ so that the symmetry of the model is broken. We can obtain new extremal GRF by the action of $S$ on $\mu^0$: let $\varphi \in S$; then

$$\mu^\varphi = \tau_\varphi \mu^0 = \lim_\Lambda \tau_\varphi \mu_\Lambda^0 = \lim_\Lambda \mu_\Lambda^\varphi \tag{4.1}$$

is an extremal GRF. Of course, $\mu^{\varphi} = \mu^{\psi}$ if and only if $\varphi - \psi \in S(\beta)$. Because all GRF are $S(\beta)$-invariant, it is natural to consider the action of $S/S(\beta)$ on the set of GRF $\Delta(J, \beta)$. All elements of the orbit of $\mu^0$ obtained by the action of $S/S(\beta)$ are different extremal GRF. Because $\mu^0$ is $Z^d$-invariant, $S(\beta)$ is $Z^d$-invariant; thus, there is a natural action of $Z^d$ on $S/S(\beta)$. Our main result is theorem 4.

*Theorem 4*: Let $J$ be a potential with properties (A1), (A2), and (A3). Let $\beta$ be regular. Then, there is a one-to-one correspondence between the $Z^d$-invariant probability measures on $S/S(\beta)$ and the $Z^d$-invariant GRF. In particular, the ergodic GRFs are in one-to-one correspondence with the ergodic probability measures on $S/S(\beta)$. The correspondence is as follows: if $\mu \in \Delta_i(J, \beta)$, then there exists a measure $\nu$ on $S/S(\beta)$ such that for all continuous functions $f$ on $\Omega$,

$$\int_{\Omega} f(\varphi)\, d\mu(\varphi) = \int_{S/S(\beta)} d\nu(\dot{\psi}) \int_{\Omega} f(\varphi)\, d\mu^{\psi}(\varphi). \tag{4.2}$$

We would also like to mention two remarks regarding this:

(1) The first result of this kind for Ising spin models was proved by Slawny.[10]
(2) Equation 4.2 is very simple if $f = \overline{\gamma} \in \Gamma(\beta)$,

$$\hat{\mu}(\gamma) = \hat{\nu}(\dot{\gamma}) \cdot \hat{\mu}^0(\gamma) \tag{4.3}$$

Let us consider the *XY* model without external magnetic field. In this model, $S(\beta)$ is either $S$ or is trivial.[11,12] Moreover, if $S = S(\beta)$, then $\beta$ is regular. When $S(\beta)$ is trivial, $S/S(\beta) = S$. All ergodic probability measures on $S$ can be parametrized by the elements $\varphi \in R/2\pi\, Z$ (see the subsection entitled LATTICE MODELS). Therefore, all ergodic GRF are parametrized by the elements $\varphi \in R/2\pi\, Z$ when there is symmetry breakdown and when $\beta$ is regular. When the dimension of the lattice is $d = 2$, there is no symmetry breakdown; thus, there is a unique translation-invariant GRF for all $\beta$. When the dimension is $d \geq 3$, there is a unique $\beta_c(d)$ such that for $\beta < \beta_c(d)$, $S(\beta) = S$, and for $\beta > \beta_c(d)$, $S(\beta)$ is trivial.

We now give several characterizations of the regular inverse temperatures. We start with theorem 5.

*Theorem 5*: Let $J$ be a potential with properties (A1), (A2), and (A3). Then, the following statements are equivalent:

(a) $\beta$ is regular;
(b) for any $\gamma \in \hat{\Omega}$, with $J(\gamma) > 0$, and for any $Z^d$-invariant GRF $\mu$, we have $\hat{\mu}(\gamma) = \hat{\mu}^0(\gamma)$;
(c) for any $\gamma \in \Gamma$ (the annihilator of the symmetry group $S$) and for any $Z^d$-invariant GRF $\mu$, we have $\hat{\mu}(\gamma) = \hat{\mu}^0(\gamma)$;
(d) there is a unique GRF that is $Z^d$-invariant and $S$-invariant.

*Remark*: When $\beta$ is not regular, only a subset of $\Delta_i(J, \beta)$ is described by theorem 4: it is the subset of all $Z^d$-invariant GRF $\mu$ such that for all $\gamma$ with $J(\gamma) > 0$, $\hat{\mu}(\gamma) = \hat{\mu}^0(\gamma)$ (see theorem 5b and the proof of theorem 4). In this case, there are at least two $Z^d$-invariant and $S$-invariant GRFs. One of these states is the state

$$\mu = \int_{S/S(\beta)} dm_{S/S(\beta)}(\dot{\psi})\, \mu^{\psi},$$

and another is the GRF $\mu^f$ that is defined with the free boundary condition (see reference 4): for any $\gamma$ with $J(\gamma) > 0$, $\hat{\mu}^0(\gamma) \geq \hat{\mu}^f(\gamma)$, and the inverse temperature $\beta$ is regular if and only if $\hat{\mu}^0(\gamma) = \hat{\mu}^f(\gamma)$ for all $\gamma$ with $J(\gamma) > 0$.

We begin by the proof of theorem 5 and then prove theorem 4.

*Proof of Theorem 5*: The equivalence of 5a and 5b is the content of theorem 2. The equivalence of 5b and 5c follows directly from theorem 3g because $\Gamma$ is generated by the elements $\gamma$ with $J(\gamma) > 0$. If $\mu$ is $S$-invariant, then $\hat{\mu}(\gamma) = 0$ for $\gamma \notin \Gamma$. Therefore, 5c and 5d are equivalent.

*Proof of Theorem 4*: We consider a subset $\Delta^*(J, \beta)$ of $\Delta(J, \beta)$:

$$\Delta^*(J, \beta) = \{\mu \in \Delta(J, \beta) : \hat{\mu}(\gamma) = \hat{\mu}^0(\gamma)\ \forall \gamma, J(\gamma) > 0\}. \tag{4.4}$$

Let $\mu$ be in $\Delta^*(J, \beta)$ and let $\mu'$ be an extremal GRF that enters into the extremal decomposition of $\mu$. By theorem 3f, we have

$$|\hat{\mu}'(\gamma)| \leq \hat{\mu}^0(\gamma) = \hat{\mu}(\gamma), \quad \text{if } J(\gamma) > 0. \tag{4.5}$$

Therefore, we must have $\hat{\mu}'(\gamma) = \hat{\mu}^0(\gamma)$ for $J(\gamma) > 0$. In other words, any element $\mu \in \Delta^*(J, \beta)$ has a unique decomposition into extremal GRFs that belong to $\Delta^*(J, \beta)$. We prove that all extremal GRFs that are in $\Delta^*(J, \beta)$ are the elements of the orbit of $\mu^0$ by the action of $S/S(\beta)$. We already know that such GRFs are extremal GRFs. Therefore, let $\mu$ be an extremal GRF in $\Delta^*(J, \beta)$. Then

$$\mu_* = \int_{S/S(\beta)} dm_{S/S(\beta)}(\varphi)\tau_\varphi \mu \tag{4.6}$$

is a GRF in $\Delta^*(J, \beta)$, and equation 4.6 is its extremal decomposition. However, this decomposition can also be written as

$$\mu_* = \int_{S/S(\beta)} dm_{S/S(\beta)}(\varphi)\mu^\varphi. \tag{4.7}$$

Because the extremal decomposition is unique, we must have $\mu = \mu^\psi$ for some $\psi \in S$. To prove the equivalence of equations 4.6 and 4.7, it is sufficient to compute the Fourier-Stieltjes transforms of the right-hand sides of equations 4.6 and 4.7. The Fourier-Stieltjes transforms vanish for all $\gamma \notin \Gamma(\beta)$ because all GRFs are $S(\beta)$-invariant. If $\gamma \in \Gamma$, then the Fourier-Stieltjes transforms coincide because both GRFs of equations 4.6 and 4.7 are in $\Delta^*(J, \beta)$ (use theorem 3g). Finally, if $\gamma \in \Gamma(\beta)\setminus\Gamma$, then the Fourier-Stieltjes transforms will again vanish. Indeed,

$$\int_{S/S(\beta)} dm_{S/S(\beta)}(\varphi)\tau_\varphi \hat{\mu}(\gamma) = \int_{S/S(\beta)} dm_{S/S(\beta)}(\varphi)\gamma(\varphi) \cdot \hat{\mu}(\gamma) = 0. \tag{4.8}$$

Therefore, there is a one-to-one correspondence between the probability measures $\nu$ on $S/S(\beta)$ and the elements $\mu$ of $\Delta^*(J, \beta)$; the correspondence is such that

$$\mu = \int_{S/S(\beta)} d\nu(\varphi)\mu^\varphi, \tag{4.9}$$

which gives the extremal decomposition of $\mu$.

If $\beta$ is regular, then $\Delta_i(J, \beta) \subset \Delta^*(J, \beta)$ by theorem 5. Because the correspondence given by equation 4.9 is linear, it is clear that there is a one-to-one correspondence between the $Z^d$-invariant (ergodic) GRF and the $Z^d$-invariant (ergodic) probability measures on $S/S(\beta)$.

## PROOF OF THEOREM 3

Let $T = R/2\pi Z$. If $G = T$, then theorem 3 is proved in reference 4. The basic tools used in the proof are correlation inequalities.[4,12,13] If we can establish these inequalities in the general case, then the proof of theorem 3 is the same as in reference 4. For example, if the next lemma is true, then theorems 3a, b, c, and f are true.

*Lemma 6*: Let $J$ be a potential satisfying the summability and positivity conditions of properties (A2) and (A3). Let $\Lambda$ be an arbitrary finite subset of $Z^d$ and let $\gamma \in \hat{\Omega}$, $[\gamma] \subset \Lambda$. Then,

$$\int \gamma(\varphi)\, d\mu_{\Lambda}^{0}(\varphi) \leq \int \gamma(\varphi)\, d\mu_{\Lambda'}^{0}(\varphi) \quad \text{if } [\gamma] \subset \Lambda' \subset \Lambda \tag{5.1}$$

and

$$\int \gamma(\varphi)\, d\mu_{\Lambda}^{0}(\varphi) \geq \cos\alpha \int \operatorname{Re}\gamma(\varphi)\, d\mu_{\Lambda}^{\theta}(\varphi) + \sin\alpha \int \operatorname{Im}\gamma(\varphi)\, d\mu_{\Lambda}^{\theta}(\varphi) \tag{5.2}$$

for all $\theta \in \Omega$ and all $\alpha \in [0, 2\pi]$.

Indeed, if $\Lambda_n$ is any sequence of finite subsets such that $\Lambda_n \subset \Lambda_{n+1}$ and if eventually any finite subset is contained in $\Lambda_n$, then the inequality in equation 5.2 shows that the measures $\mu_{\Lambda_n}^{0}$ converge weakly to a measure $\mu^0$. If the potential $J$ is $Z^d$-invariant, then $\mu^0$ is $Z^d$-invariant because the convergence is independent of the choice of the sequence $\Lambda_n$. Let us suppose that $\mu_{\Lambda_n}^{\theta}$ converges to a measure $\mu$. Let

$$\int \gamma(\varphi)\, d\mu_{\Lambda}^{\theta}(\varphi) \equiv r_{\Lambda} e^{i\alpha_{\Lambda}}. \tag{5.3}$$

Then, equation 5.2 can be written as

$$\int \gamma(\varphi)\, d\mu_{\Lambda}^{0}(\varphi) \geq r_{\Lambda} \cos(\alpha - \alpha_{\Lambda}). \tag{5.4}$$

By choosing $\alpha = \alpha_{\Lambda}$, we get

$$\int \gamma(\varphi)\, d\mu_{\Lambda}^{0}(\varphi) \geq \left| \int \gamma(\varphi)\, d\mu_{\Lambda}^{\theta}(\varphi) \right|. \tag{5.5}$$

Taking the limit $\Lambda_n \uparrow Z^d$, we get

$$\hat{\mu}^{0}(\gamma^{-1}) \geq |\hat{\mu}(\gamma^{-1})|. \tag{5.6}$$

Any extremal GRF $\mu$ can be obtained as a weak limit of measures $\mu_{\Lambda_n}^{\theta}$.[7] Thus, equation 5.6 is true for any extremal GRF, and hence it is true for any GRF. This proves theorem 3f. Clearly, equation 5.6 implies that $\mu^0$ is an extremal GRF. Because $\mu^0$ is $Z^d$-invariant, it is also an ergodic GRF. The inequality in equation 5.6 implies that $\hat{\mu}(\gamma) = 0$ for all $\gamma \notin \Gamma(\beta)$. This proves theorem 3c.

Starting from correlation inequalities valid for finite-volume Gibbs measures, the other statements of theorem 3 are proved in the same way. For details, see reference 4. We show how we can generalize lemma 6 for an arbitrary metrizable compact Abelian group. We use the next theorem, which is proved in reference 9.

*Theorem 7*: In the class of all compact Abelian groups $G$, the following three properties are equivalent:

(a) $G$ is metrizable;
(b) $\hat{G}$ is countable;
(c) $G$ is a closed subgroup of $T^{\omega}$, where $T^{\omega} = T^{N}$.

We first show that lemma 6 is valid for $G = T^\omega$ and then show that it is valid for an arbitrary closed subgroup of $T^\omega$.

Let $G = T^\omega$. An element $\varphi \in G$ is a function $\varphi : \mathrm{N} \to T$. The value of $\varphi$ at $k \in \mathrm{N}$ is $\varphi(k)$. Let $M(\mathrm{N})$ be the set of all Z-valued functions $m$ defined on N such that the support of $m$ is finite. An element $\gamma$ of $\hat{G}$ is of the form,

$$\chi_m(\varphi) = \exp i\, m \cdot \varphi, \tag{5.7}$$

with $m \cdot \varphi \equiv \Sigma_k\, m(k)\varphi(k)$. Thus, $\hat{G}$ is isomorphic to the group $M(\mathrm{N})$, whose group operation is $(m + n)(k) = m(k) + n(k)$. Let $\Omega = \Pi_x\, G_x$. An element of $\Omega$ is a function $\varphi$ defined in $\mathrm{Z}^d \times \mathrm{N}$ with values in $T$. All characters of $\Omega$ are of the form of equation 5.7 with $m \in M(\mathrm{Z}^d \times \mathrm{N})$, which is the set of Z-valued functions on $\mathrm{Z}^d \times \mathrm{N}$ that have a finite support. Here, $m \cdot \varphi = \Sigma_{(x,k)}\, m(x, k)\varphi(x, k)$, with $(x, k) \in \mathrm{Z}^d \times \mathrm{N}$.

Let $\Lambda$ be a finite subset of $\mathrm{Z}^d$. Let $H_\Lambda[\varphi(\Lambda)\,|\,\theta(\Lambda)]$ be defined as in equation 2.9. Let $L \in \mathrm{N}$. We also introduce $H^L_\Lambda[\varphi(\Lambda)\,|\,\theta(\Lambda)]$ by summing (in equation 2.9) only over the $\gamma \in \hat{\Omega}$ such that the associated $m$ satisfies the condition $m(x, k) = 0$ if $k > L$. In equation 2.9, we have absolute convergence and

$$\lim_{L\to\infty} H^L_\Lambda[\varphi(\Lambda)\,|\,\theta(\Lambda)] = H_\Lambda[\varphi(\Lambda)\,|\,\theta(\Lambda)]. \tag{5.8}$$

Let $\mu^\theta_{\Lambda,L}$ be the Gibbs measure defined by $H^L_\Lambda$. Because the Haar measure on $T^\omega$ is a product measure, lemma 6 holds with $\mu^0_{\Lambda,L}$ resp. $\mu^\theta_{\Lambda,L}$ instead of $\mu^0_\Lambda$ resp. $\mu^\theta_\Lambda$. (We simply have the case with $G = T$ on a new lattice $\mathrm{Z}^d \times \mathrm{N}$.) By taking the limit $L \to \infty$, we prove lemma 6 for $G = T^\omega$.

Let $G$ be a closed subgroup of $T^\omega$. The annihilator of $G$ in $\hat{T}^\omega$ is isomorphic to the subgroup $I$ of $M(\mathrm{N})$ defined by

$$I = \{m : \chi_m(\varphi) \equiv 1\ \forall \varphi \in G\}. \tag{5.9}$$

The dual of $G$ is therefore isomorphic to the quotient group $M(\mathrm{N})/I$. In other words, every character of $G$ is of the form of equation 5.7, and the functions $\chi_{m_1}$ and $\chi_{m_2}$ define the same character if and only if $m_1 - m_2 \in I$. Hence, we can proceed as before. The only difference is that the Haar measure on $G \subset T^\omega$ is not a product measure on $T^\omega$. In order to solve this problem, we construct a family of measures on $T^\omega$, $m_{L,\lambda}$, $\lambda \in \mathrm{R}^+$, and $L \in \mathrm{N}$, such that $m_{L,\lambda}$ is absolutely continuous with respect to the Haar measure $m$ of $T^\omega$. We prove that the Haar measure $m_G$ on $G$ is the weak limit of the measures $m_{L,\lambda}$ when $L \to \infty$ and then $\lambda \to \infty$. Moreover, for a fixed $\gamma$, lemma 6 holds for the measures defined on $\Pi_{x\in\Lambda}\, T^\omega$ by equations 2.10 and 2.11 if we replace $H_\Lambda$ by $H^L_\Lambda$ and the product measure $m(\varphi(\Lambda)) = \Pi_x m(\varphi(x))$ by the measure $\Pi_x m_{L,\lambda}(\varphi(x))$ ($L$ must be large enough). Because all functions that we have considered are continuous functions and because $m_{L,\lambda}$ converges weakly to $m_G$, we get lemma 6 in the general case by taking $L \to \infty$ and $\lambda \to \infty$.

Let $I_L = \{m \in I;\ \mathrm{supp}\, m \subset [1, L]\}$. The $I_L$ is a subgroup of $\mathrm{Z} \oplus \cdots \oplus \mathrm{Z}$ ($L$ copies of Z). In particular, $I_L$ is finitely generated. Therefore, we can find finite families $F_L$ of $I$ such that $F_{L_1} \subset F_{L_2}$ if $L_1 < L_2$, where $F_L$ is a family of generators for $I_L$. We also suppose that $m \in F_L$ implies $-m \in F_L$. Let $\lambda > 0$ and let

$$\mathcal{H}_L = -\lambda \sum_{m\in F_L} \chi_m. \tag{5.10}$$

We define $m_{L,\lambda}$ on $T^\omega$ by the formula

$$dm_{L,\lambda}(\varphi) = \frac{\exp[-\mathcal{H}_L(\varphi)]}{\int_{T^\omega} \exp[-\mathcal{H}_L(\psi)]\, dm(\psi)}\, dm(\varphi). \tag{5.11}$$

The measure $m_{L,\lambda}$ is $G$-invariant by construction and it is also invariant under the transformation of $\varphi \rightarrow -\varphi$. The $\mathcal{H}_L$ is a Hamiltonian satisfying the conditions of properties (A2) and (A3). For the measure of equation 5.11, we have Ginibre inequalities.[13] Let

$$C_m(\varphi) = \tfrac{1}{2}[\chi_m(\varphi) + \chi_{-m}(\varphi)]. \tag{5.12}$$

Ginibre inequalities give

$$\int C_{m_1}(\varphi)C_{m_2}(\varphi)\, dm_{L,\lambda}(\varphi) \geq \int C_{m_1}(\varphi)\, dm_{L,\lambda}(\varphi) \int C_{m_2}(\varphi)\, dm_{L,\lambda}(\varphi) \tag{5.13}$$

and

$$\int C_m(\varphi)\, dm_{L,\lambda}(\varphi) \geq \int C_m(\varphi)\, dm_{L',\lambda'}(\varphi), \qquad L \geq L', \quad \lambda \geq \lambda'. \tag{5.14}$$

Because $dm_{L,\lambda}$ is invariant under the transformation of $\varphi \rightarrow -\varphi$, we have

$$\int C_m(\varphi)\, dm_{L,\lambda}(\varphi) = \int \operatorname{Re} \chi_m(\varphi)\, dm_{L,\lambda}(\varphi) = \int \chi_m(\varphi)\, dm_{L,\lambda}(\varphi). \tag{5.15}$$

Therefore,

$$\hat{m}_{L,\lambda}(\chi_m) = 0, \qquad m \notin I, \tag{5.16}$$

and

$$\hat{m}_{L,\lambda}(\chi_m) \geq \hat{m}_{L',\lambda'}(\chi_m), \qquad L \geq L', \quad \lambda \geq \lambda'. \tag{5.17}$$

The inequality of equation 5.17 shows that the weak limit of $m_{L,\lambda}$, $L \rightarrow \infty$, exists and defines a measure $m_\lambda$ on $T^\omega$ that is $G$-invariant (see equation 5.16). Similarly, the limit of $m_\lambda$, $\lambda \rightarrow \infty$, exists and defines a measure $m^*$ that is $G$-invariant. We show that $m^*$ is concentrated on $G$, and therefore $m^*$ is the Haar measure $m_G$. We must show that $\hat{m}^*(\chi_m) = 0$ for $m \notin I$ and $\hat{m}^*(\chi_m) = 1$ for $m \in I$. Let $m \in I$ and let $m \in F_L$ for some $L$. Ginibre inequalities give

$$\int C_m(\varphi)\, dm_{L,\lambda}(\varphi) \geq \frac{\int C_m(\varphi) \exp[2\lambda C_m(\varphi)]\, dm(\varphi)}{\int \exp[2\lambda C_m(\varphi)]\, dm(\varphi)}. \tag{5.18}$$

Thus, for $m \in F_L$, we have

$$\lim_{\lambda\uparrow\infty} \int C_m(\varphi)\, dm_\lambda(\varphi) = \lim_{\lambda\uparrow\infty} \int \chi_m(\varphi)\, dm_\lambda(\varphi) = \hat{m}^*(\chi_{-m}) = \hat{m}^*(\chi_m) = 1. \tag{5.19}$$

Let $m_1$ and $m_2$ be defined as above. From the inequality of equation 5.13, we have

$$\begin{aligned} \hat{m}_\lambda(\chi_{m_1+m_2}) + \hat{m}_\lambda(\chi_{m_1-m_2}) &= \int [C_{m_1+m_2}(\varphi) + C_{m_1-m_2}(\varphi)]\, dm_\lambda(\varphi) \\ &= 2\int C_{m_1}(\varphi)C_{m_2}(\varphi)\, dm_\lambda(\varphi) \geq 2\int C_{m_1}(\varphi)\, dm_\lambda(\varphi) \int C_{m_2}(\varphi)\, dm_\lambda(\varphi) \\ &= 2\hat{m}_\lambda(\chi_{m_1})\hat{m}_\lambda(\chi_{m_2}). \end{aligned} \tag{5.20}$$

Because $|\hat{m}_\lambda(\chi_m)| \leq 1$, we have

$$\hat{m}^*(\chi_{m_1+m_2}) = \hat{m}^*(\chi_{m_1-m_2}) = 1$$

in the limit $\lambda \to \infty$. This finishes the proof.

## ADDITIONAL REMARKS

(1) Theorem 3 was first established in the Ising case by Lebowitz in reference 14. Our proof in the general case is directly inspired by this work (see reference 4).

(2) Theorem 4 was proved for the *XY* model in reference 15. It can be generalized to cases where $\Omega$ is not a group. See references 5 and 16.

(3) In reference 17, a version of theorem 4 (theorem 5 of reference 17) is given in a more general setting. The notion of $\beta$ regular must be defined as follows: $\beta$ is regular if (a) there exists a unique GRF that is $Z^d$-invariant and $S$-invariant; (b) this GRF is an extremal element of the convex set of all $S$-invariant GRFs. (In reference 17, the definition of $\beta$ regular was unfortunately not correctly stated.) Of course, this definition coincides with the definition given in the present article when the potential is ferromagnetic.

## REFERENCES

1. DOBRUSHIN, R. L. 1968. Funct. Anal. Appl. **2:** 31–43; **2:** 44–57; 1969. Funct. Anal. Appl. **3:** 27–35.
2. LANFORD, O. E. & D. RUELLE. 1969. Commun. Math. Phys. **13:** 194–215.
3. SIMON, B. 1984. *In* Perspectives in Mathematics. W. Jäger, J. Moser & R. Remmert, Eds.: 421–454. Birkhäuser. Basel.
4. PFISTER, C. E. 1982. Commun. Math. Phys. **86:** 375–390.
5. PFISTER, C. E. 1985. *In* Infinite Dimensional Analysis and Stochastic Processes. S. Albeverio, Ed.: 98–118. Pitman. London.
6. SINAI, Y. 1982. Theory of Phase Transitions: Rigorous Results. Pergamon. Elmsford, New York.
7. FÖLLMER, H. 1975. *In* Seminaire de Probabilités IX, p. 305–317; Lectures Notes in Mathematics, p. 465. Springer-Verlag. Berlin, Heidelberg, New York.
8. ISRAEL, R. B. 1979. Convexity in the Theory of Lattice Gases. Princeton Univ. Press. Princeton, New Jersey.
9. RUDIN, W. 1967. Fourier Analysis on Groups. Interscience. New York.
10. SLAWNY, J. 1974. Commun. Math. Phys. **35:** 297–305.
11. BRICMONT, J., J. R. FONTAINE & L. J. LANDAU. 1977. Commun. Math. Phys. **56:** 281–296.
12. MESSAGER, A., S. MIRACLE-SOLE & C. E. PFISTER. 1978. Commun. Math. Phys. **58:** 19–29.
13. GINIBRE, J. 1970. Commun. Math. Phys. **16:** 310–328.
14. LEBOWITZ, J. L. 1977. J. Stat. Phys. **16:** 463–476.
15. FRÖHLICH, J. & C. E. PFISTER. 1983. Commun. Math. Phys. **89:** 303–327.
16. BRICMONT, J., J. L. LEBOWITZ & C. E. PFISTER. 1981. J. Stat. Phys. **24:** 269–277.
17. GRUBER, C. & C. E. PFISTER. 1985. *In* Spontaneous Symmetry Breakdown and Related Subjects. L. Michel, J. Mozrzymas & A. Pekalski, Eds.: 27–60. World Scientific. Singapore.

# Global Surjection and Global Inverse Mapping Theorems in Banach Spaces

A. D. IOFFE

It is well known that a nonlinear operator that is one-to-one locally (i.e., in a neighborhood of any point at which it is defined) may fail to be one-to-one on the entire domain, even if the latter is very regular (say, a convex body; see reference 1, p. 25). Global inverse mapping theorems, therefore, are of considerable interest to us. We refer to reference 1 for a detailed discussion that is mainly concerned with the finite dimensional situation and its applications. It is probably in reference 2 that a general theorem for $C^1$-maps in Banach spaces was first proved. The assumption of the theorem is a combination of the standard criterion for local univalence and a condition of a global nature. The second principal result of this paper has a similar structure, but it deals with arbitrary continuous mappings and uses a slightly weakened form of the global condition (in this part, however, without changing essentially the techniques developed by Plastock[2]). This theorem is accompanied by a number of local univalence criteria for nondifferentiable maps.

The first result that we prove here is a global surjection theorem that offers a lower estimate for the image of a map with a closed graph (which is a guaranteed radius of a ball contained within the image). As with the inverse map theorem of which we spoke of above, this one uses a combination of a local sufficient condition (this time for surjection) and a global condition similar to that in the other theorem. We refer to references 3 and 4 for specific local surjection criteria for nondifferentiable maps.

In what follows, $X$ and $Y$ are Banach spaces, and $F$ is a map from the whole of $X$ into $Y$.

## GLOBAL SURJECTION THEOREM

We set[4]

$$\text{Sur}\,(F, x)(t) = \sup\{r \geq 0\colon B[F(x), r] \subset F[B(x, t)]\}, \qquad t > 0$$

(the modulus of surjection of $F$ at $x$), where $B(w, a)$ is the closed ball of radius $a$ around $w$. Thus, for any $t > 0$, the value of the modulus of surjection of $F$ at $x$ is the maximal radius of a ball around $y$ contained in the $F$-image of the ball of radius $t$ around $x$.

We further introduce the constant of surjection of $F$ at $x$:

$$\text{sur}\,(F, x) = \liminf_{t \to 0} t^{-1}\,\text{Sur}\,(F, x)(t).$$

Obviously, $\text{sur}\,(F, x) > 0$ is a sufficient condition for $F$ to be surjective at $x$, that is, for $\text{Sur}\,(F, x)(t)$ to be positive for small $t$.

Theorem 1: Suppose the graph of $F$ is closed and there is a positive lower semicontinuous (l.s.c.) function $m(t)$ on $[0, \infty]$ such that

$$\text{sur}\,(F, x) \geq m(\|x\|),\ \forall x. \tag{1}$$

Then,

$$\text{Sur}\,(F, 0)(t) \geq \int_0^r m(t)\,dt,\ \forall r > 0.$$

Proof: For simplicity, we suppose that $F(0) = 0$. Fix a $y \in Y$ and set

$$k_y(r) = \sup\,\{\lambda \geq 0: \mu y \in F(rB_X),\ \forall \mu \in [0, \lambda]\}.$$

It suffices to show that

$$k_y(r) \geq \int_0^r m(t)\,dt \tag{2}$$

if $\|y\| = 1$. For $r = 0$, this is obvious, so we suppose that $r > 0$.

As a function of $r$, $k_y(r)$ is nondecreasing. Therefore, to prove equation 2, it suffices to show that

$$d^-k_y(t) = \liminf_{s \to +0} s^{-1}[k_y(t + s) - k_y(t)] \geq m(t) \tag{3}$$

for all $t > 0$.

Consider a sequence $\{x_n\}$ such that $\|x_n\| \leq r$ and $F(x_n) = \mu_n y$, where $\mu_n \to k_y(r)$. We claim that $\|x_n\| \to r$.

Indeed, assuming the contrary, we find a subsequence of $\{x_n\}$ with norms separated from $r$. With no loss of generality, we assume that the subsequence coincides with the entire sequence. Thus, there is $\tau > 0$ such that $\|x_n\| \leq r - \tau$ for all $n$. Set $m = \inf\,\{m(t): 0 \leq t \leq r\}$. Then $m > 0$ because $m(t)$ is l.s.c. and positive. Fix a positive $\gamma < \tau$. Then $\|x\| \leq r$ if $\|x - x_n\| \leq \gamma$ for some $n$. By this assumption, sur $(F, x) \geq m$ for any such $x$. It follows from proposition 2 of reference 4 that Sur $(F, x_n)(\gamma) \geq m\gamma$. Therefore, for any positive $\delta < \gamma$, we have

$$\mu_n y + (m\delta)B_Y \subset F(x_n + \delta B_X) \subset F(rB_X).$$

In particular, $\lambda y \in F(rB_X)$ if $\lambda < \mu_n + m\gamma$, which contradicts the definition of $k_y(r)$ because $\mu_n + m\gamma > k_y(r)$ if $n$ is sufficiently large. The contradiction proves the claim.

Take an arbitrary $\delta > 0$ and set $m_\delta(r) = \inf\,\{m(t): |t - \tau| < \delta\}$. Clearly, $m_\delta(r) \to m(r)$ as $\delta \to 0$. Take a positive $s < \delta$ and set $s_n = -r + \|x_n\| + s$. Then,

$$r - \delta \leq \|x_n\| - s_n, \quad \|x_n\| + s_n \leq r + s \leq r + \delta.$$

By proposition 2 of reference 4,

$$F(x_n) + \gamma B_Y \subset F(x_n + s_n B_X) \subset F[(r + s)B_X]$$

if $s_n \leq m_\delta(r)$ [because sur $(F, x) \geq m_\delta(r)$ if $\|x - x_n\| \leq s_n$]. It follows that $\lambda y \in F[(r + s)B_X]$ if $\lambda < \mu_n + m_\delta(r)$, so $k_y(r + s) \geq \mu_n + s_n m_\delta(r)$. Consequently, $k_y(r + s) \geq k_y(r) + s m_\delta(r)$, which implies $d^-k_y(r) \geq m_\delta(r)$ for any $\delta > 0$; thus, equation 3 (QED).

By replacing equation 1 by one or another local surjection criterion, we obtain various specific surjection theorems as corollaries. Let $\mathcal{A}$ be a homogeneous set-valued mapping from $X$ into $Y$. We set

$$C(\mathcal{A}) = \sup_{\|y\|=1} \inf \{\|x\|: y \in \mathcal{A}(x)\},$$

$$C^*(\mathcal{A}) = \inf \{\|y\|: y \in \mathcal{A}(x), \|x\| = 1\}.$$

Corollary 1.1: Suppose $F$ has a closed graph and is everywhere Gateaux differentiable. If there is a positive l.s.c. function $m(t)$ on $[0, \infty)$ such that

$$C[F'(x)]m(\|x\|) \leq 1, \forall x,$$

then the conclusion of the theorem holds.

Proof: Because $m(\ )$ is l.s.c., it follows from corollary 1.6 of reference 4 that sur $(F, x) \geq m(\|x\|)$ for all $x$. A more general result can be obtained if we apply theorem 2 of reference 4.

Corollary 1.2: Given a subdifferential $\partial$ on the class of l.s.c. functions, if $F: X \rightarrow Y$ has a closed graph and there is a positive l.s.c. function $m(t)$ on $[0, \infty)$ such that

$$C^*(D^*[F(x)]) \geq m(\|x\|), \forall x,$$

then the conclusion of the theorem holds. Here, $D^*F(x)$ is the coderivative of $F$ associated with $\partial$,

$$D^*F(x)(y^*) = \{x^*: (x^*, y^*) \in \partial \chi_{\text{Graph}F}[x, F(x)]\},$$

and $\chi_S$ is the indicator function of $S$ (which is the function equal to 0 on $S$ and equal to $\infty$ outside of $S$). We refer to reference 4 for those (few) properties of subdifferentials that are needed for applications like corollary 1.2. In fact, every subdifferential now used has these properties. It has to be mentioned, however, that the smaller the subdifferential, the better is the result. Therefore, the $G$-subdifferential described in reference 4 and the Dini subdifferential are preferable.

We mention one more corollary. Corollary 1.3: Suppose $F$ is locally Lipschitz and $Y^*$ has an equivalent uniformly convex norm. Then, the conclusion of the theorem holds provided that

$$x^* \in \partial_A(y^* \circ F)(x) \Longrightarrow \|x^*\| \geq m(\|x\|), \forall x,$$

where $m(\ )$ is as mentioned above and $\partial_A$ is the $A$-subdifferential (see references 4 and 5, and references cited therein).

## GLOBAL INVERSE MAPPING THEOREM

Theorem 2: Suppose that $F$: $X \rightarrow Y$ is a continuous mapping that is locally one-to-one (i.e., every $x$ has a neighborhood in which $F$ is one-to-one) and there exists a positive l.s.c. function $m(t)$ on $[0, \infty)$ such that

$$\int_0^\infty m(t)\, dt = \infty \qquad \text{and} \qquad \text{sur}\,(F, x) \geq m(\|x\|), \quad \forall x.$$

Then $F$ is a homeomorphism onto $Y$, the inverse mapping $F^{-1}$ is locally Lipschitz, and, for every $y$, the Lipschitz constant of $F^{-1}$ at $y$ is not greater than $m(\|F^{-1}(y)\|)^{-1}$.

Proof: According to Plastock,[2] $F$ is a homeomorphism onto $Y$ if and only if $F$ is a local homeomorphism and the following property holds:

Property I. If $p(t)$: $[0, \alpha) \to X$ is a continuous curve such that

$$F[p(t)] = ty_1 + (1 - t)y_2,$$

then there is a sequence $t_n \to \alpha$ such that $\lim p(t_n)$ exists.

We begin by showing that for every $x$, the local inverse of $F$ (we also denote it $F^{-1}$ for simplicity) is a Lipschitz mapping with a Lipschitz constant at $y = F(x)$ not exceeding $m(\|x\|)^{-1}$.

Indeed, fix an $\epsilon \in [0, m(\|x\|)]$. Then sur $(F, u) \geq m(\|x\|) - \epsilon = m_\epsilon$ for all $u$ of a neighborhood of $x$. Using proposition 1.2 of reference 4, we can find a $t_0 > 0$ such that $F$ is one-to-one on $B(x, 2t_0)$ and Sur $(F, u)(t) \geq m_\epsilon t$ if $\|u - x\| \leq t_0$, $0 < t \leq t_0$.

Take a $u$ such that $\|u - x\| \leq t_0$, set $\mathbf{v} = F(u)$, and let $z$ be such that $\|z - \mathbf{v}\| = m_\epsilon t$, $t \leq t_0$. Then, there is a unique $w$ such that $z = F(w)$ and $\|w - u\| \leq t = \|z - \mathbf{v}\|/m$. It follows that

$$\|F^{-1}(z) - F^{-1}(y)\| \leq m_\epsilon^{-1}\|y - \mathbf{v}\|$$

if $z$ and $\mathbf{v}$ are sufficiently close to $y = F(x)$.

Thus, $F^{-1}$ is Lipschitz and $F(\quad)$ is a local homeomorphism. We also have $m_\epsilon^{-1} \to m(\|x\|)^{-1}$ when $\epsilon \to 0$.

Suppose now that $p(\quad)$ is given as in Property I. Fix a $t \in [0, \alpha)$. It follows from the just proven property that for any $\epsilon > 0$, there is a $\delta > 0$ such that

$$\begin{aligned}\|p(\tau) - p(\tau')\| &\leq (m[\|p(t)\|] - \epsilon)^{-1}\|F[p(\tau)] - F[p(\tau')]\| \\ &= (m[\|p(t)\|] - \epsilon)^{-1}|\tau - \tau'| \cdot \|y_2 - y_1\| \end{aligned} \quad \mathbf{(4)}$$

if $|\tau - t| < \delta$, $|\tau' - t| < \delta$, and $\tau, \tau' < \alpha$. This implies that the function $t \to p(t)$ is locally Lipschitz on $[0, \alpha)$ and that

$$\frac{d}{dt}\|p(t)\| \leq \frac{\|y_2 - y_1\|}{m[\|p(t)\|]}$$

almost everywhere on $[0, \alpha)$. In particular, for every positive $\bar{t} < \alpha$, we have

$$\int_{\|p(0)\|}^{\|p(\bar{t})\|} m(r)\, dr = \int_0^{\bar{t}} m[\|p(t)\|]\, d\|p(t)\| \leq \bar{t}\|y_2 - y_1\| \leq \alpha\|y_2 - y_1\|.$$

By this assumption, the integral of $m(\quad)$ between 0 and $\infty$ equals $\infty$. It follows that $\|p(t)\| \leq k < \infty$ for all $t \in [0, \alpha)$. In view of equation 4, this implies that $p(\quad)$ itself is locally Lipschitz on $[0, \alpha)$ with a Lipschitz constant nowhere greater than $c = \|y_2 - y_1\|/m$, where $m$ is the lower bound of the values of $m(\quad)$ corresponding to $r \leq \max\{\|y_2\|, \|\alpha y_1 + (1 - \alpha)y_2\|\}$. The latter, in turn, implies that $p(\quad)$ is Lipschitz on the whole of $[0, \alpha)$ with a Lipschitz constant not greater than $c$. Indeed, let $\bar{t}$ be the upper bound of those $t \in [0, \alpha)$ for which $\|p(\tau) - p(\tau')\| \leq c|\tau - \tau'|$ if $0 \leq \tau, \tau' \leq t$. We claim that $\bar{t} = \alpha$.

Next, if we assume the contrary, then, as follows from equation 4, a $\delta > 0$ exists such that $\bar{t} + \delta < \alpha$ and $\|p(\tau) - p(\tau')\| \leq c|\tau - \tau'|$ if $|\tau - \bar{t}| < \delta, |\tau' - \bar{t}| < \delta$. Suppose now that $\tau', \tau'' \leq \bar{t} + \delta$ and $\tau' < \tau''$. If $\tau'' < \bar{t}$, then

$$\|p(\tau') - p(\tau'')\| \leq c|\tau' - \tau''| \tag{5}$$

by definition of $\bar{t}$; if $\tau' > \bar{t} - \delta$, then equation 5 is valid due to the choice of $\delta$. Finally, for $\tau'' \geq \bar{t}, \tau' \leq \bar{t} - \delta$, we choose a $\tau(\bar{t} - \delta, \bar{t})$ and have

$$\begin{aligned}\|p(\tau') - p(\tau'')\| &\leq \|p(\tau') - p(\tau)\| + \|p(\tau) - p(\tau'')\| \\ &\leq c(\tau - \tau') + c(\tau'' - \tau) = c|\tau'' - \tau'|.\end{aligned}$$

Thus, $p(\ \ )$ is a Lipschitz curve and, therefore, $\lim_{t\to\alpha} p(t)$ exists. This completes the proof of the theorem.

Remark: We observe that the Lipschitz constant of the inverse map is completely defined by surjection properties of $F$, while the property of being one-to-one enters the statements in a purely qualitative way.

## CRITERIA FOR LOCAL UNIVALENCE

Suppose that $F: X \to Y$ is defined in a neighborhood of $x_0$. Recall that a homogeneous set-valued mapping $\mathcal{A}$ from $X$ into $Y$ [i.e., such that $\mathcal{A}(\lambda x) = \lambda\mathcal{A}(x)$ for $\lambda > 0$] is a strict prederivative of $F$ at $x_0$ if

$$F(x + h) \subset F(x) + \mathcal{A}(h) + r(x; h)\|h\|B, \tag{6}$$

where $r(x; h) \to 0$ as $x \to x_0$ and $h \to 0$. We refer to reference 6 for more details.

Proposition 1: Suppose $F$ has a strict prederivative $\mathcal{A}$ at $x_0$ such that $\rho[0, \mathcal{A}(h)] \geq c\|h\|$ $(c > 0)$. Then, $F$ is one-to-one in a neighborhood of $x_0$. [Here, $\rho(x, S)$ is the distance from $x$ to $S$.]

Proof: Take an $\epsilon > 0$ such that $r(x; h) < c$ for $\|x - x_0\| < \epsilon$, $\|h\| < 2\epsilon$. If $u$ and $x$ belong to the $\epsilon$-ball around $x_0$ and $h = u - x$, then $0 \in \mathcal{A}(h) + r(x; h)\|h\|B$ (otherwise, $\mathcal{A}(h)$ would contain a vector with a norm less than $c\|h\|$) and consequently $F(u) = F(x + h) \neq F(x)$.

A linear operator $A$ is a strict prederivative if and only if it is a strict derivative [i.e., if $\|h\|^{-1}\|F(x + h) - F(x) - Ah\| \to 0$ as $h \to 0, x \to x_0$] The condition $\rho[0, \mathcal{A}(h)] \geq c\|h\|$ assumes the form $\|Ah\| \geq c\|h\|$, and it means that $A$ is a linear homeomorphism onto a closed subspace of $Y$ and the norm of the inverse operator (from the subspace into $X$) is not greater than $c^{-1}$.

Thus (in view of the Banach open mapping theorem), proposition 1 contains the fact that $F$ is one-to-one near $x_0$ provided that $F$ is strictly differentiable at $x_0$, ker $F'(x_0) = \{0\}$, and Im $F'(x_0)$ is a closed subspace of $Y$.

In general, the condition $\rho[0, \mathcal{A}(h)] \geq c\|h\|$ is equivalent to $C^*(\mathcal{A}) = \inf\{\|y\|: y \in \mathcal{A}(h), \|h\| = 1\} \geq c$. If $\mathcal{A}$ is convex-valued, then

$$C^*(\mathcal{A}) = -\sup_{\|h\|=1}\inf_{\|y^*\|\leq 1} s(y^*, h),$$

where $s(y^*, h) = \sup\{\langle y^*, y\rangle : y \in \mathcal{A}(h)\}$ is the support function of $\mathcal{A}$ (see reference 6). The following proposition is obvious.

Proposition 2: If $\mathcal{A}_1$, $\mathcal{A}_2$ are strict prederivatives of $F$ at $x$, then so is $\mathcal{A}(x) = \mathcal{A}_1(x) \cap \mathcal{A}_2(x)$. If $\mathcal{A}_1(x) \subset \mathcal{A}_2(x)$ for all $x$, then $C^*(\mathcal{A}_1) \geq C^*(\mathcal{A}_2)$.

Strict prederivatives can often be represented in the form

$$\mathcal{A}(h) = \mathrm{cl}\,\{Ah: A \in \mathfrak{a}\}, \tag{7}$$

where $\mathfrak{a}$ is a collection of linear continuous operators and "cl" denotes the closure operation. This is particularly true for $D^*F(x)$ if $Y$ is reflexive and $F$ is Lipschitz near the point in question.

Proposition 3: Let $\mathcal{A}$ be a homogeneous set-valued mapping defined by equation 7. Then $C^*(\mathcal{A}) > 0$ if and only if every $A \in \mathfrak{a}$ is a homeomorphism onto a closed subspace of $Y$ (depending on $A$) and the norms of the inverse operators are bounded by the same constant for all $A \in \mathfrak{a}$. In this case,

$$C^*(\mathcal{A}) = \inf\{\|A^{-1}\|^{-1}: A \in \mathcal{A}\}.$$

Proof: We have

$$\begin{aligned} C^*(\mathcal{A}) &= \inf\{\|y\|: y \in \mathcal{A}(h),\ \|h\| = 1\} \\ &= \inf\{\|Ah\|: A \in \mathfrak{a},\ \|h\| = 1\} \\ &= \inf_{A\in\sigma} \inf\{\|Ah\|: \|h\| = 1\} = \inf_{A\in\mathfrak{a}} C^*(A). \end{aligned}$$

Thus, $C^*(\mathcal{A}) \geq k > 0$ if and only if $C^*(A) \geq k$ for all $A \in \mathfrak{a}$. The latter means that every $A$ is a homeomorphism onto a closed subspace of $Y$ and $\|A^{-1}\| \leq 1/k$.

For locally Lipschitz mappings, an important class of strict prederivatives can be defined in terms of the generalized gradients of Clarke functions $y^*\circ F$, $y^* \in Y^*$. Namely, for any $\epsilon > 0$, the function

$$s_\epsilon(y^*, h) = \sup_{\|x-x_0\|<\epsilon} d^0(y^*\circ F)(x; h)$$

[where $d^0\varphi(x; h)$ stands for Clarke's directional derivative of $\varphi$ at $x$] is the support function of a strict prederivative of $F$ at $x_0$ that will be denoted $D^0_\epsilon F(x_0)$:

$$D^0_\epsilon F(x_0)(h) = \{y: \langle x^*, h\rangle \leq s_\epsilon(y^*, h),\ \forall y^*\}.$$

This prederivative is by definition convex-valued and, moreover, it has an additional property of being a fan [a convex-valued homogeneous mapping such that $\mathcal{A}(x + u)$ belongs to the closure of $\mathcal{A}(x) + \mathcal{A}(u)$ for every $x$, $u$.[6]]

If $X$ and $Y$ are both finite dimensional and $F$ is Lipschitz near $x_0$, then

$$s(y^*, h) = \sup\{\langle x^*, h\rangle: x^* \in \partial_c(y^*\circ F)(x_0)\}$$

is the support function of the smallest strict prederivative of $F$ at $x_0$ (where $\partial_c$ denotes Clarke's generalized gradient).

By $C^*(F^*, x)$, we denote the upper bound of "dual Banach constants" $C^*(\mathcal{A})$ over the collection of all strict prederivatives $\mathcal{A}$ of $F$ at $x$. Then, proposition 1 can be reformulated in the following way: if $C^*(F, x) > 0$, then $F$ is one-to-one in a neighborhood of $x$.

If $F$ is strictly (or continuously) differentiable at $x$, then $C^*(F, x) = C^*[F'(x)]$ easily follows from proposition 2. It is difficult, however, to find a recipe to calculate $C^*(F, x)$ in a more general situation. In certain important cases, a smaller quantity is calculable: namely, the upper bound of $C^*(\mathcal{A})$ over all convex-valued strict prederivatives. We shall denote it by $c^*(F, x)$.

Proposition 4: Suppose $F$ is Lipschitz near $x_0$. Then,

$$c^*(F, x_0) = \lim_{\epsilon \to 0} C^*[D_\epsilon^0 F(x_0)]$$

$$= -\lim_{\epsilon \to 0} \sup_{\|h\|=1} \inf_{\|y^*\| \leq 1} \sup_{\|x - x_0\| < \epsilon} d^0(y^* \circ F)(x; h).$$

In addition, if $\dim X = \dim Y < \infty$, then

$$c^*(F, x_0) = C^*[D^0 F(x_0)] = -\sup_{\|h\|=1} \inf_{\|y^*\| \leq 1} d^0(y^* \circ F)(x_0; h),$$

and in cases where this quantity is positive,

$$c^*(F, x_0) = \inf\{\|A^{-1}\|^{-1}: A \in \partial_c F(x_0)\},$$

where $\partial_c F(x)$ is Clarke's generalized gradient of $F$ at $x$.

Proof: The first equality of $c^*(F, x_0) = \lim C^*[D_\epsilon^0 F(x_0)]$ follows from proposition 9.8 of reference 6; the rest follow from the discussion preceding the statement.

Proposition 5: Suppose $F$ is Gateaux differentiable everywhere in a neighborhood of $x_0$. Then,

$$c^*(F, x_0) = -\lim_{\epsilon \to 0} \sup_{\|h\|=1} \inf_{\|y^*\| \leq 1} \sup_{\|x - x_0\| < \epsilon} \langle y^*, F'(x)h \rangle. \tag{8}$$

Remark: In the case where $F$ is continuously Gateaux differentiable, proposition 5 is an immediate consequence of proposition 4 because, in this case,

$$d^0(y^* \circ F)(x; h) = \langle y^*, F'(x)h \rangle.$$

Proof: Set

$$s_\epsilon(y^*, h) = \sup\{\langle y^*, F'(x)h \rangle: \|x - x_0\| < \epsilon\}.$$

This function is sublinear and l.s.c. in each argument; hence, the set

$$\mathcal{A}_\epsilon(h) = \{y: \langle y^*, y \rangle \leq s_\epsilon(y^*, h), \forall y^*\}$$

is (obviously convex-closed and) nonempty for any $h$. It follows from the well-known equality

$$\langle y^*, F(u) - F(x) \rangle = \int_0^1 \langle y^*, F'[x + t(u - x)](u - x) \rangle \, dt$$

that $F(u) - F(x) \in \mathcal{A}_\epsilon(u - x)$ if both $u$ and $x$ belong to the $\epsilon$-ball around $x_0$; thus, $\mathcal{A}_\epsilon$ is a strict prederivative of $F$ at $x_0$.

Suppose now that $\mathcal{A}$ is an arbitrary strict prederivative of $F$ at $x_0$; that is, equation 6 holds. Fix a $\delta > 0$ and choose $\epsilon > 0$ so small that $r(x; h) < \delta$ if $\|x - x_0\| < \epsilon$, $\|h\| < \epsilon$.

For such $x$ and $h$,

$$\langle y^*, F'(x)h\rangle = \lim_{t\to 0} t^{-1}\langle y^*, F(x + th) - F(x)\rangle \leq s(y^*, h) + \delta\|h\|$$

[where $s(y^*, h)$ is the support function of $\mathcal{A}$]; hence,

$$s_\epsilon(y^*, h) \leq s(y^*, h) + \delta\|h\|.$$

It follows that $C^*(\mathcal{A}_\epsilon) \geq C^*(\mathcal{A}) - \delta$ and—as $\mathcal{A}$ is an arbitrary convex-valued strict prederivative and $\delta$ is an arbitrary positive number—that

$$\lim_{\epsilon\to 0} C^*(\mathcal{A}_\epsilon) = c^*(F, x_0).$$

It remains for us to recall that $\mathcal{A}_\epsilon$ is convex-valued and to apply the formula for $C^*$ in the case of the convex-valued mapping that was mentioned before the statement of proposition 2.

## REFERENCES

1. PARTHASARATHY, T. 1983. On global univalence theorems. Lecture Notes in Math., vol. 977. Springer-Verlag. Berlin/New York.
2. PLASTOCK, R. 1974. Homeomorphisms between Banach spaces. Trans. Am. Math. Soc. **200:** 169–183.
3. AUBIN, J-P. & I. EKELAND. 1984. Applied Nonlinear Analysis. Wiley–Interscience. New York.
4. IOFFE, A. On the local surjection property. To appear in J. Nonlinear Anal. Theory Methods Appl.
5. IOFFE, A. 1984. Approximate subdifferentials and applications. 1. The finite dimensional theory. Trans. Am. Math. Soc. **281:** 389–416.
6. IOFFE, A. 1981. Nonsmooth analysis: differential calculus of nondifferentiable mappings. Trans. Am. Math. Soc. **266:** 1–56.

# Local Structure of Convex Equilibrium Problems

R. POLYAK

Let $f_i(z)$, $i = \overline{1,p}$ be convex functions in $E^n$ and let $\Omega = \{z: f_i(z) \le 0, i = \overline{1,p}\}$ be a convex compact. The function $\phi(z; Z)$, which is continuous in $(z; Z)$ and concave in $Z$, is defined on the $\Omega \times \Omega$. Consider the following equilibrium problem:

$$z^* = \arg\max \{\phi(z^*; Z) \mid Z \in \Omega\}. \tag{1}$$

The existence of $z^*$ is implied by the Kakutani theorem. The vector $G(z) = \phi'_Z(z; Z)|_{Z=z}$ will be called the pseudogradient and the Jacobi matrix $J[G(z)] = H(z)$ will be called the pseudohessian of function $\phi(z; Z)$; also let us consider the matrix $H(z^*)$ to be nonpositive quasi-defined. If $I^* = \{i: f_i(z^*) = 0\} = \{1, r\}; f(z) \equiv f(x, y) = [f_1(z), \ldots, f_r(z)], x \in E^{N-r}, y \in E^r; f'(z) = J[f(z)] = [f'_x(z), f'_y(z)], r \operatorname{ang} f'(z^*) = r$, then there exist $U_i^* \ge 0$, $i = \overline{1,r}$, $G(z^*) = \Sigma_{i=1}^r u_i^* f_i(x^*)$, and $\det f'_y(z^*) \ne 0$. Therefore, the system $f(x, y) = 0$ determines in $S(z^*; \epsilon) = \{z: \|z - z^*\| \le \epsilon\}$ such a vector-function $y(x)$ that $y(x^*) = [y_1(x^*), \ldots, y_r(x^*)] = y^*$. Let $\varphi(x, X) = \phi[x, y(x); X, y(X)]$ be a restriction of $\phi(z; Z)$ on $\Omega_{z^*} = \{z; f(z) = 0\}$; let $g(x) = \varphi'_X(x; X)|_{X=x}$ be the pseudogradient and $J[g(x)] = h(x)$ be the pseudohessian of the restriction. Let $m_0$, $M_0$, $m_i$, $M_i$, $\mu$, $\mu$ be accordingly the minimal and maximal eigenvalues of matrices $-\frac{1}{2}[H(z^*) + H^T(z^*)]; f''_i(z^*); y'_x(x^*)[y'_x(x^*)]^T$. Then the following theorem holds true:

*Theorem* (about the spectrum of pseudohessian)—If $r \operatorname{ang} f'(z^*) = r$ and the matrices $H(z^*), f''_i(z^*)$, $i = \overline{1,r}$ exist, then for each eigenvalue $\lambda$ of pseudohessian $h(x^*)$, the inequalities

$$-(M_0 + \sum_{i=1}^{r} u_i^* M_i)(1 + \mu) \le \operatorname{Re} \lambda \le -(m_0 + \sum_{i=1}^{r} u_i m_i)(1 + \mu) \tag{2}$$

are satisfied.

If $m_0 + \Sigma_{i=1}^r u_i^* m_i > 0$, and if in a neighborhood of $V(x^*; \delta) = \{x: \|x - x^*\| \le \delta\}$, the pseudohessian $h(x)$ is continuous, then the vector function $g(x)$ is a strongly monotonous operator in $V(x^*; \delta)$, for example,

$$[g(x_1) - g(x_2), x_1 - x_2] \le -m\|x_1 - x_2\|^2, \qquad \forall [x_1, x_2 \in V(x^*; \delta)];$$

if one has $y(x)$ in an explicit form, the solution $z^* = (x^*; y^*)$ of the initial problem (equation 1) can be obtained by searching the solution of the system $g(x) = 0$.

Here, we can use the methods of solving operator equations for strongly monotonous and sufficiently smooth operators and the methods of searching for a minimum of a strongly convex $(m > 0)$ function $\|g(x)\|^2$.

In reference 1, a technique enabling us to use the above-mentioned methods without explicitly defining $y(x)$ has been developed. On the basis of this technique, relaxation operators acting in $S(z^*; \epsilon)$ have been constructed under certain conditions linearly, superlinearly, and quadratically. Usage of these operators in the computational process is realized in the framework of a general method[1] that enables us to construct processes converging to $z^*$ from any initial approximation with the above-mentioned rate for the problem (equation 1).

In conclusion, we note that the following problems are reduced to the initial problem (equation 1):

(1) the convex programming problem $z^* = \arg\max \{\varphi(z) | z \in \Omega\}$ [in this case, $\phi(z; Z) = [\varphi'(z), Z]$; $G(z) = \varphi'(z)$; $H(z) = \varphi''(z)$];
(2) the two-person convex game;
(3) the problem of finding an equilibrium point for Nash's concave $n$-person game;
(4) the Walras-Wald model and other equilibrium models of mathematical economics.

## REFERENCE

1. POLYAK, R. 1978. On finding a fixed point of a class of set-valued mappings. Sov. Math. Dokl. **19:** no. 5.

# Smooth Optimization Methods in Discrete Minimax Problems

R. POLYAK

A class of equivalent problems whose Lagrange functions have certain important properties has been built for the classical discrete minimax problem. The discovered properties are used to construct methods for solving this problem. These methods converge under ordinary assumptions at least linearly with ratio, which can be made as small as necessary at the expense of increasing the penalty factor.

Let $f_i(x)$, $i = \overline{1, m}$: $E^n \longrightarrow R$ be continuously differentiable, let $F(x) = \max \{f_i(x) | i = \overline{1, m}\}$, and let such an $x_0$ exist that $\Omega = \{x{:}F(x) \leq F(x_0)\}$ is bounded. Then

$$x^* = \arg\min \{F(x) | x \in E^n\} \tag{1}$$

exists and the Kuhn-Tucker's conditions,

$$\mathcal{L}'_x(z^*) = \mathcal{L}'_x(x^*, u^*) = \sum u_i^* f_i'(x^*) = 0,$$

$$\sum_{i=1}^{m} u_i^* = 1, \qquad u_i^* \geq 0, \quad i = \overline{1, m}; \quad u_i^*[F(x^*) - f_i(x^*)], \tag{2}$$

are satisfied.

Let $I^* = \{i{:}f_i(x^*) = F(x^*)\} = \{1, \ldots, r\}$, $r < n$, and $f(x) = [f_1(x), \ldots, f_r(x)]$. Also, let $f'(x) = J[f(x)]$ denote the Jacobi matrix of $f(x)$, $e = (1, \ldots, r) \in E^r$.

Let us assume that the ordinary strict regularity conditions for the initial problem (equation 1), that is,

$$r\,\text{ang}\,[f'(x^*), -e^T] = r, \qquad u_i^* > 0, i = \overline{1, r}, \tag{3}$$

$$f'(x^*)y = y_1 e \Longrightarrow [\mathcal{L}''_{xx}(z^*)y, y] \geq \lambda(\|y\|^2 + y_1^2), \qquad \lambda > 0, y_1 \in R, \tag{4}$$

are satisfied. Let us consider the function $\psi(t)$, which is twice differentiable and strictly increasing on $R$ with $\psi''(t) > 0$. For any $\kappa > 0$, the functions $\bar{f}_i(x, \kappa) = \kappa^{-1}\psi[\kappa f_i(x)]$ define the problem

$$x^* = \arg\min \left\{\max_{1 \leq i \leq m} \bar{f}_i(x, \kappa)/x \in E^n\right\}, \tag{1'}$$

which is equivalent to the initial problem. Let the Lagrange function $F(x, u, \kappa) = \kappa^{-1} \cdot \Sigma_{i=1}^m u_i \psi[\kappa f_i(x)]{:}E^n \times S_m \times R_+ \longrightarrow R$ be the Lagrangian of the problem, where $S_m = \{u{:} \Sigma_{i=1}^m u_i = 1; u_i \geq 0, i = 1, m\}$.

It turns out that this function has a number of remarkable properties that $\mathcal{L}(x, u)$ does not possess. Those properties are particularly displayed by the dual function $\varphi(u, \kappa) = \text{ⲙ} = \min \{F(x, u, \kappa) | x \in E^n\}$. Let $F_r(x, u, \kappa) = \kappa^{-1} \Sigma_{i=1}^r u_i \psi[\kappa f_i(x)]$.

*Theorem 1:* Let equations 2–4 be satisfied and let the function $f_i(x)$, $i = \overline{1, r}$ be sufficiently smooth. Then there exists such $\rho > 0$ and such $K_\rho$ that for $\kappa \geq K_\rho$ and for any $u = (u_1, \ldots, u_r; 0, \ldots, 0) \in S(u^*, \rho) = \{u: \|u - u^*\| \leq \rho\}$, the following is true:

(1) there exists the only point $\hat{x} \equiv \hat{x}(u, \kappa) = \arg\min\{F_r(x, u, \kappa) | x \in E^n\}$ on a neighborhood of $x^*$;
(2) $F_r(x, u, \kappa)$ is strongly convex by $x$ on a neighborhood of $\hat{x}$;
(3) for the vectors $\hat{x}$, $\hat{u} = (\hat{u}_1, \ldots, \hat{u}_r)$, and $\hat{u}_i = u_i\psi'[\kappa f_i(\hat{x})](\Sigma_{i=1}^{r} u_i\psi[\kappa f_i(\hat{x})])^{-1}$ (with $i = \overline{1, r}$), the estimates below are valid:

$$\|\hat{x} - x^*\| \leq C\kappa^{-1}\|u - u^*\|; \qquad \|\hat{u} - u^*\| \leq C\kappa^{-1}\|u - u^*\|. \tag{5}$$

Let $\hat{\psi}$ and $U^*$ be diagonal $r \times r$ matrices with elements $\psi[\kappa f_i(\hat{x})]$ and $u_i^*$, respectively, on the main diagonal, and let $\varphi_r(u, \kappa) = \min\{F_r(x, u, \kappa) | x \in E^n\}$.

*Theorem 2:* If $f_i(x)$, $i = \overline{1, r}$ are twice continuously differentiable and equations 2–4 are satisfied, then for any $\kappa \geq K_\rho$:

(1) the function $\varphi_r(u, \kappa)$ is twice continuously differentiable in $u$ on a neighborhood of $S(u^*, \rho)$ and $\varphi'_{ru}(u, \kappa) = \kappa^{-1}(\psi[\kappa f_1(\hat{x})], \ldots, \psi[\kappa f(\hat{x})])$, $\varphi''_{ruu}(u, \kappa) = -\hat{\psi}f'(\hat{x})[F''_{xx}(\hat{x}, u, \kappa)]^{-1}[\hat{\psi}f'(\hat{x})]^T$, and $\psi''_{ruu}(u^*, \kappa) = -\psi[\kappa F(x^*)]f'(x^*)[\mathcal{L}''_{xx}(z^*) + \kappa f'^T(x^*)U^*f'(x^*)]^{-1}f'^T(x^*)$;
(2) $\varphi_r(u, \kappa)$ is strongly concave;
(3) $\varphi_r(u^*, \kappa) = \max\{\varphi_r(u, \kappa) | u \in S(u^*, \rho) \cap S_r\} = \min\{F_r(x, u^*, \kappa) | x \in E^n\}$.

*Theorem 3:* Let all $f_i(x)$ be convex and twice continuously differentiable, let equation 3 be satisfied, and let there be such $i_0 \in I^*$ that $f_{i_0}(x)$ is strongly convex. Then:

(1) $\|x(\kappa) - x^*\| = 0(\kappa^{-1})$; $\|u(\kappa) - u^*\| = 0(\kappa^{-1})$, where for $x(\kappa) = \arg\min\{F(x, \kappa) = \kappa^{-1}\Sigma_{i=1}^{m}\psi[\kappa f_i(x)] | x \in E^n\}$, $u(\kappa) = u_i(\kappa) = \psi'(\kappa f_i[x(\kappa)])[\Sigma_{i=1}^{m}\psi'(\kappa f_i[x(\kappa)])^{-1}$, $i = \overline{1, m}$];
(2) for any fixed $\kappa > 0$, the estimates of equation 5 hold for $\hat{x}$ and $\hat{u}$ when $F_r(x, u, \kappa)$ is replaced by $F(x, u, \kappa)$;
(3) all the statements of theorem 2 remain valid for any $\kappa > 0$ when $\varphi_r(u, \kappa)$ is replaced by $\varphi(u, \kappa)$.

One can avoid solving the unconstrained minimization problem at each step. To that end, one has to use, instead of $\hat{x}$, the vector $\tilde{x}$ defined by the condition

$$\|F'_x(\tilde{x}, u, \kappa)\| \leq \mu\kappa^{-1}\|U(\tilde{x}, u, \kappa) - u\|, \qquad \mu > 0, \tag{6}$$

where $\tilde{u} = U(\tilde{x}, u, \kappa) = [U_i(\tilde{x}, u, \kappa) = u_i\psi'[\kappa f_i(\tilde{x})]\,(\Sigma_{i=1}^{m}u_i\psi'[\kappa f_i(\tilde{x})])^{-1}, i = \overline{1, m}]$.

*Theorem 4:* Under the assumptions of theorem 3, the following estimates are valid:

$$\|\tilde{x} - x^*\| \leq C(1 + \mu)\kappa^{-1}\|u - u^*\|; \qquad \|\tilde{u} - u^*\| \leq C(1 + \mu)\kappa^{-1}\|u - u^*\|.$$

The results obtained are used to construct a general method for solving the initial problem (equation 1). The method is based on unconstrained optimization methods for $\varphi(u, \kappa)$.

Procedures of unconstrained optimization for $\varphi(u, \kappa)$ (as a function of $u$ for fixed $\kappa$) generate some specific realizations of the general method generating sequence $\{z^s =$

$(x^s, u^s)\}_{s=1}^{\infty}$. Moreover, the convergence rate of the sequence is

(1) $\|z^s - z^*\| \leq C\kappa^{-s}q^s$, $q < 1$, or
(2) $\|z^s - z^*\| \leq C\kappa^{-s}q_1 \cdots q_s$, $q_s \rightarrow 0$, or
(3) $\|z^s - z^*\| \leq C\kappa^{-s}q^{2^s}$

if the procedure has, respectively, a geometric, superlinear, or quadratic rate of convergence.

When our method is applied, there is no need (because of certain moment) to solve the unconstrained optimization problem to find $\varphi(u, \kappa)$, $\varphi_u'(u, \kappa)$ as well as the approximations. Instead, one has to solve one step of any quadratically converging method.

Finally, both the transition from the constrained extremum problem of $x^* = \arg\min \{f_0(x) | f_i(x) \leq 0, i = \overline{1, m}\}$ to the equivalent problem of $x^* = \arg\min \{f_0(x) | [\kappa\psi'(0)]^{-1}\psi(\kappa f_i(x)) \leq \psi(0)[\kappa\psi'(0)]^{-1}\}$ and the application of the usual Lagrange function of the last problem enable us to obtain results typical for the modified Lagrange functions.[1-4] In conclusion, let us note that the function $F(x, \kappa)$ with $\psi(t) = \exp t$ was examined in reference 5, and the functions $[F(x, \kappa)]^{1/\kappa}$, $[F(x, u, \kappa)]^{1/\kappa}$ with $\psi(t) = t^{\kappa}$ were studied in references 6–8.

## REFERENCES


1. POLYAK, B. T. & N. V. TRET'JAKOV. 1973. Penalty estimates methods for the conditional extremum problems. Zh. Vychisl. Mat. Mat. Fiz. **13.**
2. KORT, B. & D. BERTSEKAS. 1976. Combined primal dual and penalty methods for convex programming SIAM. J. Control Optimization **14:** no. 2.
3. ROCKAFELLAR, R. 1973. The multiplier method of Hestenes and Powell applied to convex programming. J. Optimization Theory Appl. **12.**
4. GOL'SHTEIN, G. & N. V. TRET'JAKOV. 1980. On the modified Lagrange functions of convex programming problems. Econ. Math. Methods **XVI:** 41 (in Russian).
5. MOTSKIN, T. S. 1952. New techniques for linear inequalities and optimization in Project SCOOP. Symposium on Linear Inequalities and Programming.
6. POLYAK, R. A. 1971. On the best convex Chebushev approximation. Sov. Math. Dokl. **200:** no. 5.
7. CHARALAMBOUS, C. 1977. Nonlinear least path optimization and nonlinear programming. Math. Programming **12.**
8. CHARALAMBOUS, C. 1979. Acceleration of the least path algorithm for minimax optimization with engineering applications. Math. Programming **17.**

# Modified Barrier and Center Methods

R. POLYAK

Modifications with properties similar to the modified Lagrangians have been found for the barrier and center methods. Particularly, the barrier functions, which had been proposed by R. Frisch[1] and C. Carroll,[2] were modified. Under conventional assumptions of strict regularity (see, for example, reference 3), these modifications imply that the sequential unconstrained minimization methods converge linearly with ratio, which can be made as small as necessary at the expense of increasing the penalty factor.

Duality theorems have been proved and dual functions have been studied for the introduced modifications. While using dual functions, it is possible to obtain superlinear or quadratic convergence for the initial problem without increasing the penalty factor.

Let the function $f_i(x){:}E^n \rightarrow R$, $i = \overline{0, m}$ be sufficiently smooth. The problem is to find $x^* = \arg\min \{f_0(x) \mid x \in \Omega\}$, $\Omega = \{x{:}f_i(x) \geq 0, i = \overline{1, m}\}$. Let $I^* = \{i{:}f_i(x^*) = 0\} = \{1, \ldots, r\}$; $f(x) = [f_1(x), \ldots, f_r(x)]$. Also, let $f'(x) = J[f(x)]$ be the Jacobi matrix for the vector-function $f(x)$.

We shall assume that:

$$r\,\mathrm{ang}\, f'(x^*) = r, \quad u_i^* f_i(x^*) = 0, \quad u_i^* > 0, \quad i \in I^*; \quad \mathcal{L}'_x(x^*, u^*) = 0; \tag{1}$$

$$[\mathcal{L}''_{xx}(x^*, u^*)y, y] \geq \lambda \|y\|^2, \lambda > 0, \forall y{:}f'(x^*)y = 0, \mathcal{L}(x, u) = f_0(x) + \sum_{i=1}^{m} u_i f_i(x). \tag{2}$$

We shall call the function $C(x, u, \kappa) = f_0(x) + \kappa^{-2} \Sigma_{i=1}^m u_i [f_i(x) + \kappa^{-1}]^{-1}$ defined on the set $\Omega \times E_+^m \times R_+$ as the modified Carroll's function, and the function $F(x, u, \kappa) = f_0(x) - \kappa^{-1} \Sigma_{i=1}^m u_i \ln[f_i(x) + \kappa^{-1}]$ defined on the same set as the modified Frisch's function. Let $C_r(x, u, \kappa)$ and $F_r(x, u, \kappa)$ be the functions of the type indicated above where the sum is taken from 1 to $r$.

## THEOREM 1

Let equations 1 and 2 hold and let the functions $f_i(x)$, $i = \overline{0, r}$ be sufficiently smooth. Then there exist $\rho > 0$ and such $K_\rho$ that for any $u = (u_1, \ldots, u_r; 0, \ldots, 0) \in S(u^*, \rho) = \{u{:}\|u - u^*\| \leq \rho\}$ and $\kappa \geq K_\rho$:

(1) there exist $x_C^r$, which is a unique local minimum of $C_r(x, u, \kappa)$ in a neighborhood of $x^*$, and $x_F^r$, which is a unique local minimum of $F_r(x, u, \kappa)$ in the same neighborhood;

(2) $C_r(x, u, \kappa)$ and $F_r(x, u, \kappa)$ are strongly convex in $x$ in the neighborhoods of $x_C^r$ and $x_F^r$, respectively;

(3) for the vectors $x_C^r$, $x_F^r$, $u_C^r = u_{Ci}^r = \kappa^{-2} u_i [f_i(x_C^r) + \kappa^{-1}]^{-2}$ (with $i = \overline{1, r}$, $u_{Ci}^r = 0$, $i = \overline{r + 1, m}$), and $u_F^r = u_{Fi}^r = \kappa^{-1} u_i [f_i(x_F^r) + \kappa^{-1}]^{-1}$ (with $i = \overline{1, r}$, $u_{Fi}^r = 0$,

$i = \overline{r+1, m}$), the following estimate

$$\max\{\|x_C^F - x^*\|, \|x_F^r - x^*\|, \|u_C^r - u^*\|, \|u_F^r - u^*\|\} \leq \mu\kappa^{-1}\|u - u^*\| \quad (3)$$

is true.

## THEOREM 2

Let equations 1 and 2 hold and let the functions $f_i(x)$, $i = \overline{0, r}$ be sufficiently smooth. Then for $\kappa \geq K_\rho$:

(1) in a neighborhood of $S(u^*; \rho)$, two strongly concave functions, $\varphi_C^{(r)}(u, \kappa) = \min\{C_r(x, u, \kappa) | x \in \Omega\}$ and $\varphi_F^{(r)}(u, \kappa) = \min\{F_r(x, u, \kappa) | x \in \Omega\}$, are defined. Their smoothness is defined by the smoothness of $f_i(x)$, $i = \overline{0, r}$;

(2) the following duality relations are valid:

$$\min\{C_r(x, u^*, \kappa) | x \in \Omega\} = C_r(x^*, u^*, \kappa) = \varphi_C^{(r)}(u^*, \kappa)$$
$$= \max\{\varphi_C^{(r)}(u, \kappa) | u \in S(u^*; \rho)\};$$
$$\min\{F_r(x, u^*, \kappa) | x \in \Omega\} = F_r(x^*, u^*, \kappa) = \varphi_F^{(r)}(u^*, \kappa)$$
$$= \max\{\varphi_F^{(r)}(u, \kappa) | u \in S(u^*; \rho)\}.$$

## THEOREM 3

Let $f_0(x)$ be a convex function, let $f_i(x)$, $i = \overline{1, m}$ be concave functions, and let equation 1 hold. Thus, if $f_0(x)$ is strongly convex or one of the $f_i(x)$, $i = \overline{1, m}$ is strongly concave, then:

(1) there exist a unique vector $\bar{x}_C = \arg\min\{f_0(x) + \kappa^{-2}\Sigma_{i=1}^m [f_i(x) + \kappa^{-1}]^{-1} | x \in \Omega\}$ and a unique vector $\bar{x}_F = \arg\min\{f_0(x) - \kappa^{-1}\Sigma_{i=1}^m \ln[f_i(x) + \kappa^{-1}] | x \in \Omega\}$ such that for $\bar{x}_C$, $\bar{x}_F$, $\bar{u}_C = \bar{u}_{Ci} = \kappa^{-2}[f_i(\bar{x}_C) + \kappa^{-1}]^{-2}$ ($i = \overline{1, m}$), and $\bar{u}_F = \bar{u}_{Fi} = \kappa^{-1}[f_i(\bar{x}_F) + \kappa^{-1}]^{-1}$ ($i = 1, m$), the following estimate is valid:

$$\max\{\|\bar{x}_C + x^*\|, \|\bar{x}_F - x^*\|, \|\bar{u}_C - u^*\|, \|\bar{u}_F - u^*\|\} \leq \mu\kappa^{-1}; \quad (4)$$

(2) there exists such $\rho > 0$ that for any $u \in S(u^*; \rho)$ and for $x_C$, $x_F$, $u_C$, $u_F$, [which are defined in the same way as $x_C^r$, $x_F^r$, $u_C^r$, $u_F^r$ for $C_r(x, u, \kappa)$ and $F_r(x, u, \kappa)$ being replaced by $C(x, u, \kappa)$ and $F(x, u, \kappa)$, respectively], the estimate in equation 3 is true for any $\kappa > 0$;

(3) statements of theorem 2 are true for any $\kappa > 0$ if we write $C(x, u, \kappa)$ and $F(x, u, \kappa)$ instead of $C_r(x, u, \kappa)$ and $F_r(x, u, \kappa)$, and $\varphi_C(u, \kappa) = \min\{C(x, u, \kappa) | x \in \Omega\}$ and $\varphi_F(u, \kappa) = \min\{F(x, u, \kappa) | x \in \Omega\}$ instead of $\varphi_C^{(r)}(u, \kappa)$ and $\varphi_C^{(r)}(u, \kappa)$. The same results have been obtained for the center methods.

## CONCLUSION

As a conclusion, we would like to emphasize that there is a possibility of getting estimates of the type in equation 3 while using only one step of any unconstrained optimization method with quadratic convergence rather than solving the unconstrained optimization problem at each step.[4–7]

## REFERENCES

1. FRISCH, K. R. 1955. The logarithmic potential method of convex programming. Memorandum of May 13, 1955, University Institute of Economics, Oslo.
2. CARROLL, C. W. 1961. The created response surface technique for optimizing nonlinear restrained systems. Oper. Res. **9:** no. 2.
3. FIACCO, A. & G. MCCORMICK. 1968. Nonlinear Programming: Sequential Unconstrained Minimization Techniques. New York.
4. POLYAK, B. T. & N. V. TRET'JAKOV. 1973. Penalty estimates method for the conditional extremum problems. J. Comput. Math. Math. Phys. **13:** no. 1 (in Russian).
5. KORT, B. & D. BERTSEKAS. 1976. Combined primal dual and penalty methods for convex programming SIAM. J. Control Optimization **14:** no. 2.
6. ROCKAFELLAR, R. 1973. The multiplier method of Hestenes and Powell applied to convex programming. J. Optimization Theory Appl. **12**.
7. GOL'SHTEIN, G. & N. V. TRET'JAKOV. 1980. On the modified Lagrange functions of convex programming problems. Econ. Math. Methods **XVI:** 4 (in Russian).

# Stepwise Regression Procedures

## Overview, Problems, Results, and Suggestions

EUGENE GRECHANOVSKY

Some problems, methods, and procedures in linear regression analysis are reviewed with emphasis on least-squares estimation and subset selection. The present state of subset selection is criticized for the lack of theoretical basis. To get this field out of its present stalemate, development of the three lines of research (namely, of Interior Theory, Exterior Theory, and Comparative Theory) is recommended. The Interior Theory should investigate mathematical and statistical properties of subset selection procedures, the Exterior Theory should study relations between these procedures and the regression problems purported to be solved by them, and the Comparative Theory should deal with comparative analysis of the procedures. Unsolved problems in the Forward Selection procedure are analyzed in detail, and some solutions are suggested for two versions of this procedure. For conciseness, many abbreviations will be used in this paper. They are listed in the footnote below.[a]

## 1. INTRODUCTION

The importance of linear Regression Analysis (RA) methods for such fields as pattern recognition, economics, social sciences, etc., has been extensively documented and hardly needs elaboration. One of the major problems in these sciences involves the possible reduction in a set of predictor variables affecting the predictand (response variable of interest). This problem of selecting the "best" subset of independent variables for the regression equation has received considerable coverage in statistical literature. Popular procedures for subset selection (i.e., for obtaining the "best" regression equation) given a response variable, a set of candidate predictor variables, and a series of observations on all of them include Sequential Forward Selection (SFS),[1,2] Forward Selection (FS) [see references 3, 4, 5 (sect. 6.4), 6 (sect. 7.2.3.), and 7–11], Sequential Backward Elimination (SBE),[1,12] Backward Elimination (BE) [see references 4, 5 (sect. 6.3), and 6 (sect. 7.2.3)], Efroymson's Selection (ES) [see references 4, 5 (sect. 6.4), 6 (sect. 7.2.3), 13, 14, 15 (sect. 12.4), and 16], All Possible Regressions (APR) [see references 4, 5 (sect. 6.1), 6 (sect. 7), and 15 (sect. 12.2)], and others. Different procedures are known to select different "best" subsets when applied

[a] List of abbreviations: APR = All Possible Regressions procedure; BE = Backward Elimination procedure; cdf = cumulative distribution function; ES = Efroymson's Selection procedure; FS = Forward Selection procedure; LS = Least Squares; MB = Model Building; MSE(P) = Mean Squared Error (of Prediction); NH = Null Hypothesis; RA = Regression Analysis; RI = Regression Inference; RM = Regression Model; RP = Regression Procedure; RPb = Regression Problem; RSS = Residual Sum of Squares; SBE = Sequential Backward Elimination procedure; SFS = Sequential Forward Selection procedure; SSP = Subset Selection Procedure; SSSP = Stepwise Subset Selection Procedure; StR(s) = Stopping Rules(s).

to one data sample. The opinions regarding the advantages and disadvantages of various procedures differ, and no final word seems forthcoming (cf. reference 17).

There was noticeable tapering of interest in the topic of subset selection in the 1980s; the causes for this are still to be analyzed. One reason might be the appearance of competing methods in RA, such as robust procedures, analysis of influence, regression diagnostics, etc. Even more important, though, in our view, is the fact that subset selection failed to develop its theoretical footing and failed to provide a solid mathematical and statistical basis for studying its own procedures; in other words, it failed to solve its own problems. We will illustrate this by the example of FS. The choosing of FS does not suggest that it is any better than any other methods of deciding upon a "best" subset of variables. Rather, it has been motivated by the fact that FS is simple enough to make its analysis feasible, while at the same time, it provides enough complexity to display vast conceptual and mathematical problems associated with Stepwise Subset Selection Procedures (SSSP) involving extensive search for best regressors. Besides, it has been very popular in applied research and has received much attention in statistical literature.

At each step of FS, the predictor variable that explains more of the residual response than any other variable not yet in the equation is entered into the regression provided that the explained residual response is big enough; otherwise, FS halts. What is "big enough" is decided by a Stopping Rule (StR)—that is, by an outcome of an $F$-test performed at each step of this multistep decision-making process. To perform an $F$-test, one needs to have cutoff values, that is, $(1 - \alpha)$ quantiles, or upper critical points of the corresponding distribution of the $F$-statistic. Now, the calculation of "true" cutoff values is a major unsolved problem in the analysis of FS. The fact that the conventional method of comparing the observed $F$-value at each step of FS procedure with the table of the central $F$-distribution is not correct is a fact that has been known for years. In spite of *s* series of attempts to study the "actual" distribution of the $F$-statistic and of some related statistics in FS and to develop "true" tests,[3,18–24] no universally accepted alternatives for discredited standard tests have been discovered. Even the factors responsible for "corruption" of the $F$-distribution under extensive search have never been properly analyzed. As reported by Pinsker *et al.*,[9] there are three such factors:

Factor 1—The search for the best predictor variable at the current step.
Factor 2—Significance tests at earlier steps.
Factor 3—The search for best predictor variables at earlier steps.

Analysis of these factors runs into the following problems connected with the $F$-statistic in FS:

(1) Framing of the Null Hypothesis (NH) calls for extension of the traditional hypothesis testing theory. There exists more than one possible extension.
(2) The $F$-statistic is not defined for all observed values of $Y$. There exists more than one way to define it.
(3) The distribution of the $F$-statistic is not known and, in addition, it depends on unknown (nuisance) regression parameters for the variables already included in the subset. Alternative ways of defining the $F$-statistic and of eliminating the nuisance parameters imply different distributions of the $F$-statistic that

differ considerably from the tabled values and between themselves. This arouses the problems of choice and interpretation.

These problems are analyzed in more detail in subsection 3.3 below; partial solutions are given in references 7–11 and are summarized in subsection 4.2. The work on FS is far from complete; nevertheless, we hope that its study will facilitate progress in other SSSPs.

The organization of the paper is as follows: Section 2 addresses those facets of RA that by and large remain outside the scope of recent authoritative reviews,[4,25–29] and presents some competing viewpoints. Section 3 deals specifically with the procedures for subset selection. After expository and historical material in subsections 3.1 and 3.2, subsection 3.3 briefly describes some of the most important subset selection procedures and analyzes their problems using the FS procedure as an example. Further analysis and suggestions for future research are contained in subsection 3.4, while subsection 3.5 criticizes some recent works. Our own research on SFS and FS is summarized in section 4.

## 2. REGRESSION ANALYSIS

In this section, we sketch our views on some facets of RA that are somewhat different from those developed in standard courses and reviews. No "objectivity" is intended: the presentation is biased towards Least Squares (LS) estimation and subset selection.

### *2.1. Structural Elements of Regression Analysis*

RA, in particular linear RA, is one of the major methods of statistical data processing. Up to the 1950s, the LS-fitting was almost the sole method of RA.[27] Although the LS method is still important both as a Regression Procedure (RP) and as a constituent of other RPs (see subsection 2.4 below), there has appeared a lot of different methods within the last quarter of a century—some of these are competitors for LS, others modify or make use of LS, while still others solve alternative problems besides fitting. Very roughly, RA could be divided into Regression Design[30–32] and Regression Inference (RI) (see reference 33, sect. 1.1.1). RI can be formal or informal (see subsection 2.5 below); we will consider only the formal one. A major goal of modern RA is Model Building (MB) (see reference 33, sect. 1.1.4)—that is, obtaining a Regression Model (RM) satisfying some criterion or regression goal. Modern RA assumes that the given data may be contaminated, or nonadequate, and that direct application of LS to them is, in general, inappropriate. There are two major approaches:

(1) Robust methods (ridge, absolute deviations, etc.).
(2) Two-stage procedures including:

   (*i*) Preselection analysis that analyzes and possibly sorts out the data, or cleans them, or transforms them (regression diagnostics, analysis of residuals, transformations; e.g., see references 27, 28, 34, and 35).

(*ii*) On passing stage *i*, one might apply LS-fitting or an LS-based Subset Selection Procedure (SSP; see subsections 2.4 and 3 below) to the data for obtaining RM.

Cook and Weisberg (reference 35, chap. 1) comment:

> While it seems true that these approaches are in some way competitive, one is not likely to replace the other in the foreseeable future. As long as least squares methods are in widespread use, the need for corresponding diagnostics will exist. Indeed, the use of robust methods does not abrogate the usefulness of diagnostics in general, although it may render certain of them unnecessary.

A rough map of RA is given in FIGURE 1.

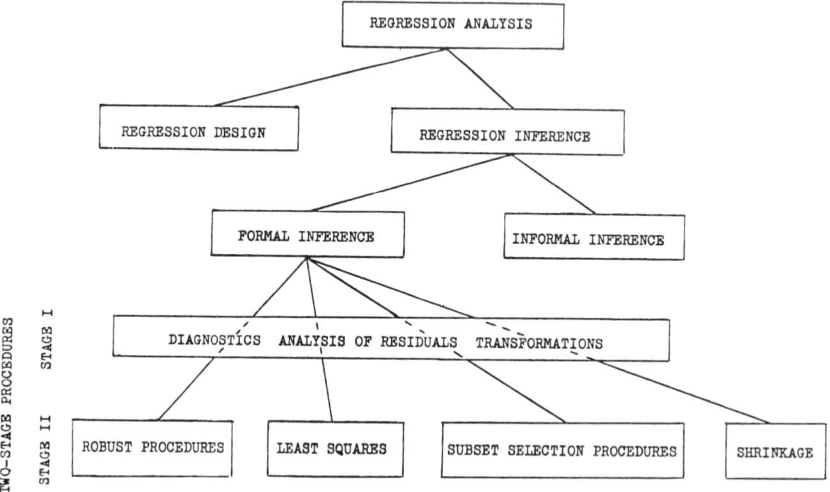


**FIGURE 1.** A rough map of Regression Analysis (RA).

### 2.2. *Regression Inference and Regression Models*

Somewhat differently, the major goal (or problem) of RA, or of RI, could be framed in the following way: It is the statistical inference about the random variable $y$ (or about some functions thereof) that is assumed to be fluctuating around an unknown fixed value $\mathring{y}$:

$$y = \mathring{y} + e, \qquad \textbf{(2.2.1)}$$

where $e$ is a random disturbance (noise).

RI will be performed under the following assumptions:

(1) Only linear models with nonrandom regressors are considered:

$$y = \sum_{i=1}^{k} X_i \mathring{b}^i, \qquad \textbf{(2.2.2)}$$

where $X_i$ are nonrandom variables and $\mathring{b}^i$ are unknown regression coefficients. Therefore, the set $\{X_i\}$ is presumed to include all relevant regressors and probably some irrelevant ones (with corresponding coefficients, $\mathring{b}^i$, equaling zero).

(2) There are $n$ observed values of the random variable $y$ that are interpreted as sample values of a random $n$-vector with independent components:

$$Y = (Y^1, Y^2, \ldots, Y^n)'.$$

(3) There are $n$ (nonrandom) values of each quantity $X_i$,

$$X_i = \{X_i^j\}, \qquad j = 1, \ldots, n, \quad i = 1, \ldots, k,$$

which make the matrix

$$X = [X_1 X_2 \ldots X_k].$$

The usual (standard) assumptions about the matrix $X$ are
(a) $k < n$,
(b) rank $X = k$.
Sometimes others are added; for instance,
(c) $X$ is orthogonal.

(4) There is an unobservable vector of disturbances,

$$e = (e^1, e^2, \ldots, e^n)',$$

so equation 2.2.1 may be written as

$$Y = \mathring{Y} + e = X\mathring{b} + e. \qquad \textbf{(2.2.3)}$$

The usual assumptions on $e$ are
(a) $Ee = 0$ (cf. equation 2.2.2),
(b) $\text{cov}(e) = \sigma^2 I_n$, where $\sigma^2$ is unknown.
Often, the normality is assumed:
(c) $e \sim N(0_n; \sigma^2 I_n)$.

(5) There may be restrictions on the vector of coefficients $\mathring{b}$; for example,

$$\mathring{b}^i \geq 0.$$

(6) There may be restrictions on the vector of estimates $b = (b^1, \ldots, b^k)$; for example,[36a]

$$b^i \geq 0.$$

The two somewhat different formulations for the major goal of RA given in this and in the preceding subsections are related to our view that MB is a part of RI. The major goal of the traditional RA was in obtaining estimates $b$, $\hat{\sigma}$, confidence intervals for $\mathring{b}$ and $\sigma$, etc., for a given RM, that is, for a given (specified a priori) set of $X_i$. The modern RA (e.g., see reference 36b) includes a philosophical approach to modeling as an aid to science in finding appropriate models. In particular, it includes model criticism, checking of normality assumptions, and analysis of outliers, along with selection of regressors, addition of new regressors, etc. Crocker[36b] criticizes Seber[15] because his "starting position is nearly always with model in hand with no awareness of a need for understanding of the physics of a system or of the relevance of goals to the modeling process." We will be applying the RI for inference in RA including MB, criticism of RMs, etc., as opposed to the traditional inference based on the a priori selected model and LS-fitting. RI is a process whose goal is the solution of some particular Regression Problem (RPb)—for instance, obtaining a model satisfying some criterion—along with computation of estimates $b$, $\hat{\sigma}$, confidence intervals, etc., connected to the process of selection of a model rather than to a fixed a priori model. Particular RPbs are arrived at by the process of specification and detailization of the main RPb in the context of the chosen assumptions. Particular RPbs are solved by using existing and newly invented RPs. RPbs and RPs are respectively considered in the two following subsections.

### 2.3. *Regression Problems (Goals or Uses)*

Lindley,[37] Mantel,[38] Crocker,[36b] and Harville,[39] among others, emphasized that the method to be employed in any analysis should be relevant to the use intended for the finally fitted regression. Different regression goals, while using different criteria, combine with different assumptions on the structure of regressors, structure of distributions, and structure of RM looked for; produce different formulations of the RPb; apply different RPs; and result in different RMs. Below, we give some examples of RPbs (or, rather, of groups of RPbs).

#### *I. Problems of Significance (Detecting the Regressors That Influence* Y*)*

(1) Is the regressor $X_i$ significant? Under the usual distributional assumptions, this problem is solved by using $t$ test (or $F$-test).

(2) Assume the regressors are ordered in decreasing importance. Find $p$ significant regressors $X_1, \ldots, X_p$ (the number $p$ is unknown) so that $X_{p+1}$ is the first insignificant one. This problem can be solved by using RPs, SFS, or SBE (see subsection 3.3 below).

(3) Find all the significant regressors, that is, break $k$ into $p$ and $r$ $(p + r = k)$ in such a way that

$$b^i \neq 0, \quad i = i1, \ldots, ip,$$

$$b^j = 0, \quad j = j1, \ldots, jr.$$

This is a problem of multiple hypotheses testing. Also, it might be solved by SSSPs (FS, BE, EA, etc.; see subsection 3.3).

(4) Find all subsets of regressors that are no worse under the given criterion than the whole set. Two approaches were given by Spjøtvoll[40] and Aitkin.[41]

#### *II. Estimation of the Regression Coefficients*

(A) Selection of the optimal (in the given sense, in the given class, under the given assumptions) vector of estimators $b$ for the vector of regression coefficients $\mathring{b}$.

(A1) In the class of linear unbiased estimators (without normality assumptions), the LS-estimator for the full model is optimal by the criterion min Var($b$) [e.g., reference 15 (chap. 3) and reference 42 (sect. 1.5)].

(A2) In the class of all unbiased estimators under the normality assumptions, the LS-estimator is optimal by the same criterion [reference 42 (chap. 1) and reference 43 (p. 319)].

(A3) Consider the class of overfitted models (i.e., models containing insignificant variables). Then, the LS-estimator for the significant model is optimal in this class by the same criterion (reference 42, sect. 2.4).

(B) Comparison of two or more estimators $b_1, b_2, \ldots$, under some criterion.

(B1) Assume the model of equation 2.2.3 has been written as

$$Y = X_p \mathring{b}_p + X_r \mathring{b}_r + e,$$

where $X_p$ is $n \times p$ matrix, $X_r$ is $n \times r$ matrix, and the LS-estimator $b$ is partitioned accordingly $b' = (b_p', b_r')$. Let $\tilde{b}_p$ denote the LS-estimator for $\mathring{b}_p$ if the variables in $X_r$ are deleted from the model. Then, $\tilde{b}_p$ is better than $b_p$ under the Mean Squared Error (MSE) criterion,

$$\text{Var}(b_p) \geq \text{MSE}\ (\tilde{b}_p),$$

provided

$$\text{Var}(b_r) - \mathring{b}_r \mathring{b}_r' \geq 0$$

(see reference 27, p. 6).

(B2) Ridge estimator has smaller MSE than the LS-estimator (reference 15, p. 89).

### *III. Prediction at the Original Points*

Here, we have two major problems:

($i$) Prediction of the "true" vector $\mathring{Y} = EY$. The usual criterion is

$$\text{MSE}(\hat{Y}) = \frac{1}{n} E(\hat{Y} - \mathring{Y})'(\hat{Y} - \mathring{Y})$$

or, more generally,

$$\text{MSE}(\hat{Y}, C) = \frac{1}{n} E(\hat{Y} - \mathring{Y})'C(\hat{Y} - \mathring{Y}),$$

where $\hat{Y}$ is the predicted value for $\mathring{Y}$ and $C$ is a positive semidefinite matrix.

($ii$) Prediction of future values of $Y$:

$$\text{MSE}(\hat{Y}) = \frac{1}{n} E(\hat{Y} - Y)'(\hat{Y} - Y)$$

or, more generally,

$$\mathrm{MSE}(\hat{Y}, C) = \frac{1}{n} E(\hat{Y} - Y)'C(\hat{Y} - Y).$$

Both problems might be complicated by additional assumptions; for example,

(1) $Xb \in L_1$, where $L_1$ is a subspace of the range space $L(X)$ (reference 15, sect. 3.9),
(2) $Xb \in L_1 \cup L_2 \cup \cdots \cup L_m$, where $L_i$ is a subspace of $L(X)$,
(3) $b^i \geq 0$ (see reference 36a),
(4) $b'b \leq h^2$.

One more criterion of prediction is

$$K = \mathrm{MSE} + S_U,$$

where $S_U$ is a loss function for using the regressors from the subset $U \subset \{X_i\}$.

Alternative criteria of goodness of fit in prediction have been suggested by references 44–53 and others (see also references 54 and 55).

### *IV. Prediction at the Target Points That Differ From the Original Ones*

For details, see references 4 and 56.

## *2.4. Regression Procedures*

For performing RI, that is, for solving RPbs like those described in subsection 2.3, different RPs are used. Some examples are: LS, SSPs like FS, BE, ES, robust procedures like shrinkage, ridge RPs, procedures of residual analysis, and procedures of regression diagnostics like rejection of outliers.

The LS method invented by Gauss and Laplace was for 150 years almost the only RP used. This fact might have been responsible for the lack of the concept of RP. Another reason might be that the concept of RP is a loose and, at the same time, rather complicated one that does not easily lend itself to mathematical refinement. In the days of the predominance of LS, RA was often identified with obtaining estimates for $\mathring{b}$ and $\sigma$. Nowadays, due to the appearance of competing methods of estimation on the one hand, and to the growing influence of various RPs not involving estimation on the other hand, LS has somewhat diminished in importance. There are four facets to LS (see below).

(1) Purely computational (approximational or descriptive) facet: It does not relate directly to statistics proper, but historically it was probably the most important one (cf. citation from Laplace's *Théorie Analytique des Probabilités* in reference 57, p. 1). The method of absolute deviations lost the battle to LS from the start just because the technique of linear equations arising in LS was superior analytically. The advent of modern computers undermined this superiority of LS, but it did not destroy it altogether.

(2) Statistical facet (LS as a method of estimation): The main stronghold of LS is the unbiased estimation. As the shares of unbiased estimation have been considerably devalued in recent years, the competitors of LS are now the methods of biased estimation like ridge and shrinkage, as well as LS-based SSPs (see section 3 below) that also give biased estimators. Goldberger[58] put forward an alternative approach to establishing the LS method of estimation: the analogy principle of estimation, which proposes that population parameters be estimated by sample statistics that have the same property in the sample as the parameters do in the population.

(3) LS as an RP: Uncritical application of LS often led to blurring out the goals of regression (subsection 2.3) and to ignoring the RPbs in which estimation is not central, or in which biased estimation is acceptable. Methodologically, the predominance of LS (it was always taken for granted) impeded the shaping of the RP concept.

(4) LS as the basis of RPs: While LS forms the basis of many RPs and, in particular, of almost all SSPs now in use, the controversy in the literature over the place of LS in RA is far from complete. Some authors develop ridge and other shrinkage procedures and advocate their use, often backing their views with simulation results.[59–63] Others, however, are cautious or even skeptical with regard to those methods[5,64–69] (also cf. discussion in reference 6, sect. 8.5.7). Thus, Draper and Smith (reference 5, p. 324) point out that the claims about superiority of ridge regression

> must be viewed with caution. Careful study typically reveals that the simulation has been done with effective restrictions on the true parameter values—precisely the situations where ridge regression is the appropriate technique theoretically. The extended inference that ridge regression is "always" better than least squares is, typically, completely unjustified.

Other widespread objections are:

—Ridge regression provides good solutions to wrong problems.

—Ridge regression is not invariant under the scale transformations. F. J. Anscombe (reference 57, p. 6) indicates conceptual advantages of LS. The situation gets especially complicated because of the proliferation of the numerous modifications of LS and of the procedures that are based on LS, but that are not LS in the strict sense.

Returning to the general concept of RP, what we mean by it is the map (or operator) from the input data to the output. In simple cases, the input is just $Y$. The output is the data resulting from the RP's performance and that computed by it: RM, estimates, confidence statements, etc. For the same input, output varies from RP to RP. For example, if an RP with extensive search selects regression subsets and estimates the regression coefficients in them, its output includes the ordered set of selected regressors, the estimates for their coefficients, the corresponding values of Residual Sum of Squares (RSS), the current estimate for $\sigma$, etc. Such entities as the form of the regression (i.e., the set of functions available for fitting), the regression goals, the criteria for achieving these goals (the structure of the loss function), and the numerical values of the significance levels (along with probably some other entities) are usually assumed more or less explicitly. It should be stressed that RP is conceived of as a separate whole, that is, as a distinct statistical entity, in spite of the fact that various nonstatistical considerations (common sense, simplicity of computations, professional intuitions for the choice of an appropriate model of the studied phenomena, or for a relevant way of data processing) may and, in fact, usually do influence the construction of RP or the choice between the existing RPs. As a result of this conceptualization, RP becomes an object of statistical study, allows statistically valid comparisons of RPs, and provides a means for statistical inference (RP-based inference) and a basis for studying the properties of non-LS estimators obtained by a given RP. Thereby, RP belongs to the sphere of formal inference (cf. next subsection). Our concept of RP basically coincides with definitions given in references 52, 63, 70, and 71 (along with others), and essentially, if not always terminologically, it agrees with the common statistical practice.

A few other examples of procedures in statistics include:

(1) Sample estimation of the population mean by the formula of the sample mean:

$$\bar{x} = \sum_i x_i/n.$$

(2) First rejecting outliers according to some rule, or to no rule at all, and then making use of the above formula.
(3) Sample estimation of the population mean by using the median of the sample values rather than the sample mean.
(4) Preliminary transformation of the data with subsequent use of one of the mentioned ways of estimation.

Introduction of the RP concept gives rise to the following lines of research:

(I) Interior Theory: Mathematical and statistical analyses of the existing RPs; investigation of their (interior) properties that enable their competent application.
(II) Exterior Theory: Investigation of when, under what conditions, and which RP is optimal for solving which RPb; supplying RPs with prescriptions indicating their potential use; constructing RPs optimal for the solution of a given RPb; modification of existing RPs in accordance with the statistical theory of RPs.
(III) Comparative Theory: Development of methods for comparison of RPs and of estimators obtained therewith both for artificial and real data.

We will consider these lines of research for SSSPs in more detail in subsection 3.4 below.

### *2.5. Formal versus Informal (Subjective) Inference*

Recently, in the statistical literature, there has been an increase in the prevalence of the views that "the fitting of equations to observational data is a risky business," that the computerized procedures cannot replace humans, and that "we would not want to leave decisions to an automated procedure" (e.g., cf. reference 27). At the same time, we are being urged to develop and supply to the user (statistician and nonstatistician alike) numerous diagnostic procedures, tools, and gadgets that he (she), the user, would have been able to use upon his (her) own judgement for processing the given data and obtaining the fitted equations. Such arguments create an impression (probably, though, more because of their unsatisfactory wording than because of the intention of the authors) that automated fitting of equations (and, more generally, automated data processing) leads into a deadlock, and that the future belongs to intuitive, semi-intuitive, subjective, and other human-driven procedures. We think, however, that competition with a competent human statistician has not been and can never be a goal of the development of automated procedures. Rather, the goals might be as follows:

(1) In analyses of routine problems, one should replace a less skilled (or statistically incompetent) human by a statistically sound automated procedure capable of selecting a good model.
(2) To a competent statistician, one should give sophisticated tools that will enable him (her) to process quickly and reliably large amounts of data and to obtain statistically valid models.

Creation of a large store of methods, tools, procedures, and recipes for data processing without development of formalized rules and criteria for their application cannot be a goal either. Such a store will hardly be of help to a nonstatistical user; it could only confuse him (her) and encourage uncritical use of the methods chosen at random. Thus, a patient without medical training would be bewildered by an abundance of patent medicines that allegedly cure any disease.

There is still another, no less serious, reason against unchecked (i.e., not guided by rigorous statistical criteria and not embedded in the framework of unified statistically valid RP) application of a large battery of methods for analyzing the data. It is the possibility of pseudofitting—of obtaining anything you like from the given data. Below, we cite some warnings against this danger.

—C. Chatfield (reference 72, p. 27):

> The dangers of over-interpretation of data are well known. I recall with amusement a paper by Armstrong (1970) where a multiple regression was carried out on a small data set. After eliminating three "obvious" outliers and several explanatory variables, a convincing model was found with $R^2 = 0.85$. Armstrong then revealed that the data were actually random numbers! If one further allows transformations, the possibility of finding a spurious model will increase.

—A. J. Miller (reference 29, p. 392):

> There are many who prefer subjective to automatic methods of variable selection. They may perhaps start by finding the simple correlations between the predictand and each of the predictors, and then look at scatter diagrams for those predictors with the largest correlations. This may show the need for a transformation, or adding polynomial terms, or the presence of outliers. After selecting one predictor, the process is repeated using the residuals from fitting this predictor, continuing until nothing more can be seen in the data. This approach is an extension of forward selection and suffers from the weaknesses of that method, though it does provide some protection against the selection of what might be considered stupid models. A formalized version of this procedure has been called "projection pursuit" by Friedman and Stuetzle (1981). Without enumeration of the family of potential models in advance, it is impossible to develop any statistical inference for such methods. The plotting and examination of residuals should of course be a part of any procedure.

—J. B. Copas (reference 73, p. 412):

> If you torture the data for long enough, in the end they will confess. The data will always confess, and the confession will usually be wrong.

The way out of this mess lies, in our view, in the development of formalized and statistically sound RPs that of course may be multistage, complex, and flexible. Any human intuitions and cookery book recipes may go into such procedures; however, they must be formalized and statistically validated, if only at later stages of their development. Therefore, any method for data processing that contains informal human-controlled elements should be viewed as a blueprint for a future automated procedure. An akin idea is expressed by S. Weisberg:[74] Useful diagnostics can often be derived by parameterizing the assumptions, thereby turning the problem of criticism, at least approximately, into one of parametric inference.

Repudiation of automated procedures for their being automated reminds me of a sentiment of a physician or a philologist (now, fortunately, only rarely met) that application of the LS method is tantamount to substituting a dumb computing device for a creative personality. Besides devising and testing the procedures, the human has other key roles to play, among them:

(1) Maintenance and supervision. In particular, interactive computer-graphic capabilities allow the statistician to supervise and control those steps of the procedure he (she) chooses.
(2) Interpretation and responsibility. Interpretation of obtained results and social responsibility for their application remain solely with the statistician.

## 3. SUBSET SELECTION

This section begins with some pros and cons of subset selection and then examines particular statistical problems in SSSPs and relevant recent research. It also puts forth some new suggestions partly developed in our own previous papers.

### *3.1. Motivations for Reducing the Number of Regressors*

There are arguments, including those of common sense, that show that a model with a smaller number of regressors is preferable to a model with a larger number of them.

Nonstatistical arguments include economic, social, practical, economy of thought, aesthetical, and computational (e.g., see references 30, 37, 39, 57, and 75). Such arguments may be either couched in formal terms and implanted into regression goals and RPs or left at the informal level and then used later when comparing models obtained by different RPs. The first way is statistically unambiguous, but it is harder to pursue. The second, when followed systematically, leads to Informal Inference (subsection 2.5).

Statistical arguments mostly turn around the two basic facts: inflation of the covariance matrix Var($b$) of the estimators for $\mathring{b}$ and poor prediction of new data when using extraregressors [references 4, 6 (sect. 7.1.2), 42 (sect. 2.4), and 76–80].

Historically, these considerations led to the inception of SSPs. It should be remembered, however, that in all the cited works, the statistical properties of the estimators are derived for just two linear models under the a priori given sets of regressors and the LS-estimation. If, on the other hand, we used two RPs with built-in significance tests to reduce the number of variables, the models obtained thereby would not be LS-models and their statistical properties might be different. This has been noted among others by McCabe (reference 81, p. 131):

> Whenever one uses a subset selection procedure to obtain an equation with a reduced number of variables, one compromises the least squares principle by setting some coefficients equal to zero. Whether or not least squares is used for the reduced problem, a deviation from the least squares estimates for the full model is present.

### *3.2. Much Ado about Subset Selection*

The arguments presented in the previous subsection stimulated the development of SSPs that are widely used and accessible to a nonstatistical user as statistical computer packages or canned programs.[82–85] Miller[29] has estimated that something of the order of $10^5$ multiple regressions are carried out per day worldwide, many of these using subset regression. At the same time, subset selection is one of the most controversial topics in statistics. Below, we list some typical viewpoints.

—F. J. Anscombe (reference 57, p. 6):

> Our aims are . . .
> (i) find a simple expression for [*equation*] 2.2.2,
> (ii) make the apparent variance of the *e*'s as small as possible compared with the gross variance of the *Y*'s,
> (iii) have the *e*'s apparently independently normally distributed with zero mean and constant variance. . . .
> Pursuit of the above aims constitutes a search. Something in the nature of a goodness-of-fit test is applied to each linear structure of the type [*in equation*] 2.2.3 that is considered, and so the whole search implies successive tests of goodness of fit applied to the same data. Curiously enough, some statisticians seem to regard successive testing as improper and unmentionable. I have heard experienced men assert that "stepwise regression" was "utterly invalid". This attitude springs from the idea that all statistical analysis should be based on a definite set of assumptions or "model", taken as "given" and so held absolutely, without verification. . . .

—W. J. Kennedy and T. A. Bancroft:[1]

> Several different procedures have been recommended for use in determining a suitable subset of independent variables for use in predicting the dependent variable of interest. These procedures involve the use of repeated tests of significance and rely upon inferences based upon the outcome of such tests. The decision rules used in these procedures were, for the most part, selected for their intuitive appeal and little consideration has been given to the consequences, with respect to the fitted model, of the effect on subsequent inferences of such repeated testing.
>
> The problem of model building in regression is one of the general class of problems called problems of incompletely specified models involving the use of repeated tests of significance. This classification serves to clarify the nature of this regression problem and to emphasize the need for more relevant theoretical development in this problem area.

—A. J. Miller:[29]

A stepwise regression is one of the most widely used of all statistical techniques; an examination of its methods and "folklore" is long overdue.

Other authors are more cautious or even skeptical.

—E. S. Page:[86]

Multiple regression . . . must be one of the most used, and misused, of standard statistical programs. . . . Certainly not all who have used the tool have realized what its limitations and dangers are.

—P. Sprent:[87]

Any paper that helps us to understand these limitations and dangers is most welcome. . . . Today the burning question thrust upon us by computer is that of choice of variables.

—J. B. Copas:[73]

. . . in many practical cases reliance on subset selection is misleading, wrong, and foolish.

—R. L. Plackett:[88]

If variable elimination has not been sorted out after two decades of work assisted by high-speed computing, then perhaps the time has come to move on to other problems.

In recent years, the center of fashions moved into robust regression and regression diagnostics with a concurrent ebbing of interest in subset regression. Fashions, however, are not always the best guides to right directions. The loss of interest in subset regression can be explained (among other things) by the fact that major problems in subset regression (see the next two subsections) remain unsolved, and as a result, subset selection is still lacking a firm theoretical basis (cf. references 3, 29, 40, and 73). On the theoretical side, subset selection is very complicated and poorly understood. As a rule, theoreticians show little interest in the topic. Thus, in a well-known monograph by Seber,[15] it is relegated to one small chapter (chap. 12), while SSSPs occupy only one subsection (subsection 12.4, which is 6 pages in a book of more than 400 pages). In a more practically oriented book by Draper and Smith,[5] this topic was extended into a big chapter (chap. 6, which is 86 pages in a book of more than 700 pages); however, theoretical development of the relevant statistical questions leaves much to be desired (see the next subsection). The majority of works on subset selection deal only with combinatorial and computational questions (reference 29, p. 424):

Most of the effort expended in this field in the past two decades has gone into computational methods and . . . only a very small number of people have been looking at the statistical problems. . . .

Very little has been done in the theory of StRs, and hardly much more in estimation and study of bias under repeated stepwise testing (cf. subsection 3.4 below).

In RPs with extensive search for best regressors, repeated testing is aggravated by the problems of stepwise multiple comparisons (or multiple hypotheses testing[89]). While in the Type I theory, very few problems have been satisfactorily solved (cf. subsection 3.4 below), the Type II theory (power of tests) is not usually even mentioned in texts on regression. In our view, the absence of a commonly accepted concept of RP or SSP, in spite of definitions given by several authors,[52,63,70,71] is indicative of an underdeveloped state of the subject. We believe that introduction of the explicit concept of RP enables one to precisely formulate, discuss, and solve methodological, statistical, and mathematical questions. For instance, the "procedural approach" allows one to clearly understand and interpret the fact that traditional estimators and test statistics get corrupted under extensive search. We have in mind distributions of test statistics like the $F$-statistic traditionally used for determining the moment of termination, or estimators like Mallows' $C_p$ [references 15 (sect. 12.2.3), 50, 54, and 55], Akaike's AIC,[45,46] Rothman's $J_p$ [reference 15 (sect. 12.2.3) and 44], or Young's Bivar.[53] This fact was repeatedly pointed out [e.g., references 3, 5 (sect. 6.5), 15 (sect. 12.4.4), 18, 23, 24, 27, 29, 40, and 90–92a], but regretfully no theory has been put forward to account for it. Many misunderstandings in the literature on subset selection resulted from the underestimation of this fact. Often, prediction estimators (like $C_p$ or AIC) derived for an a priori specified set of regressors have been used as criteria for selecting subsets, with implied false inferences about the procedures thereby obtained. Such estimators being associated with (or oriented to) one particular RP, namely, LS, cannot be used with other RPs without developing a corresponding statistical theory. Each estimator should be studied as a part of that RP with which it is used. Without the concept of RP, this problem is difficult even to formulate. However, too often, the regression people belonged to those who, in Lindley's words,[92b]

> think that to have a coherent philosophy about a subject is a luxury for cloistered academics and that operators can get by with strictly pragmatic considerations.

Numerous arguments available in the literature about superiority of ridge and shrinkage methods over subset selection take advantage of the undeveloped state of subset selection. Draper and Smith (reference 5, p. 322) give additional reasons to defend subset selection:

> Most variable selection procedures . . . would assign a value zero to a nonsignificant coefficient rather than adjusting it to a different but also nonsignificant value and such action would be no less reasonable.

However, all these arguments and counterarguments are, at best, provisional because, as pointed out by Miller:[29]

> Only when the statistical properties of subset regression methods are understood can we hope to compare the alternative methods of ridge and other shrinkage methods objectively.

### *3.3. Stepwise Subset Selection Procedures*

SSSPs form a subclass of SSPs. A partial flowchart of these procedures is given in FIGURE 2 and good reviews of them (with extensive bibliographies) may be found in references 4, 5 (chap. 6), 6, 15 (chap. 12), 25, 26, 29, and 93–95.

*Exhaustive Search* (or Global Search procedures, or All Subsets, or All Possible Regressions): This procedure requires fitting every regression equation and assessing it according to some criterion. The number of possible regressions is $2^k$. Its computational facets are described in references 29 and 96–101. Many authors point out that the procedure is cumbersome and computation-consuming, while others optimistically claim that modern computers can do anything. For the moment, however, all this is secondary because our first concern is with an array of purely statistical problems that arise in such procedures and that have nothing to do with capacities of modern computers. We will not tackle these problems in APR because we do it in connection with FS below. After discussion of FS, it will be clear that all the problems arise in APR on a larger scale. Many authors claim that APR always finds the best subset of the given size and is thereby "optimal". This claim makes sense, however, only combinatorically. The statistical "price" that must be paid for this "optimality" may be appreciated after comparison with FS in the next subsection.

*Subexhaustive Search* (Only Better Regressions): This includes searches along favorable branches [references 15 (sect. 12.3.1), 102, and 103], the directed search on *t*

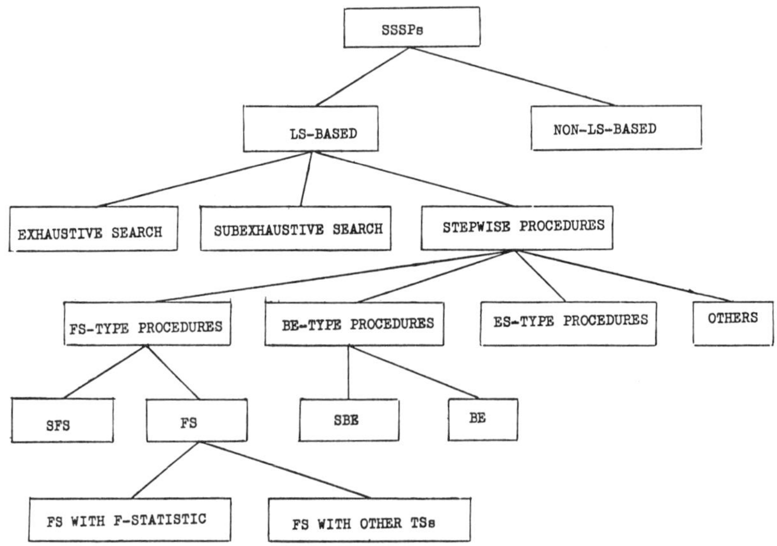


**FIGURE 2.** A partial flow chart of SSSPs.

[references 6 (sect. 7.2.2), 15 (sect. 12.3.2), and 55], breaking predictors into groups,[104] and similar methods. All the mentioned statistical problems arise here as well.

*Stepwise Procedures* (Cheap Procedures or Procedures with Extensive Search for Best Regressors) are mostly variations on the two basic ideas: forward selection (forward search) and backward elimination (backward search).

FS [references 3, 4, 5 (sect. 6.4), and 6 (sect. 7.2.3)] begins with no variables in the regression equation and brings them in one at a time. At the first step, the regressor, say, $X_{i2}$, selected for entry is the one that has the largest sample correlation with the response variable $Y$, or the one that has the largest value of the $F$-statistic for testing significance. The regressor $X_{i1}$ is entered if the value $F^1$ of the $F$-statistic exceeds a preselected $F$-value, say, $F^1_{\text{IN}}$ (or $F$-to-enter at the first step). If, on the other hand, $F^1 \leq F^1_{\text{IN}}$, FS stops. This is the StR. If there has been no stop at the first step, the regressor $X_{i2}$ chosen for entry at the second step is the one that has the largest correlation with $Y$ after adjusting for the effect on $Y$ of the first entered regressor $X_{i1}$; that is, the one that has the largest partial $F^2$ given $X_{i1}$ in the model. If $F^2 > F^2_{\text{IN}}$ (where $F^2_{\text{IN}}$ is $F$-to-enter at the second step), $X_{i2}$ is entered and the process of inclusion continues; otherwise, FS stops at the second step. The procedure terminates when either the partial $F$-statistic at some step $i$ does not exceed $F^i_{\text{IN}}$, or when all regressors have been entered.

SFS[1,2] differs from FS in that no search for best regressors is allowed. SFS applies when the regressors may be arranged at the start in the order of decreasing importance. It starts with no variables in the equation and, at each step, it brings in the next variable provided that the explained residual response due to this variable is big enough; otherwise, it stops.

We say that FS makes extensive search for best regressors at each step of its performance, while SFS is a process of sequential (stepwise) testing with no extensive search.

BE [references 4, 5 (sect. 6.3), and 6 (sect. 7.2.3)] works in the opposite direction. It begins with all $k$ variables in the model and computes the partial $F$-statistic for each regressor as if it were the last one to enter the model. The smallest of these partial $F$-statistics, $F^1$ for $X_{i1}$, say, is compared to the preselected value $F^1_{\text{OUT}}$ (or $F$-to-remove at the first step), and if $F^1 \leq F^1_{\text{OUT}}$, the regressor $X_{i1}$ is removed. If $F^1 > F^1_{\text{OUT}}$, the procedure stops and the final model includes all the variables. The process continues until either, at some step $i$, the smallest partial $F$-value $F^i$ is not less than the preselected cutoff value $F^i_{\text{OUT}}$, or when all the variables have been dropped from the model.

SBE[1,12] removes the regressors in a predetermined order and differs from BE in the same way as SFS differs from FS.

ES [references 4, 5 (sect. 6.4), 6 (sect. 7.2.3), 13, 14, 15 (sect. 12.4), and 16] combines entering and deleting variables at each step of the process. It requires two cutoff values, $F^i_{\text{IN}}$ and $F^i_{\text{OUT}}$, at each step $i$.

Numerous modifications and extensions of these procedures can be found in the literature. Thus, sets of $p$ ($p > 1$) regressors can be entered or deleted at each step; other test statistics, for example, the lack-of-fit statistic (LF),[4,29]

$$\text{LF} = \frac{(\text{RSS}_p - \text{RSS}_k)/(k - p)}{\text{RSS}_k/(n - k)},$$

can be used in StRs, etc.

While analyzing SSSRs, it is instructive to identify three basic ingredients[4] in them; however, in actual procedures, they are often intertwined:

(a) Logical (computational) scheme for selection of candidate subsets and the order of their consideration. For instance, in search-free procedures (SFS, SBE), the order of considering subsets is determined a priori, while in extensive search procedures (FS, BE, ES, APR), the order is governed by the data.
(b) Method of estimation of parameters in subsets under consideration. We consider only LS-based procedures in which the method is LS.
(c) Principles and criteria for detection of "best" subsets, which includes setting

up StRs. A natural way of constructing StRs is to use a test statistic like the $F$-statistic for testing hypotheses about some coefficient or a group of coefficients. Then, the values of $F_{IN}$ and $F_{OUT}$ at each step are computed as the quantiles (upper percentage points) of the corresponding distribution of the $F$-statistic. This approach is an extention of the traditional theory of hypotheses testing. Below, we will dwell on the problems arising here.

Another widespread approach is the comparison of subsets by some test statistic or criteria function [such as $RMS_p = RSS_p/(n - p)$, $R_p^2$, the adjusted $R^2$, $C_p$, or AIC], and then choosing the "best" subset maximizing this statistic. All these criteria, however, have been derived under assumption of no extensive search, so to use them for selecting the best subset is statistically invalid.

Now, we will outline the problems in statistical analysis of SSSPs using the setting up of StRs in FS as an illustration. Under the hypothesis testing approach, the setting up of StRs reduces the cutoff values for the test statistic to stepwise computation; that is, it determines its $(1 - \alpha)$ quantiles at each step. It gives rise to the following problems (cf. the three problems with the $F$-statistic in the INTRODUCTION):

(1) Framing the null hypothesis $NH^i$ at each step $i$;
(2) The choice of test statistic $TS^i$ at each step $i$;
(3) Computation of the distribution of $TS^i$ and obtaining its quantiles.

(1) Framing $NH^i$: If in search-free procedures like SFS and SBE, $NH^i$ is related to the significance of the $i$-th regressor $X_i$ and can be set up without reference to the data, then in FS, for different observed $Y$, different subsets of regressors are left by the $i$-th step, and it is not clear which coefficients should $NH^i$ be related to. This difficulty was first noticed by Pope and Webster,[3] who circumvented it by introducing their condition 2: The independent variables already in regression equations have been determined without reference to the data. Obviously, this condition contradicts the very logic of FS. Later authors[23,24,92,105] undertaking Monte Carlo studies of corruption of distributions of various statistics under extensive search implicitly adopted the

$$NH: \mathring{b}^j = 0, \qquad j = 1, \ldots, k,$$

at each step. This hypothesis, however, appears suitable only at the first step. For any step from the second one on, it is but self-contradictory because the very fact that the second step is being made indicates that the regressor $X_{i1}$ entered at the first step was believed significant; that is, $\mathring{b}^{i1} \neq 0$ with probability $1 - \alpha$. Therefore, the results obtained for such an hypothesis seem devoid of statistical meaning to us.

Miller[29] suggests the following formulation for $NH^i$:

> Given a subset of predictors (not chosen a priori), are the data consistent with zero regression coefficients for all remaining predictors?

This formulation catches the essentials of the problem, but Miller neither formalizes it nor develops a testing technique.

Among feasible approaches to this problem, we have chosen one suggested by a modern view that

> a realistic theory of statistics must take account of the facts that very commonly . . . the model, and sometimes indeed the questions for study, are settled in the light of the data.[106]

On this view, $\text{NH}^i$ formalizes as

$$\text{NH}^i\colon \mathring{Y}_\perp = 0, \tag{3.3.1}$$

where $\mathring{Y}_\perp$ is the projection of the vector $\mathring{Y} = EY$ onto the orthogonal complement of the space spanned by the variables selected by FS over the first $i - 1$ steps. Note that because different regressors are selected for different observed $Y$, the dependence on the data enters through the operator $\perp$. This NH has been introduced in Pinsker *et al.*[7-11] and seems to agree with Miller's suggestion. However, it is by no means the only possible extention of the traditional NH for procedures with extensive search.

(2) The choice of $\text{TS}^i$: After $\text{NH}^i$ has been framed, we face the problem of choosing $\text{TS}^i$. The traditional choice—*F*-statistic—has a number of optimal properties under definite assumptions (reference 107, sect. 4–7). Pope and Webster[3] provided justification for the use of the *F*-statistic in FS with their four conditions. However, these mostly fail in FS and other SSSPs. Therefore, the raison d'être of the *F*-statistic is cast into doubt. Alternatives (such as LF) are suggested in the literature. This question requires further study, but it seems probable that different SSSPs used for solving different RPbs will call for different TSs.

(3) Computation of the distribution of $\text{TS}^i$: Corruption of the distribution of the *F*-statistic (i.e., its deviation from the tabled *F*-distributions) under extensive search has been noticed by many authors [e.g., see references 3, 5 (sect. 6.5), 18, 23, 29, 40, 90, and 92]. The three factors responsible for it (see INTRODUCTION) were analyzed by Pinsker *et al.*[9] Draper and Smith[5] in their section 6.5 under an eloquent title, "A Drawback to Understand but Not Be Overly Concerned About", write:

> Both the backward elimination and the stepwise procedures suffer from a difficulty that is perhaps not obvious at first sight. In the stepwise procedure, for example, the partial *F*-test made on the entry variable uses the largest partial *F*-value from all those partial *F*-values that might have been selected, arising from variables that are not in the regression at that stage. The correct "null" sampling and testing distribution for this is not the ordinary *F*-distribution as we assume, and is very difficult to obtain, except in certain simple cases. Studies have shown, for example, that, in some cases where an entry *F*-test was made at the $\alpha$-level, the appropriate probability was $q\alpha$ where there were $q$ entry candidates at that stage. What can be done about this? One possibility is to find out the correct probability levels for any given case, while another is to make use of a different test statistic instead of the partial *F*. These possibilities have been discussed in recent research papers, but the overall problem has not, at the time of writing, been properly resolved to the extent that would warrant amending the procedures. Until it is resolved, we suggest that the reader use the procedures as given, not be too concerned with the actual probability levels, and simply regard the procedures as making a series of internal comparisons that will produce what appears to be the most useful set of predictors. . . .
>
> Because of the above difficulty, some programs simply ask the user to provide numbers *F* to enter and *F* to remove for making the tests (such as, for example, $F = 4$).

Other authors also seem to share this view, but they stop short of expressing it so straightforwardly. Whatever attitude one might have to such practical recommendations, as for the theoretical side, the statistical theory appears to go down the drain if this way of thinking is followed consistently. Pope and Webster[3] suggested a method for computation of distribution of the *F*-statistic in FS; the method, however, crucially uses their condition 2 (see above). Miller[29] offers several ways of approximate computation of the distribution of the *F*-statistic; all of them, though, are poorly, if at all, justified statistically. Yet, neither Pope and Webster, nor Miller, nor any other authors noticed a fundamental difficulty arising when trying to use the $F$-statistic in FS. Because FS is the process of repeated testing with possible stops at every step, the $F$-statistic is not in general defined for each observed $Y$ at all steps except for the first one. Therefore, before talking about computation of the distribution of the $F$-statistic in FS at a step $i$ ($i > 1$), it is worthwhile to rigorously define it—or, better say, to invent it. It is a fundamental difficulty for which no self-evident remedy seems to exist.

One feasible solution is the introduction of a generalized $F$-statistic $W^i$ at step $i$ by the formula

$$W^i = F^i \quad \text{if there has been no stopping over the first } i-1 \text{ steps,}$$
$$= 0 \quad \text{otherwise.}$$

This approach was developed by Pinsker *et al.*[10]

Another solution is conditioning, that is, introduction of conditional distributions of the $F$-statistic in which the realizations of $Y$ with earlier stops have zero probability.

There is an additional difficulty, though. The distribution of the $F$-statistic at step $i$ depends on the nuisance vector $(\mathring{b}^{j,1}/\sigma, \ldots, \mathring{b}^{j,i-1}/\sigma)$ determined by the regressors entered by the $i$-th step. Therefore, before using the distribution of the $F$-statistic, one should be able to somehow eliminate these nuisance vectors first (see subsection 4.2 for a summary of our results).

In order to avoid the construction of the distribution of the $F$-statistic (or, more generally, of the distribution of the chosen TS in the given SSSP) altogether, one could try to use adaptive (nonparametric) methods like the dummy variables method,[29] the permutation test,[19] the Chaotization,[91,108–111] or some empirical ready-made method. However, the questions of appropriateness and efficiency of those methods, as well as their statistical legitimacy, remain open.

Still another option is the utilization of simultaneous methods for comparing subsets.[40,41] Such procedures do not rely on particular distributions (they are SSSP-free methods) and are statistically valid, but they are crude.

Obviously, all the discussed problems arise in BE, SBE, ES, APR, and in other SSSPs.

### *3.4. Three Lines of Research in Stepwise Subset Selection Procedures*

Now we discuss some problems, solutions, and suggestions for the theory of SSSPs within the framework of subsection 2.4.

#### *I. Interior Theory*

*(A) Stopping Rules (including the theory of hypotheses testing).* We consider it a basic line of research, with all other lines presupposing a solid theory of StRs. We think that several Monte Carlo studies [see part *(D)* below] are but fruitless both methodologically and practically because they all test "wrong" hypotheses. In a similar way, we feel that the use of the "wrong" StRs mars the otherwise very good work of Kennedy and Bancroft,[1] and as a result, their findings cannot be utilized in practice. First attempts to develop a coherent theory of StRs for both SFS and FS were made by Pinsker *et al.*[2,7–11]

In principle, any method for comparing subsets[40] enables one to set up StRs: at each step of the procedure, the current subset is compared with a candidate one and the result of comparison determines continuation versus stopping. The method of Aitkin[41] for comparing the whole set of regressors with its subsets allows one to set up StRs only in the elimination-type procedures like BE and SBE.

*(B) Estimation of Regression Coefficients.* What we wish to have is the unbiased (or slightly biased) estimators $b^i$ for the regression coefficients $\mathring{b}^i$ with a small covariance matrix Var($b$). Unbiased estimators may be obtained with the split-data technique: half of the available data is used for selection, and another half is used for estimation. In some situations, this method works well, but in many fields, sample sizes are just not large enough to make it practical. When the available data do not permit splitting, it is possible that one can make use of some method of biased estimation and then one can try to eliminate the bias. Direct computation of bias in most methods of biased estimation is far too difficult, so the usual way is to estimate the bias and then try to eliminate or to reduce it. The fact that the LS-estimators are biased under the extensive search may be traced at least as far back as Miller (1962).[90] First systematic studies of bias were undertaken by Kennedy and Bancroft,[1] Bendel,[112] and Berk.[17] Miller[29] considers the study of bias to be the major line of research in subset selection and introduces three types of bias: omission bias, competition bias, and stopping rules bias. Some feasible methods of overcoming the bias are as follows:

(1) Bootstrap:[113–117] Pinsker and Trunov *et al.*[118] presented a "parametric version" of Bootstrap (the imitation method) in which the bias is estimated by Monte Carlo methods using estimates of the regression coefficients as if they were the population values. A Monte Carlo study of performance of the Bootstrap estimators of standard errors of regression coefficients is reported by Freedman and Peters.[119]
(2) Jackknife is mentioned in Miller,[29] but to our knowledge no study has been done yet.
(3) Maximum likelihood is discussed in Miller,[29] though no results are given.

It should be pointed out that a statistical study of any such method should be made: to what extent the obtained estimators are biased, what are their variances, etc.

*(C) Estimation and Confidence Intervals for MSE and MSPE.* Some possible approaches are:

(*i*) Splitting.
(*ii*) Conservative estimation.[114]
(*iii*) Various resampling plans: Bootstrap;[117,119,120] Chaotization.[91,108–111,118,121–126]
(*iv*) Asymptotic approach.[51,52]

*(D) Estimation of Other Parameters Like* $R^2$. Direct Monte Carlo estimation of the distributions of interest in various RPs under different assumptions was undertaken in references 20, 21, 23, 24, 92, 105, and 127. All these works used wrong NHs as explained in subsection 3.3 above. Negligence of theoretical matters such as a proper framing of NH at each step of the multistep procedure, the problem of nuisance parameters in TSs, etc., leads to devaluation of results. In our view, in such a complex

problem as the study of SSSPs, first things should be put first: formalizing the notion of SSSP, unambiguously defining $NH^i$ at the $i$-th step, then introducing $TS^i$, studying its distribution, and so on. Application of Monte Carlo techniques could facilitate technical problems, say, derivation of exact forms of distributions, but it can never replace the development of a conceptual framework.

### *II. Exterior Theory*

At the theoretical level, practically nothing has been done yet. This is hardly surprising because Exterior Theory presupposes a developed Interior Theory. Various recommendations and remarks, more or less intuitive and subjective, are scattered in the literature on SSSPs.[4,17,29,38,128,129] Intelligent, expert, and balanced opinions are offered by Draper and Smith (reference 5, chap. 6).

### *III. Comparative Theory*

Comparative Theory of SSSP has not been developed yet, so we can only give some examples, arguments, and references. There is a considerable number of semi-intuitive nonformalized arguments (examples) throughout the literature. They are of two kinds: nonstatistical and statistical (and sometimes mixed).

Nonstatistical arguments usually address situations in which different procedures select different regressors (or groups of regressors), as well as combinatorial properties of the selection procedures. They often manipulate with the word "optimal" invariably implying purely combinatorial properties. These arguments have helped to obtain deeper insights into properties and behaviors of various procedures and have helped to find answers to questions like: when do the subsets selected by the given procedures coincide?; or when does one procedure perform better than the others? For example, Mantel[128] and Miller[29] described the situations in which FS performs badly; the question when BE outperforms FS was clarified in the discussion of Mantel[38,128] and Beale;[93] the problem of agreement of FS and BE with APR raised by Hamaker[130] was settled by Berk.[17] It should be recalled, however, that statistics could not be reduced to combinatorics; in particular, consideration of exclusively sample (i.e., combinatorial) properties usually favors BE over FS, and APR over them both, which may prove illusory and misleading (cf. reference 17 and our analysis below). A popular opinion in the literature concerning APR is that the development of computers has made SSSPs redundant because for an ever larger number of regressors, a complete screening of all combinations of them and the choice of the "optimal" subset are possible. Indeed, exhaustive search can make available some good subsets skipped over by SSSPs for analysis. However, an exhaustive (and even fairly extensive) search is anything but innocuous: besides the often pointed out computational hurdles, it begets purely statistical problems (see subsection 3.3 above and argument 4 below). Comparison of SSSPs is anything but simple; besides a clear and explicit statement of the problem of comparison, of the regression goal, of assumptions, etc., it presupposes well-developed Interior and Exterior Theories. More than that, the development of Exterior Theory in the future will probably render superfluous many comparisons because good understanding of relations between the procedures and the problems being solved by them will enable one to assign a "right" procedure to a given problem without any need for "comparison." This shows interrelation of the Comparative Theory and the Exterior Theory: probably the former should be considered to be a subfield of the latter, but historically it has been developing independently (see below). Now, we present some arguments that might shed light on the problems of comparison.

*Argument 1.* No SSSP has absolute supremacy over all other SSSPs under all possible values of model parameters. Obviously, for any given set of values of the parameters, we could always choose a trivial procedure that picked these very values as the estimates (cf. similar argument concerning estimators in reference 43, sect. 5a).

*Argument 2.* An unambiguous problem statement and an explicit formulation of all of the relevant assumptions can often spare one the trouble of comparing procedures. Suppose, for example, we hesitate as to which of the two procedures, SFS or FS, should be applied to the given problem. If analysis of the structure of the problem shows a natural ordering of importance of the regressors, SFS should be applied and no comparison is needed.

*Argument 3.* Systematic study of properties of different procedures and taking them into consideration while solving particular problems often helps to select appropriate procedures without any comparisons. It can also show that previous arguments may be incomplete or even misleading when we have in mind a particular problem and a particular regression goal. For example, some authors (namely, Anderson[12] and Mantel[38]) claimed that FS could sometimes trigger the termination device too early and thus impair the goodness of fit. By implication, BE was favored over FS. On the intuitive grounds, however, it is perfectly clear that if the regression pool is considerable while the significant regressors are only few, BE will be no good. To untangle this paradox, the theory of the two types of errors, which is almost never used in the analysis of SSSPs, seems helpful. We define forward-type procedures as those that continue the process of selection when the NH is rejected and that stop it when it is accepted. For backward-type procedures, it is vice versa. Following Aitkin,[41] the Type I family error rate is the probability that at least one NH is incorrectly rejected. Then, we have the following theorem, which will not be proved here:

> *Theorem.* Forward-type procedures preserve the Type I family error rate, while the backward-type procedures accumulate it. With the Type II family error rate, the situation is converse.

Thus, if (as is usually the case in the Neyman-Pearson theoretical framework) the Type I error is decisive, FS should be preferred to BE.

*Argument 4.* This entails the problem of comparison of similar SSSPs with different degrees of "extensivity" of search. As mentioned above, the increased extensivity of search is traded off for the goodness of the statistical criteria used. Extensive search implies the decrease in power of the criteria used for deciding on the inclusion of regressors due to the increased probability of big sample values. Particularly, this concerns very extensive procedures like APR. It would be fairly difficult to provide complete formal analysis because, as pointed out above, it would require a developed theory of StRs. Instead, we will give a simplified comparison of the cutoff values for three SSSPs: SFS, FS, and APR for the first step under $\sigma = 1$. It should be stressed, however, that the validity of conclusions does not depend on any particular assumptions. The cutoff values for RSS must be computed. Two cases are considered: orthogonal matrix $X$ and (almost) "collapsed" matrix $X$ with the regressors $X_i$ (almost) collinear. In SFS, the first cutoff value $v$ is the $(1 - \alpha)$ quantile of $\chi_1^2$ in both cases. In FS under orthogonal $X$, the cutoff $w$ may be computed using the order statistic argument: it is the $(1 - \alpha/k)$ quantile of $\chi_1^2$. In FS under the collapsed matrix $X$, the cutoff is $u \gtrsim v$. In APR, following Aitkin,[41] the cutoff is assumed to be $kv$ for both cases. Thus, we have

| | $v_{\text{orthog}}$ | $v_{\text{collap}}$ |
|---|---|---|
| SFS | $v$ | $v$ |
| FS | $w$ | $u$ |
| APR | $kv$ | $kv$ |

where

$$v \ll w < kv,$$

$$v \lesssim u \ll kv.$$

We see that when we go from SFS to FS, the effect of search escalates sometimes dramatically ($v \rightarrow w$) and sometimes slightly ($v \rightarrow u$). As to APR, its use always implies a dramatic increase of search and the corresponding increase in the cutoff values that destroys the power of tests.

### *3.5. Comparative Studies in Literature*

Typically, in Monte Carlo comparisons, the input variables are RPs, their selection schemes, their StRs, parameters of the model, criteria of goodness of fit (usually MSE or MSEP), etc. Under the completely specified model, procedure 1 is better than procedure 2 if $\text{MSE}_1 < \text{MSE}_2$. Under the incompletely specified model, new problems arise usually for some values of parameters $\text{MSE}_1 < \text{MSE}_2$, while for others, the converse inequality holds.

(1) Kennedy and Bancroft:[1] The comparison of SFS and SBE has been made by the MSE criterion averaged over the $n$ observations for different $n$, $k$, $\lambda_i = b_i^2/\sigma^2$, and $k - r$ (with $r$ regressors in the equation). Their conclusion: SBE is preferable to SFS. We have two reservations, however:

- (i) Wrong StRs have been used;
- (ii) The results explicitly depend on the chosen values of the model parameters; under other values, the results might have been different.

Our opinion: no definite conclusion as to the relative merits of SBE against SFS could be reached at this stage.

(2) Sprevak:[131] Two procedures have been compared: one is LS, and the second is the selection of the regressors on the basis of min $C$, where

$$C = \mathrm{RSS}_p + pa^2\sigma^2.$$

Under some parameter values, LS is better; under others, the second procedure is better. Conclusion: the comparison is indecisive and cannot be used in practice.

(3) Trunov:[132] Nine StRs in polynomial regression have been compared by Monte Carlo methods under the MSEP criterion. Two among them are based on Chaotization. The variable parameters in Monte Carlo experiments have been the length $n$ of the training sample and the ratio $n\sigma^2/\Sigma_{i-1}^{n} (\mathring{b}^i)^2$. For the chosen range of parameters, no significant differences between the StRs have been detected and this might be responsible for the persistence of these StRs. Obviously, more deep and extensive experiments are needed to sustain or eliminate some of the StRs.

(4) Bendel and Afifi:[133] Probably, the most extensive comparative experiment reported. For stochastic matrix $X$, eight stopping criteria in FS have been compared under the MSEP criterion of goodness of fit for different model parameters. In their own conclusions and recommendations, the authors keep a low profile. From our point of view (Interior Theory), all their StRs are wrong. It is not clear whether such work would have been needed if the Interior and Exterior Theories had been sufficiently developed. On the other hand, under the present state of affairs, the recommendations of the authors might be of use within the limits indicated in their paper.

(5) Berk:[17] To our knowledge, it is the most relevant work on the problem of comparison of procedures. FS, BE, and APR procedures have been compared by MSE criterion on real data and on known populations. To avoid the issue of StRs, the procedures have been compared for each subset size up to the number of available predictors: FS runs all the way to the full model, BE is carried to completion, and APR finds the maximum sample $R^2$ (minimum error sum of squares) for each subset size. Thus, only stopping-free versions of FS, BE, and APR have been compared. Berk writes:

> While results based on sample data are useful, the subset selection methods really should be compared on the basis of the population parameter. Based on sample values along the all-subsets procedure is always best, but it is not necessarily best in terms of population values.

Indeed, Berk's results on real data and on known populations differ considerably. Furthermore, if on real data, APR is almost invariably better, on known populations, the difference all but vanishes, and sometimes stepwise procedures outperform APR. In our opinion, we could add that comparison is meaningful only on known populations inasmuch as on real data the purely combinatorial (as opposed to statistical) problem is being solved. The following conclusions were made by Berk himself:

> (1) Comparison on known models does not give preference to APR over FS and BE.
> (2) FS does poorly for large subsets and BE does poorly for small subsets when there are predictors which do poorly alone but predict well together.

These facts are not surprising and have been more or less known to Berk's work. Another conclusion that we cannot accept runs as follows:

> It is doubtful that APR can ever do very badly. Because it is best in the sample, the worst it can do in the population is limited by the discrepancy between sample and population values, which vanishes for large sample size. APR can do much better than the other methods and it is unlikely to do poorly, so it would be expected that, averaged over a variety of populations, APR would perform better.

All of this argument is built on combinatorics rather than on statistics, and it completely neglects the issue of StRs which is crucial here. APR has intrinsic shortcomings, of which some have been pointed out in our argument 3 above. A satisfactory theory of StRs is still lacking, and all comparisons in Berk's work, as well as in other works, look unconvincing. As to APR, our guess is that in some situations it can do as poorly as you like. Our conclusions from Berk's work are:

(1) Discussion of combinatorial examples is important and stimulating;
(2) Demonstration of the fact that superiority of APR over FS and BE on unknown models in general vanishes on known models and thereby might be illusory is very important;
(3) All Berk's conclusions about the comparisons proper, in particular, that APR is unlikely to do poorly, seem doubtful due to the absence of a viable theory of StRs.

(6) Berk:[134] Different versions of FS using RSS, PRESS,[47–49] and $C_p$ as selection criteria have been compared using Monte Carlo techniques. Because parameters have been estimated on real populations and no StRs have been used, the above criticisms apply here as well.

(7) Kempthorne[71] applied Bayesian analysis to prove that all SSPs are admissible. This result is unlikely to be helpful in advancing the theory.

## 4. RESULTS AND SUGGESTIONS

In section 2, we presented our views on the structure of RA in general. In section 3, we presented our views on the field of subset selection in particular. In subsection 3.4, we outlined the three topics for research that, when pursued, could help get the subset selection out of the mess it is presently in. Below, we sketch the research to be done in the Interior Theory in the first place and present our results related to the StRs issue.

StRs may be based on different approaches, with only the hypotheses testing approach being tackled in our works. We believe that at least six basic procedures—SFS, SBE, FS, BE, ES, and APR—should be analyzed and supplied with statistically sound StRs. For each procedure, both unconditional and conditional approaches should be developed (or rather, a number of feasible conditional approaches). Until now, we obtained some results only for SFS (both unconditional and some conditional) and for FS (unconditional approach and one version of conditional approach).

We will need the following notation: At step $p + 1$ ($p = 0, 1, \ldots, k - 1$), the vector

$\alpha_p = (\alpha^1, \ldots, \alpha^p)$ of previous significance levels and the present significance level $\alpha^{p+1}$ are given. The cumulative distribution function (cdf) of a random variable $Z$ at the point $t$ will be denoted by $P(t; Z)$. When $\sigma$ is known, the $F$-statistic at the step $p + 1$ will be denoted by $V^{p+1}$, and when $\sigma$ is unknown, it will be denoted by $W^{p+1}$. The vectors of the cutoff values at the first $p$ steps are $v_p = (v^1, \ldots, v^p)$ and $w_p = (w^1, \ldots, w^p)$, respectively. Let $Z = \{z_i\}$ be a set of random variables $z_i$, and let $F_i(t)$ be the set of their cdf's. A random variable $ZR$ with cdf $F_R(t)$ will be called the stochastic supremum (Sup) for the set $Z$ if

$$F_R(t) = \inf_i F_i(t) \quad \text{for all } t.$$

The Sup will also be called the sharp right bound or the sharp conservative bound because it furnishes a conservative test for a bound on the right family of distributions $Z$. If, in addition, $ZR \in Z$, then $ZR$ is a stochastic maximum (Max). (The definitions for Inf and Min are defined in a similar manner.) If the random variables $Y$ and $Z$ are identically distributed, we will write $Y \sim Z$. The $s^2$ will mean the estimate for $\sigma^2$ for the full fit:

$$s^2 = Y'[I_n - X(X'X)^{-1}X']Y/(n - k).$$

### *4.1. Sequential Forward Selection (reference 2)*

At step $p + 1$, the

$$\text{NH}^{p+1}\text{: } \mathring{b}^{p+1} = 0 \tag{4.1.1}$$

is tested assuming $\mathring{b}^i = 0$, $i = p + 2, \ldots, k$. Without loss of generality, we will assume $X'X = I_k$.[1]

*Unconditional Approach*

*Known $\sigma$.* Let

$$\text{NS}_p(V) = \left\{Y\text{: } \bigcap_{i=1}^{p} (V^i > v^i)\right\} \tag{4.1.2}$$

mean that there has been no stopping at the first $p$ steps. Then, the generalized $F$-statistic at step $p + 1$ is

$$\begin{aligned} V^{p+1} &= (b^{p+1})^2/\sigma^2 \qquad && \text{if NS}_p(V) \\ &= 0 && \text{otherwise.} \end{aligned} \tag{4.1.3}$$

For the given $\alpha_p$, under $\text{NH}^{p+1}$ (equation 4.1.1), the statistic $V^{p+1}$ depends on the vector of nuisance (unknown) parameters $\mathring{B}_p = (\mathring{b}_p/\sigma)^2 = [(\mathring{b}^1/\sigma)^2, \ldots, (\mathring{b}^p/\sigma)^2]$. $V\text{L}^{p+1}(\alpha_p) = V^{p+1}(\mathring{B}_p = 0, \alpha_p)$ and $VR = V^{p+1}(\mathring{B}_p, \alpha_p = 1)$ are, respectively, the sharp left and sharp right bounds for the family of null distributions

$$\{V^{p+1}(\mathring{B}_p, \alpha_p)\}\mathring{B}_p \tag{4.1.4}$$

or

$$\text{Min } (4.1.4) \sim VL^{p+1}(\alpha_p), \qquad \text{Sup } (4.1.4) \sim VR.$$

The bounds $VL^{p+1}(\alpha_p)$ and $VR$ do not depend on the nuisance vector $\mathring{B}_p$. The $VL^{p+1}(\alpha_p)$ depends on $\alpha_p$, and its cdf is

$$P(t; VL^{p+1}) = 1 - G^0(t) \prod_{i=1}^{p} G^0(v_L^i),$$

where $G^0(t) = 1 - P[t; \chi_1^2(0)]$, and $v_L^i$ is the cutoff [i.e., $(1 - \alpha^i)$ quantile] for $VL^i(\alpha_{i-1})$. The $VR$ has the central $\chi_1^2$ distribution and can be used for obtaining the conservative test.

*Unknown σ.*

$$\text{NS}_p(W) = \left\{ Y: \bigcap_{i=1}^{p} (W^i > w^i) \right\}. \tag{4.1.5}$$

The generalized $F$-statistic is

$$\begin{aligned} W^{p+1} &= (b^{p+1})^2/s^2 && \text{if NS}_p(W) \\ &= 0 && \text{otherwise.} \end{aligned}$$

For the given $\alpha_p$, under $\text{NH}^{p+1}$ (equation 4.1.1), the statistic $W^{p+1}$ depends on $\mathring{B}_p$. The statistics $WL^{p+1}(\alpha_p) = W^{p+1}(\mathring{B}_p = 0, \alpha_p)$ and $WR = W^{p+1}(\mathring{B}_p, \alpha_p = 1)$ are, respectively, the sharp left and sharp right bounds for the family of null distributions

$$\{W^{p+1}(\mathring{B}_p, \alpha_p)\}\mathring{B}_p \tag{4.1.6}$$

or

$$\text{Min } (4.1.6) \sim WL^{p+1}(\alpha_p), \qquad \text{Sup } (4.1.6) \sim WR.$$

The bounds $WL^{p+1}(\alpha_p)$ and $WR$ do not depend on the nuisance vector $\mathring{B}_p$. The $WR$ has the central $F_{1,n-k}$ distribution and can be used for obtaining the conservative test.

### *Conditional Approach*

*Known σ.* The statistic $V$ under three conditions has been considered:

(a) $\text{NS}_p(V)$;
(b) $b_p = \{Y: b_p = b_{*p}\}$, where $b_{*p}$ has been obtained for the observed $Y_*$;
(c) $V_p = \{Y: V_p = V_{*p}\}$, with $V_{*p} = (V_*^1, \ldots, V_*^p)$ obtained for the observed $Y_*$.

Under $\text{NH}^{p+1}$ (equation 4.1.1), the statistics $V^{p+1}|\text{NS}_p$, $V^{p+1}|b_p$, and $V^{p+1}|V_p$ have the central $\chi_1^2$ distribution.

*Unknown σ.* Consider the condition $\text{NS}_p(W)$. Under equation 4.1.1, for the given $\alpha_p$, the statistics $WCL = W^{p+1}[\mathring{B}_p, \alpha_p = 1|\text{NS}_p(W)]$ and $WCR^{p+1}(\alpha_p) = W^{p+1}[\mathring{B}_p = 0,$

$\alpha_p \,|\, \mathrm{NS}_p(W)]$ are, respectively, the sharp left and sharp right bounds for the family of $\mathrm{NS}_p$-conditional null distributions

$$\{W^{p+1} \,|\, \mathrm{NS}_p(W)\}\mathring{B}_p; \qquad \textbf{(4.1.7)}$$

that is,

$$\mathrm{Inf}\,(4.1.7) \sim W\mathrm{CL}, \qquad \mathrm{Max}\,(4.1.7) \sim W\mathrm{CR}^{p+1}(\alpha_p).$$

The bounds $W\mathrm{CL}$ and $W\mathrm{CR}^{p+1}(\alpha_p)$ do not depend on the nuisance vector $\mathring{B}_p$. The $W\mathrm{CL}$ has the central $F_{1,n-k}$ distribution, which could only provide liberal (that is, underestimated) cutoffs. A conservative test should use the quantiles

$$W\mathrm{CR}^{p+1}(\alpha_p) = B^{p+1}/(S \,|\, (S < \mathrm{HR}_p),$$

where the numerator and the denominator are independent, $B^{p+1} \sim \chi_1^2$ under $\mathrm{NH}^{p+1}$, $(n-k)S^2 \sim \chi_{n-k}^2$, and $\mathrm{HR}_p$ is the smallest order statistic from $p$-independent random variables having $\chi_1^2/w^i$ distributions ($i = 1, \ldots, p$).

No results have yet been obtained for any other conditional distributions.

## *4.2. Forward Selection*

*Unconditional Approach*

Under orthogonal matrix $X$, the following results have been obtained.[10]

*Known $\sigma$.* Let $\mathrm{NS}_p(V)$ be defined for FS by equation 4.1.2 and let

$$V^{p+1} = \mathrm{dRSS}^{p+1}/\sigma^2 \quad \text{if } \mathrm{NS}_p(V)$$
$$= 0 \quad \text{otherwise,} \qquad \textbf{(4.2.1)}$$

where $\mathrm{dRSS}^{p+1}$ is the decrease in RSS at the $(p+1)$ step in FS. Note that $V^{p+1}$ for FS in equation 4.2.1 differs from $V^{p+1}$ for SFS in equation 4.1.3. For the given $\alpha_p$, under $\mathrm{NH}^{p+1}$ (equation 3.3.1), the statistic $V^{p+1}$ in equation 4.2.1. depends on the vector of nuisance (unknown) parameters

$$\mathring{B}_p = [(\mathring{b}^{i1}/\sigma)^2, (\mathring{b}^{i2}/\sigma)^2, \ldots, (\mathring{b}^{ip}/\sigma)^2], \qquad \textbf{(4.2.2)}$$

where $\{i1, i2, \ldots, ip\}$ is the set of the indices of the regressors entered over the first $p$ steps. For orthogonal $X$, under $\mathrm{NH}^{p+1}$ (equation 3.3.1), the family of distributions

$$\{V^{p+1}(\mathring{B}_p, \alpha_p)\}\mathring{B}_p$$

has the sharp conservative bound $V\mathrm{R}^{p+1}$. The random variable $V\mathrm{R}^{p+1}$ is distributed as the largest order statistic from $(k-p)$-independent central $\chi_1^2$-distributed random variables and has a cdf of

$$P(t;\, V\mathrm{R}^{p+1}) = [P(t;\, \chi_1^2)]^k.$$

Thus, $V\mathrm{R}^{p+1}$ can be used for a conservative test.

*Unknown* $\sigma$. Let $\mathrm{NS}_p(W)$ be defined for FS by equation 4.1.5 and let

$$W^{p+1} = \mathrm{dRSS}^{p+1}/s^2 \qquad \text{if } \mathrm{NS}_p(W)$$
$$= 0 \qquad \text{otherwise.} \qquad \textbf{(4.2.3)}$$

For the given $\alpha_p$, under $\mathrm{NH}^{p+1}$ (equation 3.3.1), the statistic $W^{p+1}$ in equation 4.2.3 depends on the vector of nuisance parameters $\mathring{B}_p$ in equation 4.2.2. For orthogonal $X$, under $\mathrm{NH}^{p+1}$ (equation 3.3.1), the family of distributions

$$\{W^{p+1}(\mathring{B}_p, \alpha_p)\}\mathring{B}_p$$

has the sharp conservative bound $W\mathrm{R}^{p+1}$. The random variable $W\mathrm{R}^{p+1}$ is distributed as the largest order statistic from $(k - p)$-dependent (through common denominator) $F_{1,n-k}$-distributed random variables and can be used in a conservative test.

In the general case of nonorthogonal $X$, no such results could be obtained as follows from the two counterexamples in Pinsker *et al.*[10] Thus, other methods for obtaining the cutoffs should be used (conditional approach, plug-in method, etc.).

### *Conditional Approach*

*Known* $\sigma = 1$ *(see references 7–9, and 11).* We introduce the condition C that says that over the first $p$ steps, FS has selected the same set of regressors as for the observed vector $Y_*$ with the same coefficients. Then under $\mathrm{NH}^{p+1}$ (equation 3.3.1), the cdf of the C-conditional $V^{p+1}$ is

$$P(t; V^{p+1}|\mathrm{C}) = \Pr(Y_\perp \in T | Y_\perp \in S_\perp)$$
$$= \Pr(Z \in (T \cap S_\perp))/\Pr(Z \in S_\perp), \qquad \textbf{(4.2.4)}$$

where $Z \sim N(0; I_{k-p})$, $Y_\perp$ is the projection of vector $Y$ onto the orthogonal complement $L_\perp$ (in the range space of $X$) of the subspace spanned by the introduced $p$ regressors, and $S_\perp$ and $T$ are two parallelepipeds in $L_\perp$ that are completely specified by the observed $Y_*$. Numerical methods should be employed to obtain C-conditional cutoffs.

In a particular case of orthogonal matrix $X$, the null distribution of $V^{p+1}|\mathrm{C}$ is that of the largest order statistic from $(k - p)$-independent identically $P(t; \chi_1^2|\chi_1^2 < V_*^p)$-distributed random variables, and its cdf is

$$P(t; V^{p+1}|\mathrm{C}) = [P(t; \chi_1^2|\chi_1^2 < V_*^p)]^{k-p}$$
$$= [P(t; \chi_1^2)]^{k-p}/[P(V_*^p; \chi_1^2)]^{k-p} \qquad \text{if } t < V_*^p$$
$$= 1 \qquad \text{otherwise.}$$

*Unknown* $\sigma$. Under the condition C and NH of equation 3.3.1, the statistic $W^{p+1}$ depends on the nuisance parameter $\sigma$. No useful results have been obtained yet.

It seems desirable to study along these lines the other SSSPs like SBE, BE, ES, and APR. Still another line of research would be to base the theory of StRs on the prediction errors rather than on the hypotheses testing.

## 5. CONCLUDING REMARKS

We believe that progress in that part of regression analysis dealing with subset selection procedures hinges upon the success of three lines of investigation:

(1) Mathematical, statistical, and empirical study of each procedure (Interior Theory).
(2) Development of criteria for assessing performance (efficiency) of a given procedure when applied for solving a given regression problem (Exterior Theory).
(3) Comparing performances of different procedures.

Our own work thus concentrated on studying the problem of stopping rules in two forward selection procedures using the hypotheses testing framework. This research constitutes just one element of the Interior Theory. Alternative approaches might investigate different test statistics, different principles of setting up the stopping rules (based, for example, on predictive performance of models), different principles of comparing subsets, etc. By having a workable theory of stopping rules, work on performance assessment and comparison might be advanced. On the other hand, we are aware of the fact that obtaining exact distributional results for each subset selection procedure might prove far too difficult, so other alternatives (Bootstrap, Chaotization) should be tried. Such nonparametric distribution-free methods could also be directly applied to the second and third lines of investigation mentioned above.

## ACKNOWLEDGMENTS

This paper is a direct outgrowth of our common work with I. Sh. Pinsker and V. Kipnis. I. Sh. Pinsker's help was invaluable at all stages of the draft preparation. We also benefited from comments and criticisms made on our previous work by N. Mantel (American University, Bethesda, Maryland), Robert Berk (Rutgers University, New Brunswick, New Jersey), Alan Miller (CSIRO Division of Mathematics and Statistics, Australia), R. R. Hocking (Texas A&M University, College Station, Texas), and J. T. Webster (Southern Methodist University, Dallas, Texas).

## REFERENCES

1. Kennedy, W. J. & T. A. Bancroft. 1971. Model building for prediction in regression based upon repeated significance tests. Ann. Math. Stat. **42:** 1273–1284.
2. Pinsker, I. Sh., V. Kipnis & E. Grechanovsky. 1986. *F*-to-enter stopping rules in sequential forward selection. Submitted for publication.
3. Pope, P. T. & J. T. Webster. 1972. The use of an *F*-statistic in stepwise regression procedures. Technometrics **14:** 327–340.
4. Hocking, R. R. 1976. The analysis and selection of variables in linear regression. Biometrics **32:** 1–49.
5. Draper, N. R. & H. Smith. 1981. Applied Regression Analysis (2nd edition). Wiley. New York.
6. Montgomery, D. C. & E. A. Peck. 1982. Introduction to Linear Regression Analysis. Wiley. New York.
7. Pinsker, I. Sh., V. Kipnis & E. Grechanovsky. 1984. Stopping rules in forward

selection procedure. Presented at the International Scientific Seminar on Collective Phenomena, Tel Aviv.

8. PINSKER, I. SH., V. KIPNIS & E. GRECHANOVSKY. 1985. Conditional cutoffs in the forward selection algorithm (in Russian). Planirovanie Experimenta. Obshchestv. Znan. RSFSR no. 89–92. Moscow.
9. PINSKER, I. SH., V. KIPNIS & E. GRECHANOVSKY. 1985. The use of the *F*-statistic in the forward selection regression algorithm. Presented at the 145th Annual Joint Statistical Meetings of Am. Stat. Assoc., Biometric Soc., and IMS, Las Vegas. 1985 Proceedings of Stat. Comp. Sect., Am. Stat. Assoc., p. 419–423.
10. PINSKER, I. SH., V. KIPNIS & E. GRECHANOVSKY. 1985. The use of conservative cutoffs in forward selection procedure. Presented at the III[rd] International Meeting of Statistics in the Basque Country, Bilbao, Spain. To be published.
11. PINSKER, I. SH., V. KIPNIS & E. GRECHANOVSKY. 1985. The use of conditional cutoffs in forward selection procedure. Presented at the International Conference on Foundations of Statistical Inference, Tel Aviv. Revised version to appear in Commun. Stat.
12. ANDERSON, T. W. 1962. The choice of the degree of a polynomial regression as a multiple decision problem. Ann. Math. Stat. **33:** 255–265.
13. EFROYMSON, M. A. 1960. Multiple regression analysis. *In* Mathematical Methods for Digital Computers 1. A. Ralston & H. S. Wilf, Eds.: 191–203. Wiley. New York.
14. EFROYMSON, M. A. 1966. Stepwise regression—a backward and forward look. Presented at Eastern Regional Meet. Inst. Math. Stat., Florham Park, New Jersey.
15. SEBER, G. A. F. 1977. Linear Regression Analysis. Wiley. New York.
16. KIPNIS, V. M. 1979. On an algorithm for construction of quasilinear predictor (in Russian). *In* Models, Algorithms, Decision-Making. I. Sh. Pinsker, Ed.: 62–90. Nauka. Moscow.
17. BERK, K. N. 1978. Comparing subset regression procedures. Technometrics **20:** 1–6.
18. DRAPER, N. R., I. GUTTMAN & H. KANEMASU. 1971. The distribution of certain regression statistics. Biometrika **58:** 295–298.
19. FORSYTHE, A. B., L. ENGELMAN, R. JENNRICH & P. R. A. MAY. 1973. Stopping rule for variable selection in multiple regression. J. Am. Stat. Assoc. **68:** 75–77.
20. ZIRPHILE, J. 1975. Letter to the editor. Technometrics **17:** 145.
21. ZURNDORFER, E. A. & H. R. GLAHN. 1977. Significance testing of regression equations developed by screening regression. *In* 5th Conf. on Prob. and Stat. in Atmos. Sci. Am. Meteorol. Soc., p. 95–100.
22. DRAPER, N. R., I. GUTTMAN & L. LAPCZAK. 1979. Actual rejection levels in a certain stepwise test. Commun. Stat. Theory Method **A8**(2): 99–105.
23. WILKINSON, L. 1979. Tests of significance in stepwise regression. Psychol. Bull. **86:** 168–174.
24. WILKINSON, L. & G. DALLAL. 1981. Tests of significance in forward selection regression with an *F*-to-enter stopping rule. Technometrics **23:** 377–380.
25. THOMPSON, M. L. 1978. Selection of variables in multiple regression: Part I. A review and evaluation. Int. Stat. Rev. **46:** 1–19.
26. THOMPSON, M. L. 1978. Selection of variables in multiple regression: Part II. Chosen procedures, computations, and examples. Int. Stat. Rev. **46:** 129–146.
27. HOCKING, R. R. 1983. Developments in linear regression methodology: 1959–1982 (with Discussion). Technometrics **25:** 219–249.
28. HOCKING, R. R. & O. J. PENDLETON. 1983. The regression dilemma. Commun. Stat. Theory Method **12**(5): 497–527.
29. MILLER, A. J. 1984. Selection of subsets of regression variables (with Discussion). J. R. Stat. Soc. **A147:** 389–425. (The article referred to in the quoted passage in subsection 2.5 is from: FRIEDMAN, J. H. & V. STUETZLE. 1981. Projection pursuit regression. J. Am. Stat. Assoc. **75:** 817–823.)
30. BROOKS, R. J. 1972. A decision theory approach to optimal regression designs. Biometrika **59:** 563–571.
31. ATKINSON, A. C. & V. V. FEDOROV. 1975. The design of experiments for discriminating between two rival models. Biometrika **62:** 57–70.
32. ST. JOHN, R. C. & N. R. DRAPER. D-Optimality for regression designs: a review. Technometrics **17:** 15–23.

33. Box, G. E. P. & G. C. Tiao. Bayesian Inference in Statistical Analysis. Addison–Wesley. Reading, Massachusetts.
34. Atkinson, A. C. 1982. Regression diagnostics, transformations, and constructed variables (with Discussion). J. R. Stat. Soc. **B44:** 1–36.
35. Cook, R. D. & S. Weisberg. 1982. Residuals and Influence in Regression. Chapman & Hall. London.
36. (a) Waterman, M. S. 1974. A restricted least squares problem. Technometrics **16:** 135–136; (b) Crocker, D. C. 1980. Review of Seber (reference 15). Technometrics **22:** 130.
37. Lindley, D. V. 1968. The choice of variables in multiple regression. J. R. Stat. Soc. **B30:** 31–53.
38. Mantel, N. 1970. Why stepdown procedures in variable selection. Technometrics **12:** 621–625.
39. Harville, D. A. 1984. Discussion of Miller (reference 29).
40. Spjøtvoll, E. 1972. Multiple comparison of regression functions. Ann. Math. Stat. **43:** 1076–1088.
41. Aitkin, M. A. 1974. Simultaneous inference and the choice of variable subsets in multiple regression. Technometrics **16:** 221–227.
42. Demidenko, E. Z. 1981. Linear and Nonlinear Regressions (in Russian). Finansi i Statistika. Moscow.
43. Rao, C. R. 1973. Linear Statistical Inference and Its Applications (2nd edition). Wiley. New York.
44. Rothman, D. 1968. Letter to the editor. Technometrics **10:** 432.
45. Akaike, H. 1970. Statistical predictor identification. Ann. Inst. Stat. Math. **22:** 203–217.
46. Akaike, H. 1971. Information theory and an extension of the maximum likelihood principle; 1973. Second International Symposium on Information Theory. B. N. Petrov & F. Csaki, Eds.: 267–281. Akad. Kiadó. Budapest.
47. Allen, D. M. 1971. Mean square error of prediction as a criterion for selecting variables. Technometrics **13:** 469–475.
48. Allen, D. M. 1971. The prediction sum of squares as a criterion for selecting prediction variables. Dept. of Statistics, Univ. of Kentucky. Tech. report no. 23.
49. Cady, F. B. & D. M. Allen. 1972. Combining experiments to predict future yield data. Agron. J. **64:** 211–214.
50. Mallows, C. L. 1973. Some comments on $C_p$. Technometrics **15:** 661–675.
51. Shibata, R. 1976. Selection of the order of an autoregressive model by Akaike's Information Criterion. Biometrika **63:** 117–126.
52. Shibata, R. 1984. Approximate efficiency of a selection procedure for the number of regression variables. Biometrika **71:** 43–49.
53. Young, A. S. 1982. The Bivar criterion for selecting regressors. Technometrics **24:** 181–189.
54. Gorman, J. W. & R. J. Toman. 1966. Selection of variables for fitting equations to data. Technometrics **8:** 27–51.
55. Daniel, C. & F. Wood. 1980. Fitting Equations to Data (2nd edition). Wiley. New York.
56. Helms, R. W. 1974. The average estimated variance criterion for the selection-of-variables problem in general linear models. Technometrics **16:** 261–273.
57. Anscombe, F. J. 1967. Topics in the investigation of linear relations fitted by the method of least squares (with Discussion). J. R. Stat. Soc. **B29:** 1–52.
58. Goldberger, A. 1968. Topics in Regression Analysis. MacMillan Co. New York.
59. Marquardt, D. W. & R. D. Snee. 1975. Ridge regression in practice. Am. Stat. **29:** 3–20.
60. Smith, A. F. M. & M. Goldstein. 1975. Ridge regression: some comments on a paper of Conniffe and Stone. Statistics **24:** 61–66.
61. Hill, R. W. & P. W. Holland. 1977. Two robust alternatives to least-squares regression. J. Am. Stat. Assoc. **72:** 828–833.
62. Dempster, A. P., M. Schatzoff & N. Wermuth. 1977. Simulation study of alternatives to ordinary least squares (with Discussion). J. Am. Stat. Assoc. **72:** 77–106.

63. COPAS, J. B. 1983. Regression, prediction, and shrinkage. J. R. Stat. Soc. **B45:** 311–354.
64. CONNIFFE, D. & J. STONE. 1973. A critical view of ridge regression. Statistics **22:** 181–187.
65. CONNIFFE, D. & J. STONE. 1975. A reply to Smith and Goldstein. Statistics **24:** 67–68.
66. THISTED, R. 1977. Comment on Dempster *et al.* (reference 62). J. Am. Stat. Assoc. **72:** 102–103.
67. DRAPER, N. R. & R. C. VAN NOSTRAND. 1977. Shrinkage estimators: review and comments. Dept. of Statistics, Univ. of Wisconsin–Madison. Tech. report no. 500.
68. DRAPER, N. R. & R. C. VAN NOSTRAND. 1977b. Ridge regression: is it worthwhile? Dept. of Statistics, Univ. of Wisconsin–Madison. Tech. report no. 501.
69. DRAPER, N. R. & R. C. VAN NOSTRAND. 1979. Ridge regression and James-Stein estimators: review and comments. Technometrics **21:** 451–466.
70. HJORTH, U. 1982. Model selection and forward validation. Scand. J. Stat. **9:** 95–105.
71. KEMPTHORNE, P. 1984. Admissible variable-selection procedures when fitting regression models by least squares for prediction. Biometrika **71:** 593–597.
72. CHATFIELD, C. 1982. Discussion of Atkinson (reference 34). (The article referred to in the quoted passage is from: ARMSTRONG, J. S. 1970. How to avoid exploratory research. J. Advert. Res. **10:** 27–30.)
73. COPAS, J. B. 1984. Discussion of Miller (reference 29).
74. WEISBERG, S. 1983. Some principles for regression diagnostics and influence analysis (Discussion of Hocking, reference 27). Technometrics **25:** 240–244.
75. BROSS, I. D. J. 1982. Simplicity and credibility: A counterstrategy. Stat. Prob. Lett. **1:** 79–83.
76. WALLS, R. E. & D. L. WEEKS. 1969. A note on the variance of a predicted response in regression. Am. Stat. **23:** 24–26.
77. RAO, P. 1971. Some notes on misspecification in regression. Am. Stat. **25:** 37–39.
78. NARULA, S. & J. S. RAMBERG. 1972. Letter to the editor. Am. Stat. **26:** 4.
79. ROSENBERG, S. H. & P. S. LEVY. 1972. A characterization on misspecification in the general linear regression model. Biometrics **28:** 1129–1132.
80. HOCKING, R. R. 1974. Misspecification in regression. Am. Stat. **28:** 39–40.
81. MCCABE, G. P., JR. 1978. Evaluation of regression coefficient estimates using $\alpha$-acceptability. Technometrics **20:** 131–139.
82. NIE, N. H., C. H. HULL, J. G. JENKINS, K. STEINBRENNER & D. BENT. 1975. SPSS: Statistical Package for the Social Sciences (2nd edition). McGraw–Hill. New York.
83. BARR, A. J., J. H. GOODNIGHT, J. P. SALL & J. T. HILWIG. 1976. A User's Guide to SAS. SAS Institute. Raleigh, North Carolina.
84. DIXON, W. J. & M. BROWN. 1978. BMDP: Biomedical Computer Programs (2nd edition). Univ. of California Press. Berkeley.
85. DIXON, W. J., Ed. 1979. BMDP: Biomedical Computer Programs. Univ. of California Press. Berkeley.
86. PAGE, E. S. 1967. A note on generating random permutations. Appl. Stat. **16:** 133–134.
87. SPRENT, P. 1968. Discussion of Lindley (reference 37).
88. PLACKETT, R. L. 1984. Discussion of Miller (reference 29).
89. MILLER, R. G. 1981. Simultaneous Statistical Inference (2nd edition). Springer Pub. New York.
90. MILLER, R. G. 1962. Statistical prediction by discriminant analysis. Meteorol. Monogr. (Am. Meteorol. Soc.) **4:** no. 25.
91. PINSKER, I. SH. 1973. The choice of structure and estimation of parameters of a decision rule for small samples (in Russian). *In* Computer Simulation and Automated Analysis of Electrocardiograms. I. Sh. Pinsker, Ed.: 24–34. Nauka. Moscow.
92. (a) DIEHR, G. & D. R. HOFLIN. 1974. Approximating the distribution of the sample $R^2$ in best subset regressions. Technometrics **16:** 317–320; (b) LINDLEY, D. V. 1984. Discussion of Miller (reference 29).
93. BEALE, E. M. L. 1970. Note on procedures for variable selection in multiple regression. Technometrics **12:** 909–914.
94. COX, D. R. & E. J. SNELL. 1974. The choice of variables in observational studies. Appl. Stat. **23:** 51–59.

95. Draper, N. R. 1982. Applied regression analysis bibliography update 1982. Dept. of Statistics, Univ. of Wisconsin–Madison. Tech. report no. 667.
96. Garside, M. J. 1965. The best subset in multiple regression analysis. Appl. Stat. **14:** 196–200.
97. Garside, M. J. 1971. Some computational procedures for the best subset problem. Appl. Stat. **20:** 8–15.
98. Schatzoff, M., R. Tsao & S. Fienberg. 1968. Efficient calculation of all possible regressions. Technometrics **10:** 769–779.
99. Furnival, G. M. 1971. All possible regressions with less computation. Technometrics **13:** 403–408.
100. Furnival, G. M. & R. W. M. Wilson, Jr. 1974. Regression by leaps and bounds. Technometrics **16:** 499–511.
101. Morgan, J. A. & J. F. Tatar. 1972. Calculation of the residual sum of squares for all possible regressions. Technometrics **14:** 317–325.
102. Beale, E. M. L., M. G. Kendall & D. W. Mann. The discarding of variables in multivariate analysis. Biometrika **54:** 357–366.
103. Hocking, R. R. & R. N. Leslie. 1967. Selection of the best subset in regression analysis. Technometrics **9:** 531–540.
104. Gabriel, K. R. & F. C. Pun. 1979. Binary prediction of weather events with several predictors. Sixth Conference on Prob. and Stat. in Atmos. Sci., Am. Meteorol. Soc., p. 248–253.
105. Rencher, A. C. & F. C. Pun. 1980. Inflation of $R^2$ in best subset regression. Technometrics **22:** 49–53.
106. Cox, D. R. 1977. The role of significance tests. Scand. J. Stat. **4:** 49–70.
107. Lehmann, E. L. 1959. Testing Statistical Hypotheses. Wiley. New York.
108. Pinsker, I. Sh. 1978. A chaotization based diagnostic algorithm for handling small samples. *In* Modern Electrocardiology. Proceedings of the IV International Congress on Electrocardiology, Balatonfured, Hungary. A. Antaloczy, Ed.: 35–40. Akad. Kiadó. Budapest. Excerpta Medica. Amsterdam.
109. Pinsker, I. Sh. 1979. The chaotization principle and its applications in data processing (in Russian). *In* Models, Algorithms, Decision-Making. I. Sh. Pinsker, Ed.: 5–38. Nauka. Moscow.
110. Pinsker, I. Sh. 1980. Estimating validity of prediction models on the basis of the chaotization principle. Report Sent to the Task Force Meeting on Model Validity and Credibility, International Institute for Applied Systems Analysis, Laxenburg, Austria.
111. Pinsker, I. Sh. & V. G. Trunov. 1979. Comparing some criteria for effectiveness of training in estimating the relationship in empirical data (in Russian). *In* Models, Algorithms, Decision-Making. I. Sh. Pinsker, Ed.: 90–100. Nauka. Moscow.
112. Bendel, R. B. 1973. Stopping rules in forward stepwise regression. Ph.D. dissertation, Univ. of California at Los Angeles. Thesis no. 74-11496. University Microfilms. Ann Arbor, Michigan.
113. Efron, B. 1979. Bootstrap methods: Another look at the Jackknife. Ann. Stat. **7:** 1–26.
114. Efron, B. 1979. Computers and the theory of statistics: Thinking the unthinkable. SIAM Rev. **21:** 460–480.
115. Efron, B. 1982. The Jackknife, the Bootstrap and Other Resampling Plans. SIAM. Philadelphia.
116. Platt, C. A. 1982. Bootstrap stepwise regression. Proc. Bus. and Econ. Sect. Am. Stat. Assoc., p. 586–589.
117. Bunke, O. & B. Droge. 1984. Bootstrap and cross-validation estimates of the prediction error for linear regression models. Ann. Stat. **12:** 1400–1424.
118. Pinsker, I. Sh., V. G. Trunov, V. Kipnis & E. A. Aidu. 1985. Imitation estimators for quality of decisions (in Russian). *In* Search for a Relationship and Estimation of the Error. I. Sh. Pinsker, Ed.: 14–32. Nauka. Moscow.
119. Freedman, D. A. & S. C. Peters. 1984. Bootstrapping a regression equation: Some empirical results. J. Am. Stat. Assoc. **79:** 97–106.
120. Stine, R. A. 1985. Bootstrap prediction intervals for regression. J. Am. Stat. Assoc. **80:** 1026–1031.

121. PINSKER, I. SH. 1973. Assessment of a training algorithm and a training sample (in Russian). *In* Computer Simulation and Automated Analysis of Electrocardiograms. I. Sh. Pinsker, Ed.: 13–23. Nauka. Moscow.
122. PINSKER, I. SH. 1976. Representing the function in many variables as a sum of products of functions in one variable (in Russian). *In* Mathematical Processing of Medical-Biological Information. I. Sh. Pinsker, Ed.: 7–29. Nauka. Moscow.
123. PINSKER, I. SH. & V. G. TRUNOV. 1973. Developing decision rules in medical diagnostic problems (in Russian). *In* Computer Simulation and Automated Analysis of Electrocardiograms. I. Sh. Pinsker, Ed.: 151–164. Nauka. Moscow.
124. KIPNIS, V. M. 1977. An approach to assessing accuracy of prediction by means of the chaotization principle (in Russian). *In* Questions of Mathematical-Statistical Analysis of Short-Term Economical Processes. CEMI. Moscow.
125. KIPNIS, V. M. & I. SH. PINSKER. 1979. Prediction in short time series based on the chaotization principle (in Russian). *In* Models, Algorithms, Decision-Making. I. Sh. Pinsker, Ed.: 38–61. Nauka. Moscow.
126. TRUNOV, V. G. 1985. Estimating the prediction errors under the variable selection in linear regression (in Russian). *In* Search for a Relationship and Estimation of the Error. I. Sh. Pinsker, Ed.: 57–68. Nauka. Moscow.
127. LAWRENCE, M. B., C. J. NEUMANN & E. L. CASO. 1975. Monte Carlo significance testing as applied to the development of statistical prediction of tropical cyclone motion. Fourth Conf. on Prob. and Stat. in Atmos. Sci., Am. Meteorol. Soc., p. 21–24.
128. MANTEL, N. 1971. Letter to the editor. Technometrics **13:** no. 2.
129. GUGEL, H. W. 1972. SELECT—a computer program for isolating "best" regressions. Gen. Mot. Corp. Res. Lab. Res. Publ.
130. HAMAKER, H. C. 1962. On multiple regression analysis. Stat. Ned. **16:** 31–56.
131. SPREVAK, D. 1976. Statistical properties of estimates of linear models. Technometrics **18:** 283–289.
132. TRUNOV, V. G. 1976. Comparing some stopping rules while developing a decision rule for a small sample (in Russian). *In* Mathematical Processing of Medical-Biological Information. I. Sh. Pinsker, Ed.: 58–64. Nauka. Moscow.
133. BENDEL, R. B. & A. A. AFIFI. 1977. Comparison of stopping rules in forward "stepwise" regression. J. Am. Stat. Assoc. **72:** 46–53.
134. BERK, K. N. 1978. Sequential PRESS, forward selection, and the full regression model. Proc. Stat. Comput. Sect. Am. Stat. Assoc., p. 309.

# A Predictive Probabilistic Estimate for Selecting Subsets of Regressor Variables

V. L. BRAILOVSKY

## INTRODUCTION

There is an extensive literature on selecting subsets of regressor variables.[1–3] The problem may be briefly described as follows. One has, for experiment $i$, the value of a set of explanatory variables, $x_i$, and the value of a response function, $y_i$, which may be subject to some additive error, $\epsilon_i$. The problem is to find a formula, depending on the variables $x$, to predict the values of the response function. We specify a set of prospective regressors, that is, a set of known functions $\{\varphi_j(x)\}$, that we believe may be useful as components of the desired regression formula. The usual procedure for obtaining the regression formula is to examine different subsets of the set $\{\varphi_j(x)\}$ and to calculate an estimate of prediction quality for each one. The subset of regressors with the best estimate of quality is then selected.

In regression analysis, the properties of different procedures and estimates are studied under the assumption that the subset of variables is selected without reference to the data. In practice, however, one typically uses the same data for calculating regression coefficients, for estimating the prediction quality of a given subset of regressors, and for selecting the best subset on the basis of this estimate. In this case, the estimate of prediction quality for the selected regression formula will be biased by a competition effect.

This phenomenon was described long ago (e.g., see references 4 and 5), but it is almost completely overlooked by many existing procedures for estimating regressions by selecting subsets of regressors. The phenomenon of competition bias seems to be very important and some unreliable predictions, obtained with the help of an apparently good regression formula, may be ascribed to it.[11,15] When a large search is conducted, the final regression formula will typically fit the observed data quite well and will thus have a high estimate of prediction quality; however, it may in fact predict $y$ quite poorly for $x$ values not observed in the data set.

One way to deal with competition bias is to study the bias per se, so that estimates of prediction quality can be corrected accordingly. Unfortunately, the state of such studies is still far from results that can be applied to many practical problems.[6]

Another approach is to use some additional estimates, which are not subjected to competition bias, or to account for the bias in an indirect way. In reference 7, some conditions are established under which minimization of a sample estimate of prediction quality in the course of a search leads to a subset of regressors with rather good genuine prediction quality. The estimate of prediction quality used there is a corrected sample estimate, which, with high probability, works as an upper bound for the averaged deviation of the regression from the response function. This estimate depends on some characteristics of the size of the search among the prospective regressors, as well as on some assumptions concerning the probability distribution of the explanatory variables $x$ and the form of the response function.

In this article, a probabilistic estimate of prediction quality of subsets of regressor variables is suggested. This estimate is not subject to competition bias and is not a monotone function of model complexity. It should provide an effective method for selecting the best subsets of regressor variables and an unbiased assessment of the prediction quality of the selected regression formula.

## A PROBABILISTIC ESTIMATE OF PREDICTION QUALITY FOR SELECTING SUBSETS OF REGRESSOR VARIABLES AND ITS PROPERTIES

### *Part 1*

Let $X$ ($x \in X$) be a finite-dimensional space of explanatory variables and let $y$ be the response function. Assume we have data $\{y_i, x_i\}_{i=1}^N$ of the form $y_i = f(x_i) + \epsilon_i$. For definiteness, let $\epsilon_i$ be independent random variables with zero mean and common variance.

Denote the set of prospective regressors by $\phi = \{\varphi_1(x), \varphi_2(x), \ldots, \varphi_m(x)\}$ and let $\varphi_{r1}(x), \varphi_{r2}(x), \ldots, \varphi_{rs}(x)$ be a subset of $\phi$. Assume the regression function, $\Sigma c_j \varphi_{rj}(x)$, is fit by least squares and assess the quality of the fit by

$$\Delta_y = 1 - \frac{\sum_{i=1}^{N}\left[y_i - \sum_{j=1}^{s} \hat{c}_j \varphi_{rj}(x_i)\right]^2}{\sum_{i=1}^{N} y_i^2} = 1 - \frac{RSS}{\sum_{i=1}^{N} y_i^2}, \tag{1}$$

where $\hat{c}_j$ are the least-squares parameter estimates.[8] It is easy to see that $0 \leq \Delta_y \leq 1$.

As described in the previous section, if one selects from among a collection of subsets of $\phi$ that which maximizes equation 1, this value for the selected subset is subject to competition bias and will not provide a good estimate of prediction quality for the selected formula. No correction coefficient can change the situation when the main source of the bias is competition and selection. This applies to well-known estimates such as Mallows' $C_p$, Akaike's Information Criterion, etc., as well as Jackknife methods.[6]

### *Part 2*

Let us now introduce the probabilistic estimate of prediction quality. The basic idea of the estimate is to compare the observed "best fit" (in terms of equation 1) for various subsets of $\phi$ to the best fit that occurs when, in fact, there is no relation whatsoever between the response function and the explanatory variables. This provides a reference distribution for assessing the best fit that is specifically designed to account for the competition bias.

To define the estimate, define a noise function $\nu(x)$ at the $N$ sample points $x_1, x_2, \ldots, x_N$ by generating $N$-independent $N(0, 1)$ random variables $\xi_1, \xi_2, \ldots, \xi_N$. Now for the noise function $\nu(x)$, perform the same kind of search among different subsets of the set of regressors $\phi$ as was described in PART 1 for the response function $y$. For a given subset of regressors $\varphi_{r1}(x), \varphi_{r2}(x), \ldots, \varphi_{rs}(x)$, assess the fit to the noise function by

$$\Delta_\nu = 1 - \frac{\sum_{i=1}^{N} \left[ \nu(x_i) - \sum_{j=1}^{s} \hat{c}_j \varphi_{rj}(x_i) \right]^2}{\sum_{i=1}^{N} \nu^2(x_i)} . \quad (2)$$

Suppose we divide the search into stages, while considering at stage $G_s$, all subsets with $s$ regressors. Let $\Delta_{\nu\max}$ be the maximum value of equation 2 obtained at stage $G_s$.

Consider now an ensemble of all possible noise functions $\{\nu(x)\}$ as follows. Represent a function, defined on the sample set $x_1, x_2, \ldots, x_N$, as a vector in $R_N$. Functions whose vectors have the same directions are equivalent with respect to their potential for approximation (see equations 1 and 2). Thus, one considers one representative for each direction, where each direction corresponds to a point on the surface of the unit sphere in $R_N$. From the above-described way of generating of a noise function $\nu(x)$, it follows that all directions in $R_N$ are equally possible. Therefore, one can calculate the probability of obtaining the value of $\Delta_{\nu\max}$ more than or equal to $x$ in the course of the stage $G_s$, $F_{G_s}(x) = \Pr(\Delta_{\nu\max} \geq x/G_s)$, by reference to the uniform distribution on the surface of the sphere.

Now perform the regression search for the response function $y$. At stage $G_s$ of the search, one considers a regression formula, based on a subset of regressors, with the sample estimate of fit $\Delta_y$ (equation 1). We then define the predictive probabilistic estimate (PPE) at stage $G_s$ by comparing $\Delta_y$ to the reference distribution of $\Delta_{\nu\max}$:

$$F_{G_s}(\Delta_y) = \Pr(\Delta_{\nu\max} \geq \Delta_y/G_s). \quad (3)$$

Naturally, the smaller the value of PPE (equation 3), the better the indicated prediction of the regression (based on a given subset of regressors).

The basic idea of PPE (equation 3) is as follows. It is clear that an arbitrary function such as $\nu(x)$ cannot be successfully approximated outside of sample values of arguments. If our search is so extensive that an arbitrary function has a good chance of obtaining a regression with the same or better sample estimate of fit (equation 2) as the response function, then the set of prospective regressors $\phi$ and formulas, analyzed in the course of the search, are not specific to represent the response function. They can fit a broad class of arbitrary functions that are defined on the sample set of arguments $x_1$, $x_2, \ldots, x_N$. The response function is not distinguished anyhow within the class. Therefore, the chances that the regression formula will be an effective predictor outside of the sample set are as poor for the response function as for an arbitrary one.

On the other hand, our search is organized in such a way that an arbitrary function has almost zero probability of obtaining a regression with the same or better sample estimate of fit as the response function. It indicates that the regression for the response function really reflects the dependence of interest and may be effectively used for prediction.

## *Part 3*

Let us consider some properties of PPE (equation 3). From the fact that $0 \leq \Delta_j \leq 1$, the probability distribution of $\Delta_{\nu\max}$, $F_{G_s}(x) = \Pr(\Delta_{\nu\max} \geq x/G_s)$, is concentrated in the interval [0, 1]. Due to the monotonicity of the probability distribution function, for any

$0 \leq x_1 < x_2 \leq 1$,

$$\Pr(\Delta_{\nu\max} \geq x_1/G_s) \geq \Pr(\Delta_{\nu\max} \geq x_2/G_s). \quad (4)$$

From equation 4, it follows that at a given stage $G_s$ of the search, the regression with maximal $\Delta_y$ minimizes PPE (equation 3).

If one compares the prediction qualities of regression formulas obtained for subsets of regressors at different stages of the search, $G_1, G_2, \ldots, G_l$, the subset with maximal $\Delta_y$ will not necessarily minimize PPE. Suppose, for example, stage $G_s$ involves a search among all possible combinations of $s$ regressors from the set $\phi$, and suppose that we denote the maximal values of equation 1 for the response function by $\Delta^1_{y\max}, \Delta^2_{y\max}, \ldots, \Delta^l_{y\max}$. The corresponding PPE (equation 3) take minimal values at their stages. Although it is easy to see that $\Delta^1_{y\max} \leq \Delta^2_{y\max} \leq \ldots \leq \Delta^l_{y\max}$, it is impossible to say a priori how the values of $\Pr(\Delta_{\nu\max} \geq \Delta^1_{y\max}/G_1), \ldots, \Pr(\Delta_{\nu\max} \geq \Delta^l_{y\max}/G_l)$ will be ordered.

## *Part 4*

Let us consider now the form of the probability distribution of $\Delta_\nu$ (equation 2) for a specific subset of regressors. Consider the sample set of values of arguments $x_1, x_2, \ldots, x_N$, which were introduced in PART 1. Any function, defined on the sample set, may be represented as a vector in $R_N$. Let $\varphi_1(x), \varphi_2(x), \ldots, \varphi_s(x)$ be a linearly independent subset of regressors on the sample set $x_1, x_2, \ldots, x_N$. Then, the least-squares approximation of a noise function $\nu(x)$ by this subset of regressors reduces to a projection of $\nu(x)$ into the subspace $R_S$ spanned by the functions.

Consider an orthonormal basis in $R_N$ that consists of $N$ vectors, such that the first $S$ vectors form a basis in the space $R_S$: $\eta_1(x), \eta_2(x), \ldots, \eta_s(x), \eta_{s+1}(x), \ldots, \eta_N(x)$;

$$\nu(x) = \sum_{i-1}^{N} \xi_i \delta_i(x) = \sum_{i-1}^{N} \alpha_i \eta_i(x). \quad (5)$$

Here

$$\delta_i(x) = \left.\begin{matrix} 1 \text{ if } x = x_G \\ \\ 0 \text{ otherwise} \end{matrix}\right\} \quad (6)$$

and $\xi_1, \xi_2, \ldots, \xi_N$ are independent $N(0, 1)$ random variables (as introduced in PART 2). The functions $\delta_1(x), \delta_2(x), \ldots, \delta_N(x)$ form an orthonormal basis in $R_N$ and equation 5 induces a transformation from one orthonormal basis to the other by

$$\alpha = C\xi. \quad (7)$$

Here, $\alpha$ and $\xi$ are $(N \times 1)$ column vectors (see equation 5) and $C$ is an $(N \times N)$ orthogonal matrix. Thus, from the properties of $\xi_1, \xi_2, \ldots, \xi_N$ given above, $\alpha_1, \alpha_2, \ldots, \alpha_N$ are also independent $N(0, 1)$ random variables.

The quality of fit for a noise function $\nu(x)$ by $\varphi_1(x), \varphi_2(x), \ldots, \varphi_s(x)$ may thus be estimated by

$$\Delta_\nu = \frac{\sum_{i-1}^{S} \alpha_i^2}{\sum_{i-1}^{S} \alpha_i^2 + \sum_{i-s+1}^{N} \alpha_i^2}. \quad (8)$$

From geometric considerations, it is clear that the estimates of equations 2 and 8 are identical. At the same time, the value of equation 8 has a beta-probability distribution, $B_{ab}$,[9] with parameters, $a = S/2$ and $b = (N - S)/2$:

$$\Pr(\Delta_\nu \geq x) = 1 - \Pr(\Delta_\nu < x) = 1 - I_x(a, b), \tag{9}$$

$$I_x(a, b) = \frac{\Gamma(a+b)}{\Gamma(a) \cdot \Gamma(b)} \int_0^x z^{a-1}(1-z)^{b-1}\, dz, \tag{10}$$

$0 \leq x \leq 1$, $a, b > 0$, $0 \leq I_x \leq 1$; $\Gamma(\ \ )$ stands for the gamma function. This probability distribution may be tabulated directly or may be easily obtained through tables of the $F$-distribution (e.g., see reference 10).

The beta-probability distribution is concentrated in the interval [0, 1] and has a mean value of $E_{\Delta_\nu} = a/(a + b) = S/N$. By Chebyshev's inequality,

$$\Pr(\Delta_\nu \geq x) \leq \frac{E_{\Delta_\nu}}{x} = \frac{S}{xN}. \tag{11}$$

From equation 11, it follows that for any fixed $x > 0$,

$$\lim_{N\to\infty} \Pr(\Delta_\nu \geq x) = 0 \tag{12}$$

provided that $S = O(N^\gamma)$, $\gamma < 1$.

Now consider projections of a noise function $\nu(x)$ into $k$ subspaces of $R_N$ that are associated with $k$ subsets of regressors, where each subspace has dimension $S$: $R_S^1$, $R_S^2, \ldots, R_S^k$. It is easy to see that the subspace $R_P$ associated with all of the $Q = k \cdot S$ regressors has dimension $P \leq Q$. Assume further that $P < N$. Because the subspaces $R_S^1, R_S^2, \ldots, R_S^k$ are all subspaces of $R_P$, projecting a noise function $\nu(x)$ into any of the subspaces may be represented by two sequential operations—projecting into the subspace $R_P$ and then projecting the result of the first operation $\nu_P(x)$ into the subspace of interest. The quality of fit of the noise function $\nu(x)$ by the subset of regressors associated with subspace $R_S^i$, $1 \leq i \leq k$, may be estimated according to equation 8 by

$$\Delta_\nu^i = \frac{\sum_{j=1}^{S} \alpha_j^2}{\sum_{j=1}^{N} \alpha_j^2} = \frac{\sum_{j=1}^{S} \alpha_j^2}{\sum_{j=1}^{P} \alpha_j^2} \cdot \frac{\sum_{j=1}^{P} \alpha_j^2}{\sum_{j=1}^{N} \alpha_j^2} = \Delta_\nu^P \cdot \Delta_{\nu_P}^i. \tag{13}$$

The coefficients $\alpha_j$ in equation 13 are associated with an orthonormal basis in $R_N$ for which the first $S$ vectors span $R_S^i$ and the first $P$ vectors span $R_P$. $\Delta_\nu^P$ indicates the quality of the fit of $\nu(x)$ by the set of regressors associated with $R_P$ and $\Delta_{\nu_P}^i$ indicates the fit of $\nu_P(x)$ by the regressors associated with $R_S^i$.

It is clear that

$$\Delta_{\nu\max} = \Delta_\nu^P \cdot \Delta_{\nu_P\max}. \tag{14}$$

Here, $\Delta_{\nu\max} = \max\{\Delta_\nu^1, \Delta_\nu^2, \ldots, \Delta_\nu^k\}$; $\Delta_{\nu_P\max} = \max\{\Delta_{\nu_P}^1, \Delta_{\nu_P}^2, \ldots, \Delta_{\nu_P}^k\}$; $\Delta_{\nu\max} \leq \Delta_\nu^P$ for $0 \leq \Delta_{\nu_P\max} \leq 1$. From this, it follows that $\Pr(\Delta_{\nu\max} \geq x) \leq \Pr(\Delta_\nu^P \geq x)$. According to equation 12, for any $x > 0$, $\lim_{N\to\infty} \Pr(\Delta_\nu^P \geq x) = 0$ for any fixed value $P$ as well as for $P \to \infty$ with the order being $N^\gamma$, $\gamma < 1$. Therefore, under the same conditions,

$$\lim_{N\to\infty} \Pr(\Delta_{\nu\max} \geq x) = 0.$$

This asymptotic relation suggests how one can increase the size of the search asymptotically, but while still ensuring that the PPE converges to zero. The size of the search may be increased by increasing either $k$ or $S$, or both of them, just so long as $P = O(N^{\gamma})$, $\gamma = 1$.

### *Part 5*

Consider now the implications of increasing the size of a search on the PPE (equation 3) provided that the sample size $N$ is fixed. Suppose the search stage $G$ involves maximizing $\Delta_{\nu}$ over a collection $S$ of subsets of $\phi$; the corresponding reference distribution is Pr $(\Delta_{\nu\max} \geq x/G)$. Now, suppose an additional collection $S'$ of subsets is included; denote the expanded search stage by $G \cup G'$ and the reference distribution by Pr $(\Delta_{\nu\max} \geq x/G \cup G')$. Because $S \subset S \cup S'$, it is clear that for all $0 \leq x \leq 1$,

$$\Pr (\Delta_{\nu\max} \geq x/G \cup G') \geq \Pr (\Delta_{\nu\max} \geq x/G).$$

Thus, increasing the search at any stage will increase the measure of the fit of equation 1, but it will also result in a stochastically larger reference distribution.

### *Part 6*

The PPE (equation 3) is based on the sample estimate of the fit of equations 1 and 2. This is not necessary—the same approach could be applied with any proper sample estimate of fit; different estimates of fit, generally speaking, induce different PPE. However, the following theorem takes place. When the estimates of fit are linked by monotone functional dependence, their PPE are identical.

Let there be two estimates of fit, $\Delta_1$ and $\Delta_2$, and let $\Delta_1 = \theta(\Delta_2)$, where $\theta(\ \ )$ is a strictly monotone increasing function. At the stage $G_s$, let a regression formula obtain the estimate of fit $\Delta_{1y}$ and $\Delta_{2y}$. The former induces the PPE:

$$F_{G_s}(\Delta_{1y}) = \Pr (\Delta_{1\nu\max} \geq \Delta_{1y}/G_s) = \Pr \{[\theta(\Delta_2)]_{\nu\max} \geq \theta(\Delta_{2y})/G_s\}.$$

Due to the monotonicity, the maximal $\theta(\Delta_2)$ is reached for $\Delta_{2\nu\max}$; as a result, the inequalities, $\Delta_{1\nu\max} \geq \Delta_{1y}$ and $\Delta_{2\nu\max} \geq \Delta_{2y}$, are correct simultaneously for the same set in $R_N$. It follows that

$$\Pr (\Delta_{1\nu\max} \geq \Delta_{1y}/G_s) = \Pr (\Delta_{2\nu\max} \geq \Delta_{2y}/G_s).$$

Note that the properties of PPE, derived in PART 3 and PART 5, are correct not only for the estimate of fit in equations 1 and 2, but for a broad class of such estimates. Later, we shall consider the estimate of fit in equations 1 and 2 only.

### *Part 7—Some Additional Remarks*

(a) In PART 2 of this section, decomposition of the search into stages was introduced. In PART 3, it was shown that at a given stage $G_s$ of the search, the regression with maximal $\Delta_y$ (equation 1) minimizes PPE (equation 3). If one compares PPE for regressions obtained at different stages, the subset with maximal $\Delta_y$ will not necessarily minimize PPE (see PART 5). Therefore, only proper decomposition of the search process into stages leads to new ways for estimating the predictive quality of subsets of regressors.

In fact, the stages of a search may be defined in a different way. The decomposition (described in PART 2) is surely not the only possibility; there may be other interesting and useful ways to divide up the subsets of regressor variables into stages. When a search includes procedures like transformations of the set of regressors $\phi$, different transformations of the response function with the purpose of finding a better fit, and so on, these procedures should also be introduced as stages of the search and their corresponding probability distributions should be used for PPE.[11]

(b) The property described in PART 5 demonstrates how PPE (equation 3) behaves as the size of a search grows. The following example demonstrates how the estimate (equation 3) corrects the tendency of the sample estimate of the fit of equation 1 to indicate better results for more complex regression formulas.

Suppose one analyzes a subset of $N$ linearly independent regressors on the sample set $x_1, x_2, \ldots, x_N$. It is clear that these regressors will fit the response function, $y_1, y_2, \ldots, y_N$, with $RSS = 0$, that is, with a perfect sample estimate (equation 1) $\Delta_y = 1$. The same, however, is true for any noise function $\nu(x)$, so Pr $(\Delta_\nu = 1) = 1$; thus, PPE (equation 3) of the given regression formula is maximal, thereby correctly reflecting the fact that such a formula cannot be used for prediction.

(c) In principle, the proposed PPE (equation 3) would have to be produced by Monte Carlo methods. It makes the method quite computer intensive and, as such, it might be compared to similarly computer intense methods such as Bootstrap.[12] Modern computers render such methods practical. There are, however, some problems associated with the computations.

If the regressors provide a good fit to the response function, the corresponding PPE is likely to be far out in the right tail of the distribution of $\Delta_{\nu\max}$. The straightforward Monte Carlo approach may require a very large number of generated noise functions to provide an accurate estimate of extreme quantiles. Perhaps some alternative Monte Carlo approaches that alleviate the problem should be considered.

Therefore, in defining the sample measure of fit in equations 1 and 2, it is possible to use adjusted (for the sample average) sums of squares. Because empirical models often include a constant term, the value of the adjusted measure of fit of the response function as a rule would be less than equation 1, and, as a result, the not-so-extreme quantiles of the distribution of $\Delta_{\nu\max}$ should be estimated. This then alleviates the computational problems.

The above idea may be developed in such a way that by analogy to stepwise regression at each step, the relevant measure of fit is based on the adjusted sums of squares for all terms already in the model.

It is clear that some numerical examples would be very much important for solving the above problems, as well as for studying the proposed scheme as a whole. Unfortunately, I have no access to the computers now. I hope this work will be possible in the nearest future.

(d) The search process may be organized as follows. Decompose the search into stages $G_1, G_2, \ldots, G_l$ and find the subset of regressors that minimizes PPE: Pr $(\Delta_{\nu\max} \geq \Delta^i_{y\max}/G_i)$. When the stages are reduced for examination of some subsets of regressors, one may find the best subset for each stage according to the estimate of equation 1 and then one can perform the final selection according to the above rule. The above rule may also be applied when the search process is more complex and includes transformations of regressors, of the response function, and so on.

In practice, instead of PPE, one has to deal with its statistical estimate, and the precision of it should be taken into account.

The approach to estimating the prediction quality of regression described here has some predecessors. One example[6] is to augment the set of regressors with noise variables (produced by a random-number generator) and use them in a stepwise procedure to select a good subset of regressors. The selection is done according to the rule: "when the first artificial variable is selected, stop the procedure."

## BEHAVIOR OF PPE (EQUATION 3) FOR SOME MODELS FOR SELECTING SUBSETS OF REGRESSORS

### *Part 1*

In PART 5 of the previous section, it was shown that the increasing of the search leads to the increasing (i.e., deteriorating) of PPE (equation 3) for any fixed value $\Delta_y$ and sample size $N$. To estimate how PPE increases, we consider here two models that demonstrate different characters of the increase under different conditions. To deal with models that allow for appropriate results to be easily calculated in closed form, we should introduce some assumptions that hardly seem to be fulfilled in applications. In the meantime, the study of the models seems to be helpful in understanding the behavior of PPE under different circumstances.

We now return to the expansion in equation 5 of a noise function $\nu(x)$ and its properties (discussed in PART 4 of the previous section). Suppose one generates independently $k$ noise functions, $\nu_1(x), \nu_2(x), \ldots, \nu_k(x)$, with the parameters of the expansion in equation 5 for a given set of regressors as follows: $\alpha_1^1, \alpha_2^1, \ldots, \alpha_N^1; \alpha_1^2, \alpha_2^2, \ldots, \alpha_N^2; \ldots; \alpha_1^k, \alpha_2^k, \ldots, \alpha_N^k$; this is a set of $k \cdot N$ independent normal (0, 1) random variables. For each of the $k$ noise functions and a given set of regressors, one may compute the values $\Delta_{\nu_1}, \Delta_{\nu_2}, \ldots, \Delta_{\nu_k}$ (equation 8). It follows that the latter are independent random variables as well, and from equation 9, it is evident that

$$\Pr\,(\Delta_{\nu\max} \geq x) = 1 - [I_x(a, b)]^k, \tag{15}$$

where $\Delta_{\nu\max} = \max\,\{\Delta_{\nu_1}, \Delta_{\nu_2}, \ldots, \Delta_{\nu_k}\}$.

Also, consider another interpretation of equation 15. We can find a linear orthogonal transformation in $R_N$ that maps $\nu_2(x)$ into a vector $\nu_2^*(x)$ that coincides with $\nu_1(x)$ by direction. This transformation induces a mapping of the subset of regressors $\{\varphi_1(x), \varphi_2(x), \ldots, \varphi_s(x)\}$ into another subset $\{\varphi_1'(x), \varphi_2'(x), \ldots, \varphi_s'(x)\}$.

From the fact that equation 8 does not depend on the norm of the vector $\alpha_1, \alpha_2, \ldots, \alpha_N$ and from the orthogonality of the above transformation, then (a) the values $\Delta_{\nu_1}'$ and $\Delta_{\nu_2}^{*\prime}$ obtained by equation 8 for $\nu_1(x)$ and $\nu_2^*(x)$, respectively, with respect to the subspace of $R_N$, based on the subset of functions $\{\varphi_1'(x), \varphi_2'(x), \ldots, \varphi_s'(x)\}$, are equal: $\Delta_{\nu_1}'$ and $\Delta_{\nu_2}^{*\prime} = \Delta_\nu'$; (b) this value $\Delta_\nu'$ coincides with $\Delta_{\nu_2}$ obtained for the noise function $\nu_2(x)$ with respect to the subspace of $R_N$, based on the functions $\{\varphi_1(x), \varphi_2(x), \ldots, \varphi_s(x)\}$, that is, $\Delta_\nu' = \Delta_{\nu_2}$.

Therefore, instead of projections of two independent noise functions, $\nu_1(x)$ and $\nu_2(x)$, into the subspace spanned by $\varphi_1(x), \varphi_2(x), \ldots, \varphi_s(x)$ with the estimates $\Delta_{\nu_1}$ and $\Delta_{\nu_2}$, respectively, one can consider projections of the function $\nu_1(x)$ into two subspaces—the first one based on the functions $\varphi_1(x), \varphi_2(x), \ldots, \varphi_s(x)$ and the second based on the functions $\varphi_1'(x), \varphi_2'(x), \ldots, \varphi_s'(x)$ with the same respective estimates, $\Delta_{\nu_1}$ and $\Delta_{\nu_2}$. As mentioned above, the random variables $\Delta_{\nu_1}$ and $\Delta_{\nu_2}$ are independent. One calls the two sets of functions random independent subsets of regressors. Because of this, we now consider that the random independent subsets of regressors and the results of this part will be obtained just for this case.

Equation 15 may now be interpreted as follows. Project a noise function $\nu(x)$ into $k$ subspaces of $R_N$, with the subspaces being spanned by random independent subsets of regressors $\varphi_1^i(x), \varphi_2^i(x), \ldots, \varphi_s^i(x)$; $1 \leq i \leq k$. The fit of each projection to the noise function $\nu_i(x)$ may be described by $\Delta_{\nu_i}$ (equation 8). The probability that the maximum of these values exceeds $x$ is given by equation 15 with $a = S/2$ and $b = (N - S)/2$.

Now, we return to the problem of approximating the response function. Suppose a search among $k$ subsets of regressors, with each consisting of $S$ terms, yields a best subset with respect to the estimate in equation 1 whose value is $\Delta_{y\max} = x$. If these $k$ subsets of regressors may be considered as random and independent, one may find the value $I(a, b) = I_{\Delta_{y\max}}[S/2, (N - S)/2]$ (equation 10) with the help of tables like in reference 10, and one can calculate PPE of the selected subset of regressors by equation 15, taking into account the size $k$ of the search.

One can see that as the size of the search $k$ increases, Pr $(\Delta_{\nu\max} \geq x) \rightarrow 1$ for any $x < 1$; that is, the PPE of a selected subset of regressors whose sample estimate (equation 1) equals $x$ and is increasing (i.e., worsening) up to the worst value equals 1. The fact that the PPE increases as the size of a search grows is quite general (as proved in PART 5 of the previous section). For this model, equation 15 provides a precise rate of increase and a limit value as well.

### *Part 2*

In the above, one considered projecting a noise function $\nu(x)$ into $k$ subspaces associated with $k$ independent subsets of regressors. Now, we shall study the case where all prospective regressors are situated in a subspace $R_Q$ of dimension $Q < N$; this is, in fact, the case where subsets of regressors are obtained as a result of random sampling from the subspace $R_Q$.

If a noise function $\nu(x)$ is projected into the subspace $R_Q$, the quality of fit may be estimated by

$$\Delta_\nu^Q = \frac{\sum_{i-1}^{Q} \alpha_i^2}{\sum_{i-1}^{Q} \alpha_i^2 + \sum_{i-Q+1}^{N} \alpha_i^2} = \frac{\| \nu_Q(x) \|^2}{\| \nu(x) \|^2}. \tag{16}$$

Here, $\nu_Q(x)$ stands for the projection of $\nu(x)$ into $R_Q$.

Now, consider a subset of $S$ regressors that spans the space $R_S$ such that $S < Q$ and $R_S \subset R_Q$. Analogous to equation 13, one obtains the sample estimate of fit for $\nu_Q(x)$

as

$$\Delta^S_{\nu Q} = \frac{\sum_{i=1}^{S} \alpha_i^2}{\sum_{i=1}^{S} \alpha_i^2 + \sum_{i=S+1}^{Q} \alpha_i^2}. \quad \textbf{(17)}$$

The quality of fit of the noise function $\nu(x)$ by the set of $S$ regressors has the form,

$$\Delta_\nu = \frac{\sum_{i=1}^{S} \alpha_i^2}{\sum_{i=1}^{S} \alpha_i^2 + \sum_{i=S+1}^{N} \alpha_i^2} = \Delta^S_{\nu Q} \cdot \Delta^Q_\nu. \quad \textbf{(18)}$$

The random variables $\Delta_\nu$, $\Delta^Q_\nu$, and $\Delta^S_{\nu Q}$ have beta-distributions with parameters, $a_{\Delta_\nu} = S/2$ and $b_{\Delta_\nu} = (N - S)/2$; $a_{\Delta^Q_\nu} = Q/2$ and $b_{\Delta^Q_\nu} = (N - Q)/2$; $a_{\Delta^S_{\nu Q}} = S/2$ and $b_{\Delta^S_{\nu Q}} = (Q - S)/2$; here, $\Delta^S_{\nu Q}$ and $\Delta^Q_\nu$ are stochastically independent.

In $R_Q$, consider $k$ independent subsets of regressors, each consisting of $S$ terms. The meaning of the term "independent" is the same as in PART 1 of this section. From equations 16–18, one obtains an expression analogous to equation 15:

$$\Pr\,(\Delta^S_{\nu Q\max} \geq x) = 1 - \left[I_x\left(\frac{S}{2}, \frac{Q - S}{2}\right)\right]^k. \quad \textbf{(19)}$$

Here

$$\Delta^S_{\nu Q\max} = \max\,\{\Delta^{S1}_{\nu Q}, \Delta^{S2}_{\nu Q}, \ldots, \Delta^{Sk}_{\nu Q}\}. \quad \textbf{(20)}$$

Now consider the random variables

$$\Delta_{\nu\max} = \Delta^Q_\nu \cdot \Delta^S_{\nu Q\max}. \quad \textbf{(21)}$$

From the above remark, $\Delta^Q_\nu$, $\Delta^{S1}_{\nu Q}$; $\Delta^Q_\nu$, $\Delta^{S2}_{\nu Q}$; . . . ; $\Delta^Q_\nu$, $\Delta^{Sk}_{\nu Q}$ are pairs of stochastically independent variables, which implies that $\Delta^Q_\nu$ and $\Delta^S_{\nu Q\max}$ are stochastically independent.

Consider what happens when $k$ grows. From arguments similar to those presented at the end of PART 1 of this section, one can prove that for any $x < 1$, $\lim_{k\to\infty} \Pr\,(\Delta^S_{\nu Q\max} \leq x) = 0$; naturally, $\Pr\,(\Delta^S_{\nu Q\max} \leq 1) = 1$. Thus, the random variable $\Delta^S_{\nu Q\max}$ converges in distribution to a point mass at $x = 1$. Therefore, for large $k$, $\Pr\,(\Delta_{\nu\max} < x) \approx \Pr\,(\Delta^Q_\nu < x)$. Thus, as the size of the search $k$ grows, for any $x < 1$, $\Pr\,(\Delta_{\nu\max} \geq x) \to \Pr\,(\Delta^Q_\nu \geq x)$. This result indicates the difference between the model considered here and that discussed in PART 1 of this section.

## CONCLUSIONS

(1) In this paper, a probabilistic measure of the prediction quality (PPE) of selected subsets of regressors is introduced. For a given estimate of fit to the response function, the probabilistic measure is defined as the probability of obtaining the same

or better quality fit for a noise function as a result of the same kind of search that was performed for the response function. This value may be used effectively for selecting a best subset of regressors if the search process is properly decomposed into stages.

Some properties [such as the use of normal distributions to generate noise functions $\nu(x)$] were introduced to simplify the derivation of some results and, in fact, are not necessary. The procedure could also be applied to $\nu(x)$ generated from other distributions.

(2) In principle, PPE (equation 3) may be approximated using Monte Carlo methods; however, for some complicated problems, it may require considerable computation. That is why some alternative Monte Carlo approaches that alleviate the problem should be considered. Some models for which PPE may be easily calculated may be helpful as well. Two such models are considered in the previous section.

(3) The approach used here to obtain PPE of a selected subset of regressors is of a rather general nature and may be useful for other problems involving search, selection, and estimates based on the selected set.

### *Remark*

Though this work has something in common with references 13 and 14, the approach here seems to be quite different. The uniform probability distribution on the surface of the unit sphere in $R_N$ (mentioned in PART 2 of the second section) is in fact a kind of a priori probability distribution. If one has some a priori information concerning the form of a response function (e.g., monotonicity, limitations on rate of change, and so on), this should lead to an appropriate change of this a priori probability distribution. In regard to this point, the approach developed here is similar to a Bayesian one.

It should be noted that the idea to compare the observed "best fit" (equation 1) to the best fit that occurs when there is no relation whatsoever between a response function and the explanatory variables was developed in another form by I. Sh. Pinsker.[15]

## ACKNOWLEDGMENTS

The comments of D. Steinberg regarding this paper were very helpful. I want to especially mention his comments concerning the use of Monte Carlo methods and the decomposition of a search process into stages, as well as his help in improving the language. I want to thank D. Steinberg for similar contributions to my previous work[16] as well.

## REFERENCES

1. HOCKING, R. R. 1976. The analysis and selection of variables in linear regression. Biometrics **32:** 1–49.
2. HOCKING, R. R. 1983. Developments in linear regression methodology 1959–1982. Technometrics **25:** 219–230.

3. THOMPSON, M. L. 1978. Selection of variables in multiple regression. Parts I and II. Int. Stat. Rev. **46:** 1–19, 129–146.
4. MILLER, R. G. 1962. Statistical prediction by discriminant analysis. Meteorol. Monog. (Am. Meteorol. Soc.) **4:** 25.
5. LUNTZ, A. L. & V. L. BRAILOVSKY. 1967. On the estimation of features in statistical decision rules. Izv. Akad. Nauk. SSSR (Proc. Acad. Sci. USSR) Tekh. Kibern. (Eng. Cybern.) (in Russian) **3:** 99–110.
6. MILLER, A. J. 1984. Selection of subsets of regression variables. J. R. Stat. Soc. **A147:** part 2.
7. VAPNIK, V. N. 1982. Estimation of Dependences Based on Empirical Data. Springer-Verlag. New York.
8. SEBER, G. A. F. 1977. Linear Regression Analysis. Wiley. New York.
9. CRAMER, H. 1946. Mathematical Methods of Statistics. Princeton Univ. Press. Princeton, New Jersey.
10. MARDIA, K. V. & R. G. ZEMROCH. 1978. Tables of the F and Related Distributions with Algorithms. Academic Press. New York.
11. FREEDMAN, D. A. 1983. A note on screening regression equations. Am. Stat. **37:** 152–155.
12. EFRON, B. 1982. The Jackknife, the Bootstrap and Other Resampling Plans. SIAM. Philadelphia.
13. DIEHR, G. & D. R. HOFLIN. 1974. Approximating the distribution of the sample $R^2$ in the best subset regressions. Technometrics **16:** 317–320.
14. RENCHER, A. C. & F.C. PUN. 1980. Inflation of $R^2$ in best subset regression. Technometrics **22:** 49–53.
15. See, for example: PINSKER, I. SH. 1979. The chaotization principle and its application in data analysis. *In* Models, Algorithms, Decision Making. (In Russian.) I. Sh. Pinsker, Ed.: 5–38. Nauka. Moscow.
16. BRAILOVSKY, V. L. 1985. On an incompletely determined model for function approximation by experimental data. Ann. N.Y. Acad. Sci. **452:** 316–333.

# A New Approach to Decision Making in Multiple-Objective Linear Programming (MOLP) Problems

D. SOLOVEICHIK

## INTRODUCTION

Multiple-Objective Linear Programming (MOLP) has been analyzed from various viewpoints.[1-3] Multiobjective linear programs are defined as follows:

$$\text{Maximize } C \cdot x \text{ such that } A \cdot x = b, x \geq 0. \tag{1}$$

Here, we take $x \in E^n$ and $b \in E^m$, so $A$ is $m \times n$ and $C$ is $r \times n$.

The following MOLP problem is also considered:

$$\text{Maximize } \lambda' \cdot C \cdot x \text{ such that } A \cdot x = b, x \geq 0, \tag{2}$$

where the vector of weights

$$\lambda \in E^r, \qquad \lambda > 0, \qquad \sum_{j-1}^{r} \lambda_j = 1.$$

It is well known that the set of nondominated solutions (NS) (also known as efficient, or pareto-optimal) for MOLP is obtained by solving equation 2. The weight coefficients, $\lambda_j$, are parameters that are varied to locate NS points. The space of weights is represented by a regular simplex.

The representative subset of NS may be obtained by means of a simplex lattice (SL) design.[5] Points $\lambda^1 = (\lambda_1^1, \lambda_2^1, \ldots, \lambda_r^1), \ldots, \lambda^t = (\lambda_1^t, \lambda_2^t, \ldots, \lambda_r^t)$ of SL[4,5] are considered and corresponding $t$ LP problems (equation 2) are solved.

Let us consider a matrix of objective function values for these solutions; let us call it $\{Z\}$. The rows of this matrix represent solutions, while the columns of this matrix represent objectives.

The basic strategy of the approach is as follows: It is expected that the final solution can be chosen by the decision maker (DM) after analysis of the zone of compromise solutions. Such a zone as a simplex in the space of weights is considered. This zone is constructed by means of a SL design and a system of polynomial equations is obtained.

For this purpose:

(1) The matrix $\{Z\}$ is constructed after solving $t$ LP problems (equation 2) for predetermined points, $\lambda^1, \lambda^2, \ldots, \lambda^t$, of SL.
(2) The system of polynomial equations (models)

$$Z_p = f_p(\lambda_1, \lambda_2, \ldots, \lambda_r), \qquad p = 1, 2, \ldots, r,$$

is constructed. Coefficients in each model are represented by functions of the values of corresponding columns of the matrix $\{Z\}$.

(3) If the model fits the process adequately, the row ($\ell$) from matrix $\{Z\}$ is selected by means of the methods of multivariate statistical analysis. This row and the corresponding weights, $\lambda^{\ell}(\lambda_1^{\ell}, \lambda_2^{\ell}, \ldots, \lambda_r^{\ell})$, are considered to be representatives of the zone of compromise solutions. A model of higher degree is determined if the model fits the process inadequately.

(4) The vertices of the new SL are determined by means of the system of polynomial equations. The steps from point $\lambda^{\ell}$ in the directions of the vertices of SL are constructed. The objective values in the points of the simplex are obtained by means of the system of polynomial equations. The steps in the directions of some vertices are stopped if the points obtained are satisfactory for DM. The steps in the directions of other vertices, though, are continued. The value of the step is changed if it is needed at every iteration of the procedure.

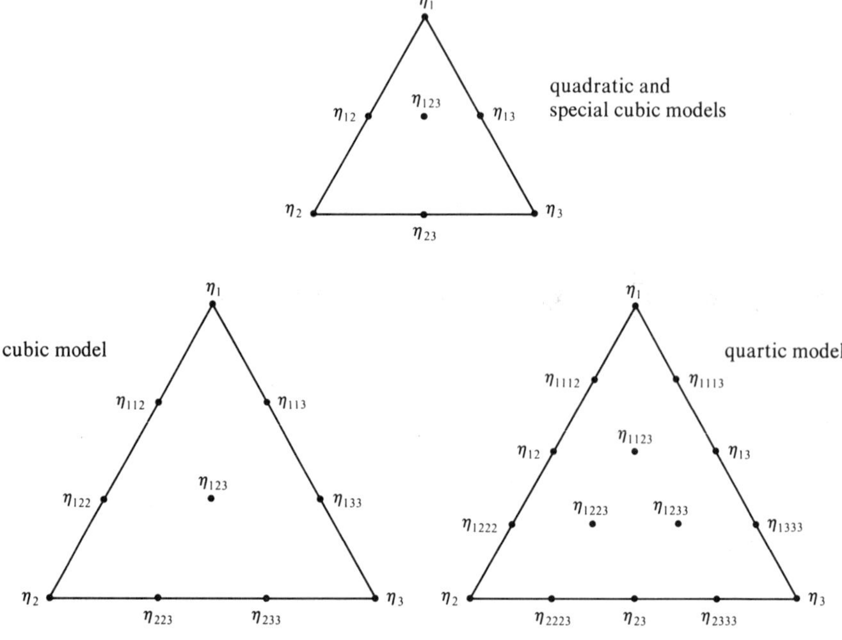


**FIGURE 1.** Weight nomenclature for three objectives.

The procedure is finite because it takes place in the direction of the vertices where the extreme values of the objectives are achieved. It requires less computational efforts because LP problems are not solved.

Selected points $\{\lambda^q\}$ are considered as vertices of the new SL. The procedure is reiterated. The objectives are calculated by means of the models obtained as polynomial values of corresponding weights.

It is of interest to examine quadratic, special cubic, cubic, and quartic models according to the number of points of the SL. Models of higher degree give approximations with higher accuracy, but they require more points of SL for the initial objective value calculation.

## METHOD OF RESEARCH

Models in the case of $r$ objectives are considered as follows:

*Quadratic model:*

$$Z = \sum_{1 \le i \le r} \beta_i x_i + \sum_{1 \le i < j \le r} \beta_{ij} x_i x_j;$$

*Special cubic model:*

$$Z = \sum_{1 \le i \le r} \beta_i x_i + \sum_{1 \le i < j \le r} \beta_{ij} x_i x_j + \sum_{1 \le i < j < k \le r} \beta_{ijk} x_i x_j x_k;$$

*Cubic model:*

$$Z = \sum_{1 \le i \le r} \beta_i x_i + \sum_{1 \le i < j \le r} \beta_{ij} x_i x_j + \sum_{1 \le i < j \le r} \gamma_{ij} x_i x_j (x_i - x_j) + \sum_{1 \le i < j < k \le r} \beta_{ijk} x_i x_j x_k;$$

*Quartic model:*

$$\begin{aligned} Z = \sum_{1 \le i \le r} \beta_i x_i + \sum_{1 \le i < j \le r} \beta_{ij} x_i x_j + \sum_{1 \le i < j \le r} \gamma_{ij} x_i x_j (x_i - x_j) \\ + \sum_{1 \le i < j \le r} \delta_{ij} x_i x_j (x_i - x_j)^2 + \sum_{1 \le i < j < k \le r} \beta_{iijk} x_i^2 x_j x_k + \sum_{1 \le i < j < k \le r} \beta_{ijjk} x_i x_j^2 x_k \\ + \sum_{1 \le i < j < k \le r} \beta_{ijkk} x_i x_j x_k^2 + \sum_{1 \le i < j < k \le r} \beta_{ijk\ell} x_i x_j x_k x_\ell . \end{aligned}$$

All coefficients in the models are simply calculated. For example, weight nomenclature for the three objectives ($r = 3$) is represented in FIGURE 1.

The coefficients of quadratic and special cubic models in this case are as follows:

*Quadratic model:*

$$\beta_1 = \eta_1, \beta_2 = \eta_2, \beta_3 = \eta_3, \tag{3}$$

$$\beta_{12} = 4\eta_{12} - 2\eta_1 - 2\eta_2,$$

$$\beta_{13} = 4\eta_{13} - 2\eta_1 - 2\eta_3,$$

$$\beta_{23} = 4\eta_{23} - 2\eta_2 - 2\eta_3, \tag{4}$$

*Special cubic model:*

$$\beta_{123} = 27\eta_{123} - 12(\eta_{12} + \eta_{13} + \eta_{23}) + 3(\eta_1 + \eta_2 + \eta_3),$$

TABLE 1. Objective Values for the Marginal Solutions[a]

| Solutions | Objectives[b] | | | | | |
|---|---|---|---|---|---|---|
| $N$ | 1 | 2 | 3 | 4 | 5 | 6 |
| 1 | **2612** | 2415 | 52 | 2364 | 4036 | 1652 |
| 2 | 2600 | **2416** | 60 | 2356 | 4022 | 1667 |
| 3 | 2532 | 2378 | **94** | 2283 | 3948 | 1690 |
| 4 | 2300 | 2140 | 84 | **2056** | 3587 | 1520 |
| 5 | 2300 | 2140 | 53 | 2087 | **3430** | 1585 |
| 6 | 2300 | 2140 | 55 | 2085 | 3663 | **1374** |

[a]Matrix $\{Z\}$ for the marginal solution.
[b]Boldface numbers represent extremal values.

where coefficients $\beta_1$, $\beta_2$, $\beta_3$, $\beta_{12}$, $\beta_{13}$, and $\beta_{23}$ from equations 3 and 4 are derived. The coefficients of the models considered are presented in the APPENDIX.

The step-by-step description of the procedure is:

Step 1. The first SL (quadratic, for example) and the appropriate model are constructed. Matrix $\{Z\}$ and their respective points $\lambda^1, \ldots, \lambda^Z$ are considered. If the models fit the process adequately,[4] go to step 2. Otherwise, the adequate model of higher degree should be formed.

Step 2. The methods of multivariate statistical analysis are used for the selection of the solution that is close to the "ideal" solution as a result of an iteration process.[6] The polynomial equations can be clearly presented to DM as contour curves of equal objective values on the triangle diagrams (FIGURE

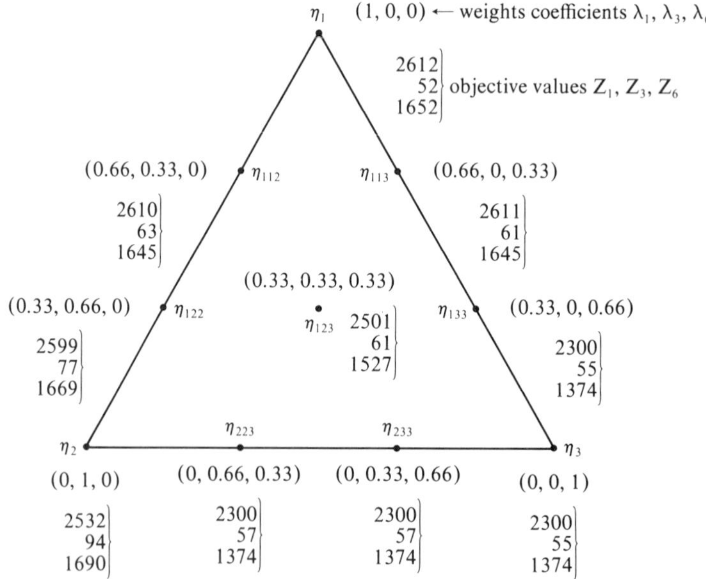


FIGURE 2. Cubic SL (objectives 1, 3, 6).

**TABLE 2.** Coefficients of Special Cubic and Quadratic Models

| | $\beta_1$ | $\beta_2$ | $\beta_3$ | $\beta_{12}$ | $\beta_{13}$ | $\beta_{23}$ | $\beta_{123}$ |
|---|---|---|---|---|---|---|---|
| Objective 1 ($Z_1$) | 2612 | 2532 | 2300 | 112 | 180 | −464 | 1047 |
| Objective 3 ($Z_3$) | 52 | 94 | 55 | 16 | 26 | −70 | −78 |
| Objective 6 ($Z_6$) | 1652 | 1690 | 1374 | −8 | 56 | −632 | 537 |

1). It is impossible to construct such diagrams if there are more than three objectives (weights) in the MOLP problem. In this case, triangle diagrams are formed by discrete values of rest components. The sections of multidimensional simplex are obtained.

The solution is recommended for the final choice if it is acceptable. In this case, the procedure is finished. Otherwise, the point of SL $\lambda^\ell(\lambda_1^\ell, \ldots, \lambda_r^\ell)$ and the corresponding solution that is close to the ideal solution (and, therefore, the most preferable for DM) are considered. We must then find a different solution to DM for the final choice. In such a case, we use the system of polynomial equations as an aid in the decision-making process. Now go to step 3.

Step 3. Let $R^1, R^2, \ldots, R^r$ be the vertices of SL. Let $\lambda^\ell(\lambda_1^\ell, \ldots, \lambda_r^\ell)$ be the selected point of SL (step 2). The point $\lambda^\ell$ then belongs to the compromise zone of solutions for the final choice.

The new points $\lambda^q(h, R)$, $q = 1, \ldots, r$, are constructed by performing steps from $\lambda^\ell$ in the directions of the vertices of SL:

$$\lambda^q = \lambda^\ell + h(R^q - \lambda^\ell), \qquad q = 1, \ldots, r,$$

where $h$ is the step (iteration), $\lambda^\ell$ is the initial point selected on step 2, and $R^q$ is the vertex of SL. The objective values for these points are obtained by means of polynomial equations (step 1) without solving any LP problem. These points are considered as vertices of the new SL if objectives are preferable to DM. Otherwise, we replace $h \rightarrow 0.5h$ or $h \rightarrow 1.5h$, and construct new points $\lambda^q(h, R)$. New objective values by means of polynomial equations are constructed. The procedure is reiterated until one receives preferable objective values. Then one can go to step 1.

This approach to decision making in several MOLP problems was used and the results have been favorable.

**TABLE 3.** Weights and Values of Objectives

| | $\lambda_1$ | $\lambda_3$ | $\lambda_6$ | $Z_1$ | $Z_3$ | $Z_6$ |
|---|---|---|---|---|---|---|
| Point 1 | 0.125 | 0.875 | 0 | 2554 | 90.5 | 1684 |
| Point 2 | 0 | 0.875 | 0.125 | 2452 | 81 | 1581 |
| Point 3 | 0.125 | 0.75 | 0.125 | 2495 | 78 | 1593 |

TABLE 4. Examination of Special Cubic Model by Three Control Points

| | ① | | | ② | | | ③ | | |
|---|---|---|---|---|---|---|---|---|---|
| | $\lambda_1, \lambda_3, \lambda_6$ / $\lambda_2, \lambda_4, \lambda_5$ 0, 0.25, 0.75 | | | $\lambda_1, \lambda_3, \lambda_6$ / $\lambda_2, \lambda_4, \lambda_5$ 0.25, 0, 0.75 | | | $\lambda_1, \lambda_3, \lambda_6$ / $\lambda_2, \lambda_4, \lambda_5$ 0.25, 0.25, 0.5 | | |
| | Calculated Values | Facts | % | Calculated Values | Facts | % | Calculated Values | Facts | % |
| $Z_1$ | 2336 | 2358 | 0.9 | 2461 | 2358 | 4.3 | 2503 | 2478 | 1.0 |
| $Z_2$ | 2149.5 | 2196 | 2.1 | 2297 | 2196 | 4.6 | 2314 | 2314 | — |
| $Z_3$ | 87 | 89 | 2.2 | 90 | 89 | 1.1 | 92 | 92 | — |
| $Z_4$ | 2086 | 2108 | 1.0 | 2207 | 2108 | 4.7 | 2245.5 | 2123 | 5.7 |
| $Z_5$ | 3681 | 3707 | 0.7 | 3851 | 3707 | 3.9 | 3901 | 3873 | 0.7 |
| $Z_6$ | 1506 | 1532 | 1.7 | 1602 | 1532 | 4.6 | 1651 | 1608 | 2.7 |

## NUMERICAL EXAMPLE

A numerical example with 41 decision variables ($n = 41$), 6 objective functions ($r = 6$), and 30 constraints ($m = 30$) is considered.

The objectives 1, 2, and 3 are maximized and the objectives 4, 5, and 6 are minimized. Two groups of objectives are considered. In TABLE 1, the objective values for the marginal solutions are presented (matrix $Z$). The analysis of the marginal solutions testifies to the close dependence inside of every group of objectives (1, 2, 3 and 4, 5, 6).

As a result, the three most interesting objectives (1, 3, 6) are considered by DM. A cubic SL is shown in FIGURE 2. The quadratic and special cubic models for every objective are considered (see TABLE 2). For every model, three control points were selected. The distinctions between the calculated and actual values are acceptable in those control points. These models are adequate and are used for the search for the compromise solution.

The "ideal" vector that is more preferable to DM is:

$$2351, 2174, 85, 2262, 3774, 1511.$$

The values of the objectives in the points of SL show that the point $\eta_2$ (FIGURE 2) may be used as the selected point that is close to this "ideal" vector. The values of the objectives in two points (1, 2) that are close to the selected point confirm our supposition (see TABLE 3). Obviously, the compromise solution can be found as a result

TABLE 5. Coefficients of Special Cubic and Quadratic Models (Compromise Simplex)

| | $\beta_1$ | $\beta_2$ | $\beta_3$ | $\beta_{12}$ | $\beta_{13}$ | $\beta_{23}$ | $\beta_{123}$ |
|---|---|---|---|---|---|---|---|
| $Z_1$ | 2548 | 2532 | 2358 | 32 | 296 | −348 | 1860 |
| $Z_2$ | 2382 | 2283 | 2196 | 8 | 292 | −364 | 1875 |
| $Z_3$ | 94 | 94 | 86 | 0 | 12 | −4 | 48 |
| $Z_4$ | 2288 | 2283 | 2110 | 14 | 280 | −358 | 1812 |
| $Z_5$ | 3962 | 3948 | 3722 | 28 | 368 | −520 | 2550 |
| $Z_6$ | 1692 | 1690 | 1515 | 4 | 230 | −282 | 1728 |

of analysis of SL with vertices in points $\eta_2(0, 1, 0)$, $1(0.125, 0.875, 0)$, $2(0, 0.875, 0.125)$.

The cubic SL is constructed for the new simplex (compromise simplex). Corresponding models for every objective are also formed. Examination of the models by three control points shows that the models obtained are adequate and that they can be used for searching for the compromise solution (TABLE 4). Special cubic models for the other three objectives (2, 4, 5) by using this SL are also constructed (see TABLE 5).

DM can find the compromise solution with the help of contour curves that correspond to every objective. The analysis of such contour curves gives the DM the possibility of deriving conclusions about the interconnections between objective values in the MOLP problem.

## REFERENCES

1. ZELENY, M. 1974. Linear Multiobjective Programming. Springer-Verlag. New York.
2. ZIONTS, S. 1980. Multiple criteria decision making: an overview and several approaches. SUNY at Buffalo working paper no. 454.
3. EVANS, J. P. & R. E. STEUER. 1973. A revised simplex method for linear multiple objective programs. Math. Programming **5:** 54–72.
4. SCHEFFE, H. 1958. Experiments with mixtures. J. R. Stat. Soc. **B20:** 344–360.
5. GORMAN, G. & G. HINMAN. 1962. Technometrics **4:** 463–487.
6. SOLOVEICHIK, D. 1983. A method for solving the multiple-objective linear programming problem. Ann. N.Y. Acad. Sci. **410:** 213.

## APPENDIX

*Quadratic model:*

$$\beta_i = \eta_i;$$

$$\beta_{ij} = 4\eta_{ij} - 2\eta_i - 2\eta_j.$$

*Special cubic model:*

$$\beta_i = \eta_i;$$

$$\beta_{ij} = 4\eta_{ij} - 2\eta_i - 2\eta_j;$$

$$\beta_{ijk} = 27\eta_{ijk} - 12(\eta_{ij} + \eta_{ik} + \eta_{jk}) + 3(\eta_i + \eta_j + \eta_k).$$

*Cubic model:*

$$\beta_i = \eta_i;$$

$$\beta_{ij} = (9/4)(\eta_{iij} + \eta_{ijj} - \eta_i - \eta_j);$$

$$\gamma_{ij} = (9/4)(3\eta_{iij} - 3\eta_{ijj} - \eta_i + \eta_j);$$

$$\beta_{ijk} = 27\eta_{ijk} - (27/4)(\eta_{iij} + \eta_{ijj} + \eta_{ikk} + \eta_{jjk} + \eta_{iik} + \eta_{jkk}) + (9/2)(\eta_i + \eta_j + \eta_k).$$

*Quartic model:*

$$\beta_i = \eta_i; \beta_{ij} = 4\eta_{ij} - 2\eta_i - 2\eta_j;$$

$$\gamma_{ij} = (8/3)(-\eta_i + 2\eta_{iiij} - 2\eta_{ijjj} + \eta_j);$$

$$\delta_{ij} = (8/3)(-\eta_i + 4\eta_{iiij} - 6\eta_{ij} + 4\eta_{ijjj} - \eta_j);$$

$$\beta_{iijk} = 32(3\eta_{iijk} - \eta_{ijjk} - \eta_{ijkk}) + (8/3)(6\eta_i - \eta_j - \eta_k) - 16(\eta_{ij} + \eta_{ik})$$
$$- (16/3)(5\eta_{iii} + 5\eta_{iiik} - 3\eta_{ijjj} - 3\eta_{ikkk} - \eta_{jjjk} - \eta_{jkkk});$$

$$\beta_{ijjk} = 32(3\eta_{ijjk} - \eta_{iijk} - \eta_{ijkk}) + (8/3)(6\eta_j - \eta_i - \eta_k) - 16(\eta_{ij} + \eta_{jk})$$
$$- (16/3)(5\eta_{ijjj} + 5\eta_{jjjk} - 3\eta_{iiij} - 3\eta_{jkkk} - \eta_{iiik} - \eta_{ikkk});$$

$$\beta_{ijkk} = 32(3\eta_{ijkk} - \eta_{ijjk} - \eta_{iijk}) + (8/3)(6\eta_k - \eta_i - \eta_j) - 16(\eta_{ik} + \eta_{jk})$$
$$- (16/3)(5\eta_{ikkk} + 5\eta_{jkkk} - 3\eta_{iiik} - 3\eta_{jjjk} - \eta_{iiij} - \eta_{ijjj});$$

$$\beta_{ijk\ell} = 256\eta_{ijk\ell} - 32(\eta_{iijk} + \eta_{iij\ell} + \eta_{iik\ell} + \eta_{ijjk} + \eta_{ijj\ell} + \eta_{jjk\ell}$$
$$+ \eta_{ijkk} + \eta_{ikk\ell} + \eta_{ij\ell\ell} + \eta_{jk\ell\ell} + \eta_{ik\ell\ell} + \eta_{jkk\ell})$$
$$+ (32/3)(\eta_{iiij} + \eta_{iiik} + \eta_{iii\ell} + \eta_{ijjj} + \eta_{jjjk} + \eta_{jjj\ell}$$
$$+ \eta_{ikkk} + \eta_{jkkk} + \eta_{kkk\ell} + \eta_{i\ell\ell\ell} + \eta_{j\ell\ell\ell} + \eta_{k\ell\ell\ell}).$$

# Using Reduction Techniques in Multiple-Objective Linear Programming (MOLP)

D. SOLOVEICHIK

## INTRODUCTION

The purpose of this paper is to discuss the results obtained from applications of new recently developed MOLP reduction techniques[1,2] to several MOLP problems. We also discuss this approach by using comparisons with the results obtained from applications in some other MOLP techniques to these problems.

Four MOLP problems are examined. Using reduction techniques, we first obtain the representative subset of nondominated solutions (NS) with the help of the simplex lattice. Then, the factor analysis and the method of multidimensional classifying for determination of the more detailed structure of criteria dependence and grouping homogeneous solutions are used.

The paper is organized as follows. In the next section, we give a brief summary of the reduction techniques in MOLP (for a more detailed description, see references 1 and 2). In the third section, the data set is discussed, and in the final section, the outcome of our calculation is presented.

## REDUCTION TECHNIQUES

The MOLP problem is considered:

$$\text{Maximize } C \cdot x \text{ such that } A \cdot x \leq b, x \geq 0, \tag{1}$$

where $C$ is $(l \times n)$, $A$ is $(m \times n)$, $b$ is $(m \times 1)$, and $x \geq 0$ is an $n$-dimensional decision variable vector.

To determine the most interesting nondominated solutions (NS) for MOLP, we recapitulate some of the basis notions in reduction techniques. These techniques include factor analysis, pattern search, and function aggregation, and they consist of the following steps.

Step I: The model gives the decision maker (DM) the subset of all NS by means of a simplex lattice design.

The MOLP problem:

$$\text{Maximize } \sum_{i-1}^{n} \lambda_i c_i x \text{ such that } Ax \leq b, x \geq 0, \tag{2}$$

where $\lambda_i \geq 0$ and $\Sigma_i \lambda_i = 1$ are the weighting coefficients that are considered. All NS could be computed by solving equation 2 with various $\lambda$. There are five values of weights for the quartic model: 0, ¼, ½, ¾, and 1. The number of NS in the case of the quartic model is $\frac{l}{2}(l^2 + 1)$, where $l$ is the number of objectives.[3]

**TABLE 1.** Productivity and Resources of Plants

| Rolling Mill | Kind of Product 1 | 2 | 3 | 4 | 5 | 6 | 7 | Resource (up time) |
|---|---|---|---|---|---|---|---|---|
| I | 115.1 | 148.1 | | | 169.7 | 169.7 | | 6092 |
| II | | 27.7 | | | 27.1 | 25.2 | 29.2 | 6461 |
| III | | 37.7 | 20.1 | 15.5 | | 25.8 | 28.1 | 7021 |
| IV | 21.5 | 29.8 | 21.7 | 22.9 | | | | 6794 |
| V | | | 18.0 | 16.1 | | 16.9 | 17.5 | 7488 |

It is well known that there is no need to solve each linear programming (LP) problem (equation 2) from an initial phase 1 of simplex tableau. The maximal extreme point of one of the LPs can be used as a starting extreme point of the next.

Step II: We use factor analysis (FA) to reduce the dimensionality of the space of objectives. The method of principal components is used. The factor loading analysis and the graphic representation of the location of the objectives in the space of the orthogonal factors give information for the control grouping of the objectives and pick out the groups of objectives.

Step III: In order to reduce the number of NS considered, we aggregate them into small numbers of groups of similar NS for the DM. These solutions are grouped in the space of factors described earlier. Several cluster analysis techniques are used for this purpose,[4] with favorable results being obtained when the number of factors is not too large.

The DM can choose one set of NS on the basis of the information obtained. If the number of NS in the preferable set is not too large and the set is homogeneous, the DM can choose one of the NS after comparison.

There may be some cases where additional constructions are needed. The investigation of the neighborhood of the required vector of objectives is proposed in this situation. It is expected that the final solution can be chosen by DM after analysis of the zone of compromise solutions.[5] The information obtained could be given to DM for

**TABLE 2.** Planned Requirement in Kinds of Products

| Industrial Consumer | Kind of Product | Quantity | Industrial Consumer | Kind of Product | Quantity |
|---|---|---|---|---|---|
| 1 | | 140,000 | 13 | | 45,000 |
| 2 | ① | 39,000 | 14 | ④ | 25,000 |
| 3 | | 78,000 | 15 | | 7600 |
| 4 | | 300,000 | 16 | | 60,000 |
| 5 | | 90,000 | 17 | ⑤ | 12,000 |
| 6 | ② | 85,000 | 18 | | 8000 |
| 7 | | 20,000 | 19 | | 320,000 |
| 8 | | 15,000 | 20 | ⑥ | 120,000 |
| 9 | | 47,400 | 21 | | 34,000 |
| 10 | | 40,000 | 22 | | 3000 |
| 11 | ③ | 30,000 | 23 | ⑦ | 2000 |
| 12 | | 17,400 | 24 | | 1000 |

**TABLE 3.** Matrix of Objectives (Matrix $C$)

| | Objectives | | | | | Objectives | | | | | Objectives | | | | | Objectives | | | |
|---|---|---|---|---|---|---|---|---|---|---|---|---|---|---|---|---|---|---|---|
| Variables | 1 | 4 | 2 | 3 | Variables | 1 | 4 | 2 | 3 | Variables | 1 | 4 | 2 | 3 | Variables | 1 | 4 | 2 | 3 |
| 1 | 1200 | 0.00869 | 0 | 1 | 19 | 1300 | 0.02652 | 0 | 1 | 37 | 980 | 0.05556 | 0 | 1 | 55 | 760 | 0.00589 | 0 | 1 |
| 2 | 800 | 0.00869 | 1 | 1 | 20 | 1400 | 0.02652 | 0 | 1 | 38 | 1020 | 0.05556 | 0 | 1 | 56 | 810 | 0.00589 | 0 | 1 |
| 3 | 650 | 0.00869 | 0 | 1 | 21 | 1200 | 0.02652 | 0 | 1 | 39 | 730 | 0.05556 | 0 | 1 | 57 | 360 | 0.00589 | 0 | 1 |
| 4 | 540 | 0.04651 | 0 | 1 | 22 | 1100 | 0.02652 | 0 | 1 | 40 | 320 | 0.06431 | 0 | 1 | 58 | 480 | 0.03968 | 1 | 1 |
| 5 | 500 | 0.04651 | 0 | 1 | 23 | 790 | 0.02652 | 0 | 1 | 41 | 490 | 0.06431 | 1 | 1 | 59 | 520 | 0.03968 | 0 | 1 |
| 6 | 720 | 0.04651 | 1 | 1 | 24 | 590 | 0.02652 | 1 | 1 | 42 | 920 | 0.06431 | 0 | 1 | 60 | 490 | 0.03968 | 0 | 1 |
| 7 | 600 | 0.00675 | 0 | 1 | 25 | 720 | 0.03356 | 1 | 1 | 43 | 980 | 0.04367 | 0 | 1 | 61 | 200 | 0.03876 | 0 | 1 |
| 8 | 400 | 0.00675 | 1 | 1 | 26 | 600 | 0.03356 | 0 | 1 | 44 | 560 | 0.04367 | 0 | 1 | 62 | 320 | 0.03876 | 1 | 1 |
| 9 | 900 | 0.00675 | 0 | 1 | 27 | 1100 | 0.03356 | 1 | 1 | 45 | 710 | 0.04367 | 0 | 1 | 63 | 610 | 0.03876 | 0 | 1 |
| 10 | 800 | 0.00675 | 0 | 1 | 28 | 1200 | 0.03356 | 0 | 1 | 46 | 870 | 0.06211 | 1 | 1 | 64 | 230 | 0.05917 | 0 | 1 |
| 11 | 1400 | 0.00675 | 0 | 1 | 29 | 1300 | 0.03356 | 0 | 1 | 47 | 710 | 0.06211 | 0 | 1 | 65 | 270 | 0.05917 | 0 | 1 |
| 12 | 430 | 0.00675 | 0 | 1 | 30 | 1300 | 0.03356 | 0 | 1 | 48 | 690 | 0.06211 | 0 | 1 | 66 | 340 | 0.05917 | 0 | 1 |
| 13 | 960 | 0.0361 | 1 | 1 | 31 | 260 | 0.04975 | 0 | 1 | 49 | 1500 | 0.00589 | 0 | 1 | 67 | 980 | 0.03425 | 0 | 1 |
| 14 | 500 | 0.0361 | 0 | 1 | 32 | 830 | 0.04975 | 1 | 1 | 50 | 1300 | 0.00589 | 0 | 1 | 68 | 810 | 0.03425 | 0 | 1 |
| 15 | 820 | 0.0361 | 0 | 1 | 33 | 890 | 0.04975 | 0 | 1 | 51 | 600 | 0.00589 | 0 | 1 | 69 | 690 | 0.03425 | 0 | 1 |
| 16 | 940 | 0.0361 | 0 | 1 | 34 | 790 | 0.04608 | 1 | 1 | 52 | 800 | 0.03690 | 1 | 1 | 70 | 690 | 0.03559 | 1 | 1 |
| 17 | 650 | 0.0361 | 0 | 1 | 35 | 360 | 0.04608 | 0 | 1 | 53 | 720 | 0.03690 | 0 | 1 | 71 | 730 | 0.03559 | 0 | 1 |
| 18 | 480 | 0.0361 | 0 | 1 | 36 | 590 | 0.04608 | 0 | 1 | 54 | 920 | 0.03690 | 0 | 1 | 72 | 920 | 0.03559 | 0 | 1 |
| | | | | | | | | | | | | | | | 73 | 420 | 0.05714 | 0 | 1 |
| | | | | | | | | | | | | | | | 74 | 360 | 0.05714 | 0 | 1 |
| | | | | | | | | | | | | | | | 75 | 960 | 0.05714 | 0 | 1 |

the final decision choice. The selection of the solution could then be the result of an iterative process.

## THE DATA

The data set used consists of four MOLP problems. The first example is the MOLP problem with 75 decision variables ($n = 75$), 4 objective functions ($l = 4$), and 29 constraints ($m = 29$). The development of a new product is examined. There are five plants (Rolling Mill) that produce seven kinds of products with different productivities and expenses. Resources are limited and there are restrictions on the amount of each product that can be produced (see TABLES 1–3).

The problem is modeled with 4 objective functions and 2 categories of constraints:

$$\min \text{ (objectives 1,4)},$$

$$\max \text{ (objectives 2,3)}$$

**TABLE 4.** Factor Loading Matrix

| Criteria (Objectives) | Factors | | Commonalities |
|---|---|---|---|
| | $F_1$ | $F_2$ | |
| I | 0.65 | 0.37 | 0.74 |
| II | 0.84 | 0.006 | 0.70 |
| III | −0.17 | 0.94 | 0.91 |
| IV | 0.95 | −0.26 | 0.96 |
| Eigenvalues | 2.05 | 1.27 | — |
| Factor and total percentage of variance | 0.51 | 0.32 | 0.83 |

such that

$$\text{[resources constraints } (\leq)] \text{ } 5,$$

$$\text{[products constraints } (\geq)] \text{ } 24.$$

The second example with 31 decision variables, 5 objective functions, and 13 constraints is presented (see reference 6). The third example consists of 4 objectives, 7 decision variables, and 7 constraints (reference 7, p. 26, example no. 7310). The last example with 5 objective functions, 8 decision variables, and 8 constraints (reference 7, p. 11, example no. 74) is also considered.

## EMPIRICAL FINDINGS

It is well known that there are five values of weights for the quartic model: 0, ¼, ½, ¾, and 1. A quartic lattice consists of $\frac{1}{2}(l^2 + 1)$ points (vector of weights). In the first

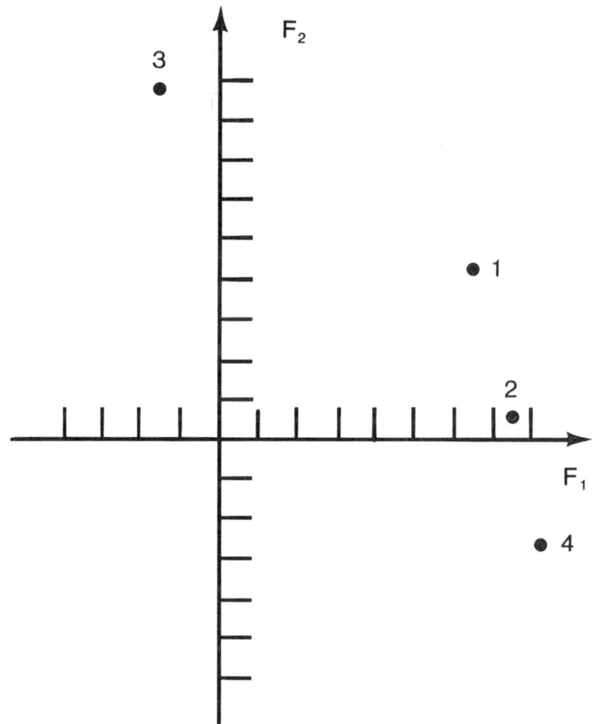


**FIGURE 1.** The four objectives in the plane of two factors.

**TABLE 5.** Classification of the Set on NS by Two Factors[a]

| Groups of Solutions by the Second Factor | 1 | 2 | 3 | 4 | 5 | 6 | 7 | 8 | 9 | 10 | 11 |
|---|---|---|---|---|---|---|---|---|---|---|---|
| 10 | | | | | | | | | 1 | | 1 |
| 9 | | | | | | | | | | | 1 |
| 8 | | | | 3 | | | | | | | |
| 7 | 3 | 1 | | 3 | | | | | | | |
| 6 | | | 6 | 3 | | | 1 | | | | 2 |
| 5 | | | | | | | | | | | 3 |
| 4 | | | | | | | | | | 1 | |
| 3 | | | | | | | | | | 1 | |
| 2 | | | | | | 1 | | 1 | | | |
| 1 | | | | | 2 | | | | | | |

Groups of Solutions by the First Factor (columns 1–11)

[a]The number of the solutions in the cells is given.

**TABLE 6.** Objective Function Values

| Nondominated Solutions | Objectives | | | |
|---|---|---|---|---|
| | I | II | III | IV |
| 8 | 1569.1 | 753.1 | 1175 | 33,857 |
| 9 | 1566 | 762.8 | 1189 | 33,856 |
| 10 | 1539 | 634 | 1043 | 33,856 |
| 23 | 1566 | 766 | 1192 | 33,855 |
| 24 | 1539 | 711 | 1083 | 33,855 |
| 25 | 1539 | 475 | 970 | 33,502 |

example, 34 NS ($l = 4$) are determined after solving 34 LP problems (equation 2) with the help of the vector of weights. Then, 65, 34, and 65 LP problems (equation 2) are solved for the second, third, and fourth examples, respectively. Correspondingly, 65, 34, and 65 NS are determined. In each example, the matrices of the values of all objective functions for every NS are constructed.

The factor loading matrix can be determined on the basis of the correlation coefficient matrix using the method of principal components. The matrix of factor loading is presented in TABLE 4. The three loadings of the first factor are positive, and two of them are rather large and close to 1. This factor can be regarded as the most important. The second factor is connected with objective function 3. Two factors account for about 83% of the total variance. The consideration of the rest of the factors is not necessary.

The grouping of criteria is presented in FIGURE 1 in the plane of two major factors. These four objectives can be partitioned into three or two groups. (Objective functions 2 and 4 could form one group, while criteria 1 and 3 could form the second and third groups, respectively. Objectives 1, 2, and 4 as one group and criteria 3 as the second group of objective functions are also considered.)

The solutions were grouped along the first as well as the second factor. The solutions along both factors were partitioned into 11–13 groups. TABLE 5 is the result of dividing the subset of NS along two factors.

Four cells are especially interesting for the DM. Solutions from these cells are presented in TABLE 6 for the next investigation to the DM. If some solutions from this set are most preferable to DM, the final solution can be chosen by DM after analysis of

**TABLE 7.** Factor Loading Matrix

| Criteria (Objectives) | Factors | | Commonalities |
|---|---|---|---|
| | $F_1$ | $F_2$ | |
| I | 0.61 | 0.52 | 0.64 |
| II | −0.93 | 0.27 | 0.94 |
| III | −0.23 | 0.92 | 0.89 |
| IV | −0.91 | 0.27 | 0.91 |
| V | −0.88 | −0.08 | 0.78 |
| Eigenvalues | 2.96 | 1.20 | — |
| Factor and total percentage of variance | 0.59 | 0.24 | |

the zone of compromise solutions. Such a zone as a simplex in the space of weights is considered. This zone is constructed by means of a simplex lattice (SL) design and a system of polynomial equations is obtained.[5]

Using SL design for a simplex of weights coefficients, we can approximately calculate objective values of NS as polynomial values of corresponding weights. For each objective polynomial equation, a function of λ is constructed. Polynomial coefficients in such equations are calculated using objective values in special predeter-

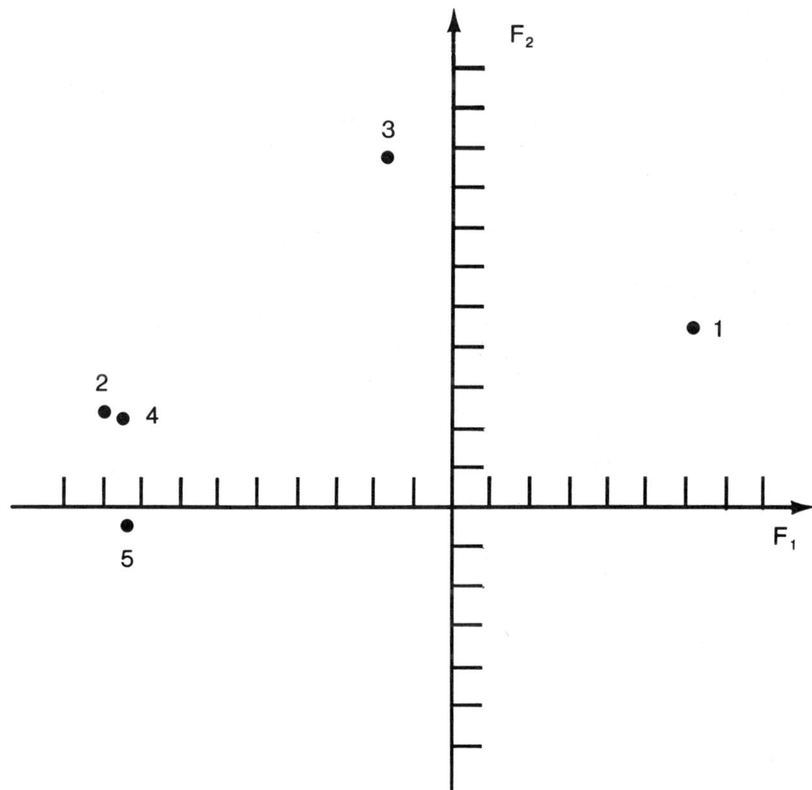


**FIGURE 2.** The five objectives in the plane of two factors.

mined SL points. The polynomial equations that correspond to the points of SL may be successfully used in the decision analysis of the MOLP problem.

In the second example, the matrices of the values of the objective functions for every one of the 65 NS are computed. These solutions are considered to be a representative selection from the general totality (set) of all NS.

The matrix of factor loading is presented in TABLE 7. Two factors account for about 83% of the total variance. The factor loadings are considered to be correlation

**TABLE 8.** Classification of the Set on NS by Two Factors[a]

| Groups of Solutions by the Second Factor | | | | | | | | | | | |
|---|---|---|---|---|---|---|---|---|---|---|---|
| | 10 | | | | | | 1 | 2 | | 3 | 5 |
| | 9 | | | | 12 | 7 | | | | | |
| | 8 | | 7 | 4 | | | | | | | |
| | 7 | | 7 | | | | | | | | |
| | 6 | | | | | | 3 | | | | |
| | 5 | 1 | 1 | | | | | | | | |
| | 4 | | | | | | | | | | |
| | 3 | | | | | | | | | | |
| | 2 | 10 | | | | | | | | | |
| | 1 | | | | | | | | | | |
| | | 1 | 2 | 3 | 4 | 5 | 6 | 7 | 8 | 9 | 10 |

Groups of Solutions by the First Factor

[a]The number of the solutions in the cells is given.

coefficients between the factors and the objectives. Analysis of the factor loadings and the graphic representation of the location of the objectives in the space of the orthogonal factors (FIGURE 2) give important information to DM. Objectives 2 and 4 as one group of objectives may be considered. In this group, objective function 5 may be included too. The DM, though, has important arguments in order to reduce the number of objectives considered.

Sixty-five solutions are grouped in the space of the factors obtained (described above). The graphic representation of the projections of the solutions on two-dimensional shears is presented in TABLE 8.

On the basis of the information obtained, DM chooses four classes of NS. Four typical solutions can be recommended for choice when it is acceptable (TABLE 9).

This example is also considered in reference 4 and the Interactive MOLP approach is used. In this approach, 11 single-objective LP problems (equation 2) were solved in the first step (after computing 11 convex combination trial gradients). The list of 11 plans to DM is also presented, and one of them was selected as being the most preferred.

The second gradient cone was computed from the previous one. The 6 single-objective LP problems (equation 2) were solved. The best plan from this was selected

**TABLE 9.** Objective Function Values

| | Objectives | | | | |
|---|---|---|---|---|---|
| NS | I | II | III | IV | V |
| 1 | 39.5 | 26.96 | 18.0 | 4.88 | 1.25 |
| 2 | 39.5 | 28.44 | 17.32 | 5.75 | 2.27 |
| 3 | 39.5 | 28.62 | 17.02 | 6.12 | 2.8 |
| 4 | 24.27 | 28.3 | 18.1 | 4.86 | 1.25 |

**TABLE 10.** Factor Loading Matrix

| Criteria | Factors | | | Commonalities |
|---|---|---|---|---|
| | $F_1$ | $F_2$ | $F_3$ | |
| I | 0.966 | −0.16 | −0.17 | 0.99 |
| II | −0.04 | −0.05 | 0.99 | 0.99 |
| III | −0.87 | −0.46 | −0.16 | 0.98 |
| IV | 0.04 | 0.99 | −0.06 | 0.99 |
| Factor and total percentage of variance | 0.422 | 0.308 | 0.263 | — |
| Eigenvalues | 1.83 | 1.10 | 1.03 | — |

and the third gradient cone was formed. The next 6 LP problems (equation 2) were then solved and one plan was selected as the best from this group. The fourth gradient cone thus was constructed. This case produced 30 efficient extreme points. Using the filter in the "forward" direction, the 30 points were reduced to the 6 least redundant extreme points. The best plan was selected and the filter was used in the "reverse" direction for the best iteration. The list of plans obtained from the reverse filter for the fifth iteration consisted of 9 extreme points. The best solution was then selected. This solution looks like solution 2 of our approach (TABLE 9).

Our approach, like the Interactive MOLP approach, does not require criterion weights of any kind. In reference 4, the five iterations are calculated and a lot of alternatives are presented to DM for analysis. Overall, the DM examined more than thirty solutions.

Only one iteration in our approach is considered and only four solutions are presented to DM. These solutions are close to the solutions selected in reference 4. In

**TABLE 11.** Classification of the Set on NS by Two Factors[a]

| Groups of Solutions by the Second Factor | | | | | | | | | | | | |
|---|---|---|---|---|---|---|---|---|---|---|---|---|
| | 10 | | | | 1 | 2 | | | | | | |
| | 9 | | | | | | 1 | | | | | |
| | 8 | | | | | | | | | 1 | 3 | |
| | 7 | | | 7 | | | | | | | | |
| | 6 | 1 | | | | | | | | | | |
| | 5 | | | 1 | | | | | | | | 4 |
| | 4 | | 7 | | | | | | 3 | | | |
| | 3 | | | | | | 1 | | | | | |
| | 2 | | | | | | | | | 1 | | |
| | 1 | | | | | | | 1 | | | | |
| | | 1 | 2 | 3 | 4 | 5 | 6 | 7 | 8 | 9 | 10 | 11 |

Groups of Solutions by the First Factor

[a]The number of the solutions in the set is given.

**TABLE 12.** Objective Function Values

| NS | Objectives I | II | III | IV |
|---|---|---|---|---|
| 1 | 39.09 | 7.82 | 44.86 | −13.58 |
| 2 | 37.04 | 7.41 | −25.93 | 55.56 |
| 3 | 32.22 | 6.67 | −20.0 | 58.89 |
| 4 | 37.04 | 40.74 | 40.74 | −44.44 |
| 5 | 33.33 | 66.67 | −33.33 | 0 |
| 6 | −80.00 | 100.00 | 60.00 | 40.00 |

our approach, the four best solutions are obtained without consulting with DM. Our approach gives the possibility to the DM to reduce the number of objectives and to form small groups of similar NS.

In the third example, 35 LP problems (equation 2) are solved. The matrix of factor loadings is presented in TABLE 10. Three factors account for about 99% of the total variance. The four objectives in the space of two major factors, $F_1$ and $F_2$, are presented (see FIGURE 1). TABLE 11 is the result of dividing the subset of NS by ten-by-eleven groups along factors $F_1$ and $F_2$. The projection or the most interesting solution on the factor $F_3$ gives DM information to choose the NS after comparison (TABLE 12).

In the last example, only one factor, which accounts for about 85% of the total variance, is calculated (TABLE 13). Graphic representation of the five objectives in the space of this factor is presented in FIGURE 2. DM could group the objectives considered into two groups, and select representatives could form the group of dependence criteria.

NS are grouped in the space of the factor obtained. A one-dimensional partition for several groups is constructed (TABLE 14). A one-dimensional classification can be obtained without using any additional apparatus. The DM can thus use a solution from the group no. 9 on the basis of the information obtained.

## CONCLUDING REMARKS

Two modifications of our approach may be considered. The first modification could be employed to reduce the number of solved LP problems. The most simple models

**TABLE 13.** Factor Loading Matrix

| Criteria (objectives) | Factor $F_1$ |
|---|---|
| I | 0.99 |
| II | 0.98 |
| III | 0.97 |
| IV | −0.61 |
| V | −0.99 |
| Eigenvalues | 4.27 |
| Factor and total percentage of variance | 0.853 |

**TABLE 14.** Classification of the Set on NS by One Factor[a]

| (2) | (1) | (2) | (1) | (3) | (1) | (5) | (4) | (46) |
|---|---|---|---|---|---|---|---|---|
| 1 | 2 | 3 | 4 | 5 | 6 | 7 | 8 | 9 |
| | | | | Groups of Solutions by the First Factor | | | | |

[a]The number of the solutions is given.

(quadratic or cubic) should first be considered when the methods of simplex lattice are used. The second modification is connected with using the methods of analysis of the zone of compromise solutions by means of simplex lattice design.

## ACKNOWLEDGMENTS

I wish to extend many thanks to Jonathan Kornbluth of Jerusalem University for his attention to this research and for his useful advice.

## REFERENCES

1. SOLOVEICHIK, D. 1983. A method for solving the multiple-objective linear programming problem. Ann. N.Y. Acad. Sci. **410:** 213–225.
2. SOLOVEICHIK, D. 1983. The multiple criteria decision-making process. Using the methods of multivariate statistical analysis. Ann. N.Y. Acad. Sci. **410:** 205–212.
3. SCHEFFE, H. 1958. Experiments with mixtures. J. R. Stat. Soc. **B20:** 344–360.
4. BRAVERMAN, E. M. 1974. A linguistic approach to processing large data arrays. Avtom. Telemekh. **11:** 73–88 (in Russian).
5. SOLOVEICHIK, D. 1984. A new approach to the decision-making in MOLP. To be published.
6. STEUER, R. E. & A. T. SCHULER. 1978. An interactive multiple-objective linear programming approach to a problem in Forest Management. Oper. Res. **26**(no. 2): 254–269.
7. STEUER, R. E. 1978. Repertoire of multiple-objective linear programming test problems. Univ. of Kentucky. Working paper in business administration, no. BA 23.

# A New Approach to the Flexible Planning Problem (FPP)

D. SOLOVEICHIK

## INTRODUCTION

The concept of De Novo Programming as it was formulated by Zeleny in reference 1 reads as follows:

> One should design in a satisfactory way a new system of available resources rather than optimize a system based on a potentially inferior set of a priori fixed resources.

This concept is quite similar to the FPP that was introduced by Negoita:[2]

$$\sum_{j-1}^{n} a_{ij}\bar{x}_j \in [\underline{d}^i, \bar{d}^i], \qquad x_j \geq 0. \tag{1.1}$$

Zimmerman[3] has proposed the following model for solving this problem:

$$\text{Maximize } X_{n+1} \text{ such that } \sum_{j-1}^{n} a_{ij}x_j + (\bar{d}_i - \underline{d}_i)x_{n+1} \leq \bar{d}_i, \qquad x_j \geq 0. \tag{1.2}$$

Another approach to FPP was proposed by Chernikova in reference 4. FPP may also be considered as a Multiple-Objective Linear Programming (MOLP) problem. The decision maker (DM) must obtain answers to the following questions:

—how to find the level of resources when the level of objectives is given?
—how to find the level of objectives when the level of resources is given?
—how to simultaneously find the most satisfactory levels of resources and objectives?

In the MOLP problem, objectives

$$C_k = \sum_{j-1}^{n} C_{kj}x_j, \qquad k = 1, \ldots, \ell, \tag{1.3}$$

should be "maximized" by being subjected to the constraints

$$\sum_{j-1}^{n} a_{ij}x_j \leq d_i, \qquad x_j \geq 0. \tag{1.4}$$

Zeleny[1] reduces this problem to the following one:

$$\text{Maximize } C_k \equiv \sum_{j-1}^{n} C_{kj}x_j, \qquad k = 1, \ldots, \ell, \tag{1.5}$$

and

$$\text{Minimize } d_i \equiv \sum_{j-1}^{n} a_{ij}x_j, \qquad x_j \geq 0, \tag{1.6}$$

$$\text{such that} \sum_{i=1}^{m} p_i d_i = W, \qquad j = 1, \ldots, n, \tag{1.7}$$

in which $m + \ell$ objective functions are subjected to one parametric constraint where $p_i$ ($i = 1, \ldots, m$) are the prices of $m$ resources.

The following fuzzy LP problem is considered:

$$\begin{array}{rl} \underset{\sim}{\max}\ c^T x & c, x \in R^n \\ a_i^T x \lesssim b_i & b \in R^m, (A)_{m,n} \\ d_j^T x \lesssim b'_j & i = 1, \ldots, m_1 \\ x \geq 0 & j = m_1 + 1, \ldots, m_1 + m_2 \\ & m_1 + m_2 = m. \end{array}$$

The first interpretation of this model is the following: Make a decision $x \geq 0$ in such a way that

—the value of the objective function, $c^T x$, "exceeds at least" the predetermined aspiration level $b_0$, and
—the restrictions $a_i^T x \leq b_i$, $i = 1, \ldots, m$, are satisfied "as well as possible", while
—the restrictions $d_j^T x \leq b'_j$, $j = m + 1, \ldots, m_1 + m_2$, are satisfied strictly.

Fuzzy LP (FLP) and MOLP have also been analyzed from various viewpoints.[3,5,6]

In this paper, we will consider another approach to the problem. This approach is based on the description of the set of constraint vectors $D_s$,

$$D_s = \{d \mid x \geq 0, Ax \in [\underline{d}, \bar{d}], \underline{d} \leq d \leq \bar{d}\}, \tag{1.8}$$

which we call simultaneous or $s$-vectors, instead of being based on the determination of only one vector $d$ as was done by Zimmerman.[3]

This set includes not only constraint vectors, but also the objective function value vector,

$$C = (\max C_1, \ldots, \max C_\ell), \tag{1.9}$$

which is defined in equation 1.5.

In practice, it is assumed that the DM will face the problem with unsimultaneous vectors (or $\bar{s}$-vectors) of constraints $d_0$ being given a priori ($d_0 \in D_{\bar{s}}$). The problem to find $s$-vectors (which lie in the nearest proximity to $d_0$) is considered. The real-valued function $f(d - d_0)$ can be introduced as a measure of the proximity between the points $d_0$ and $d$ in $E_m$. It may be, for example, Euclidean or any other norm.

The problem of the optimal changing $\bar{s}$-vector of constraints $d_0$ used for the obtainment of the near-lying $s$-vector $d$ in FPP can be presented in the following analytical form:

$$\min \{f(d - d_0) \mid d \in D\}. \tag{1.10}$$

The result of solving equation 1.10 is the $s$-vector $d$ nearest to the $\bar{s}$-vector $d_0$. The main difficulty in solving equation 1.10 is the necessity of the efficient description of the set $D_s$. It follows from equation 1.8 that $D_s$ is polyhedron in the space $E_m$. For most practical problems, the DM would not be able to use an analytic description, even if it is simplified, for example, by an algorithm that decreases the number of inequalities.[4]

As an alternative, we recall that $D_s$ implicitly includes all alternatives of the DM's alterations of initial constraints. A very effective description of this set can be achieved by using the decision rules $f_p = dy_p^*$ $(p = 1, \ldots, P)$ that are considered in pattern recognition (PR). On the one hand, this description sufficiently reduces the amount of redundant information. On the other hand, though, it is given in a form that is convenient for the decision-making process.

Three kinds of vectors are examined: $s$-vector, $\bar{s}$-vectors, and "suspicious" (or $\bar{\bar{s}}$-vectors). If we do not know if the vector considered is an $s$- or $\bar{s}$-vector, we name such a vector as an $\bar{\bar{s}}$-vector. The $\bar{\bar{s}}$-vectors are considered as inputs on every iteration of our algorithm.

To describe the set of $s$-vectors with the help of decision rules, it is necessary to develop the procedure by which the initial data sets of the $s$-, $\bar{s}$-, and $\bar{\bar{s}}$-vectors are defined. Such a procedure significantly reduces the computing time required.

The first step of this method is the statistical simulation of possible constraints. It is suggested that DM should give well-grounded intervals of permissible alterations of some constraints for this purpose.

At the second step, for every random computer-formed vector of model constraints, we should determine whether this vector is simultaneous or not. The set of all random computer-formed vectors is

$$D_{\bar{\bar{s}}} = \{d_s | \underline{d} \leq d_s \leq \bar{d}, s = 1, \ldots, R\}. \quad \textbf{(1.11)}$$

$D_{\bar{\bar{s}}}$ consists of $s$ or $\bar{s}$ vectors:

$$D_{\bar{\bar{s}}} = D_s \cup D_{\bar{s}}. \quad \textbf{(1.12)}$$

In this paper, we present a new computing timesaving procedure by eliminating $s$- and $\bar{s}$-vectors from the $D_s$. To make it more efficient, we use the matrix of reduced cost and the simplex method with a special rule of introducing vectors into the basis. As a result, we derive the sufficiently reliable procedure for the determination of $s$-vectors without solving the linear programming (LP) problem.

The above-mentioned procedure is illustrated in a numerical example later in this paper.

## METHODS OF RESEARCH

It is well known[7] that only one of the following alternatives is true:

—inequality

$$Ax \leq d \quad \textbf{(2.1)}$$

has a nonnegative solution;

—there are nonnegative solutions of the following system of inequalities:

$$yA \geq 0, y \geq 0 \quad \textbf{(2.2)}$$

$$dy < 0. \quad \textbf{(2.3)}$$

The corresponding LP problem can be formulated as follows:

$$\text{Minimize } dy \tag{2.4}$$

$$\text{such that } yA \geq 0, y \geq 0, \tag{2.5}$$

$$ye = 1, \tag{2.6}$$

or

$$yA' \geq \ell, y \geq 0, \tag{2.7}$$

where

$$A' = \begin{pmatrix} A \\ \ell \end{pmatrix}, \ell = \begin{pmatrix} 0 \\ 1 \end{pmatrix}.$$

Equation 2.6 is taken into account so as to eliminate the case when the minimum (equation 2.4) is $-\infty$. If we find vectors $y^* \geq 0$ (which is optimal) and $d_t y^* < 0$, then we consider vector $d_t$ as an $\bar{s}$-vector. For a given solution $y^*$, it is simple to calculate the set $D_{\bar{s}}(y^*)$ of $\bar{s}$-vectors by observing

$$f = d_t y^* < 0, \qquad t = 1, \ldots, \bar{s}_1, \tag{2.8}$$

$$D_{\bar{s}}(y_1^*) = \{d_t \mid d_t y^* < 0, \qquad t = 1, \ldots, \bar{s}_1\}. \tag{2.9}$$

We note that every solution $y^*$ is associated with a subset of $\bar{s}$-vectors $D_{\bar{s}}(y^*)$. Our purpose is to construct such a set $\{y_i^*\}_{i=1}^{p}$, along with the corresponding set $\{D_s(y_i^*)\}$. The inequality (equation 2.8) as a first decision rule is considered. At the end of the algorithm, $p$ decision rules and $D_s(y_i^*)$ and $D_{\bar{s}}(y_i^*)$ sets of $s$ and $\bar{s}$-vectors, respectively, are obtained. Finally, we note that

$$D_s = D_s(y_1^*) \cup D_s(y_2^*) \cup \cdots \cup D_s(y_p^*) \tag{2.10}$$

and

$$D_{\bar{s}} = D_{\bar{s}}(y_1^*) \cup D_{\bar{s}}(y_2^*) \cup \cdots \cup D_{\bar{s}}(y_p^*) \tag{2.11}$$

are considered.

Let $A' = (B{:}N)$ be a partition of $A'$ in equation 2.7. At the $i$-th step of elimination of $s$-vectors, the procedure proposed is based on the matrix of reduced costs $\{\Delta^i\}$ of all vectors $d_t \in D_{\bar{s}}^i$.

$D_{\bar{s}}^i$ is the matrix of $\bar{s}$-vectors on the $i$-th step of our algorithm:

$$D_{\bar{s}}^i = \left\{ d \mid d \in \left( \bigcup_{k=1}^{i-1} D_{\bar{s}}(y_k^*) \right) \cup \left( \bigcup_{k=1}^{i-1} D_s(y_k^*) \right) \right\}, \tag{2.12}$$

$$\Delta^i = D_{\bar{s}}^i - D_{\bar{s}(b)}^i \cdot B^{-1} \cdot N; \tag{2.13}$$

$\Delta^i$ is the matrix of reduced costs, where $D_{\bar{s}(b)}^i$ is the submatrix of the $D_{\bar{s}}^i$ corresponding to the basis $B$.

It is well known that a column of $\Delta$, say $\Delta_j$, gives the marginal rates of the change of the function $dy$ in equations 2.4 and 2.7 that was caused by the introduction of the variable $y_j$ into the basis. The simplex method with the special rule of introducing variable $y_j$ into the basis is considered. The variable $y_j$ is introduced into the basis if the number of negative components of the vector $\Delta_j$ is maximal. Solution $y_2^*$ is obtained as a result. The second decision rule is formed:

$$f_2 = dy_2^*.$$

It is well known that if $\Delta_{tj} \geq 0$ for all $j$, then the solution obtained in the LP problems in equations 2.4 and 2.7 is optimal for a given $d_t$. If $d_t y_i^* > 0$ for any $d_t \in D_{\bar{s}}$ and if $\Delta_{tj} \geq 0$ for all $j$, then the vector $d_t$ is considered as a $s$-vector. $\Delta^i \geq 0$ as a group of decision rules connected with vector $y_i^*$ is also examined.

This procedure is reiterated for some steps. The algorithm is continued when $\Delta_{tj} < 0$ for some $d_t$. If $\Delta_{tj} \geq 0$ for all $t$, $j$, it means that $D_s$ and $D_{\bar{s}}$ are determined finally without any errors.

The algorithm is terminated if the number of $s$-vectors obtained during the previous steps is enough for the investigation. The interactive process is organized to determine the finishing of our procedure.

The LP model (equations 2.4 and 2.7) as a linear max-min programming model is

$$\min z(y) = \max_{z=1,\ldots,R} \{d_t(y)\}, \qquad \textbf{(2.14)}$$

$$yA' \geq \ell, \; y \geq 0. \qquad \textbf{(2.15)}$$

An algorithm of linear max-min programming[8] for finding $s$- and $\bar{s}$-vectors may be considered. In our opinion, using the simplex method with the special rule of introducing a variable into the basis is more effective in our case than the algorithm.[8]

The set of $s$-vectors $d \in D_s$ is obtained by using the following two models:

$$\text{Minimize} \sum_{i=1}^{m} \beta_i' y_i \text{ such that} \sum_{j=1}^{n} a_{ij}x_j - y_i = \underline{d}_i, \qquad x_j, y_i \geq 0, \qquad \textbf{(2.16)}$$

$$\text{Maximize} \sum_{i=1}^{m} \beta_i'' z_i \text{ such that} \sum_{j=1}^{n} a_{ij}x_j + z_i = \bar{d}_i, \qquad x_j, z_i \geq 0. \qquad \textbf{(2.17)}$$

Let $\bar{x}_j^{(1)}$, $\bar{y}_i$ and $\bar{x}_j^{(2)}$, $\bar{z}_i$ be optimal solutions for equations 2.16 and 2.17, respectively. The model of equation 1.1 is inconsistent with the given system of weights, $\beta_i'$, $\beta_i''$, if for some $i$, $\underline{d}_i + \bar{y}_i \geq d_i$ or $\bar{d}_i - \bar{z}_i \leq \underline{d}_i$. Otherwise, the elements of the set of $s$-vectors,

$$d \in D_s,$$

$$d = \underline{d}_i + \bar{y}_i, \text{ or}$$

$$d = \bar{d}_i - \bar{z}_i, \qquad \textbf{(2.18)}$$

are received. Weights $\beta_i'$, $\beta_i''$ may be formed randomly in the models of equations 2.16 and 2.17.

The step-by-step description of the procedure for forming $D_s$ and $D_{\bar{s}}$ is:

Step 1. Obtain some intervals of permissible alterations for constraints $\underline{d}$ and $\bar{d}$ in equation 1.11.

Step 2. Normal (Gaussian) or uniform probability distributions are used to determine random computer-formed vectors of constraints. $R$ vectors $d_t \in D_{\bar{s}}$ are considered ($t = 1, \ldots, R$).

Step 3. The arithmetic mean of vectors $d_t \in D_{\bar{\bar{s}}}$ is determined:

$$\bar{d} = \frac{\sum_{t=1}^{R} d_t}{R}. \tag{2.19}$$

Step 4. Solve the following LP problem:

$$\text{Minimize } \bar{d} \cdot y \text{ such that } y \cdot A' \geq 1, y \geq 0.$$

Let $y_1^*$ be an optimal solution of this problem.

Step 5. Calculate

$$f_1^t = d_t y_1^*, \qquad t = 1, \ldots, R, \tag{2.20}$$

for each $t$.

Step 6. Vectors $d_t \in D_{\bar{s}}(y_1^*)$ if $f_1^t < 0$ for these vectors. The subset $D_{\bar{s}}(y_1^*)$ is formed.

Step 7. Vectors $d_t \in D_{\bar{\bar{s}}} \backslash D_{\bar{s}}(y_1^*)$ are considered. If for some vectors $d_t$, $\Delta_{tj} \geq 0$ for all $j$, then such vectors $d_t$ are determined as $s$-vectors. The subset $D_s(y_1^*)$ is formed.

Step 8. Subset

$$D_{\bar{\bar{s}}}^1 = D_{\bar{\bar{s}}} \backslash [D_s(y_1^*) \cup D_{\bar{s}}(y_1^*)] \tag{2.21}$$

is considered as input for the next step.

Step 9. Compute

$$\Delta_{tj} = D_{\bar{\bar{s}}}^1 - D_{\bar{\bar{s}}(b)} \cdot B^{-1} \cdot N.$$

Vector $y_j$ is introduced into the basis if the number of negative components of vector $\Delta_j$ for all $t$ is maximal. Solution $y_2^*$ is obtained at this step. Then go back to step 5.

Steps 5–9 are reiterated until the number of eliminated $s$- or $\bar{s}$-vectors are decreased to a great extent. In this case, the set of $\bar{\bar{s}}$-vectors remaining is considered to be the input for step 3.

Equation 2.18 or any other formula for the determination of the vector $\bar{d}$ in the LP problem (equation 2.19) is used. For example, some vectors from the set of weight vectors may be utilized to determine the vector $\bar{d}$.

The $s$-vectors $d \in D_s$ were formed by using the models of equation 2.16 or 2.17 and by pattern recognition methods. We must exclude $s$-vectors that are dominated by a convex combination of other $s$-vectors. An $s$-vector $d_k = (d_{k1}, \ldots, d_{km}) \in D$ is dominated by a convex combination of other vectors in $D$ if and only if there exists, by definition, a set of multipliers $\beta_i$, $i = 1, \ldots, t$, $i \neq k$, such that

$$\sum_{i \neq k} \beta_i d_{ij} \geq d_{kj}, \qquad j = 1, \ldots, m,$$

$$\sum_{i \neq k} \beta_i = 1$$

$$\beta_i \geq 0.$$

An efficient procedure for excluding $s$-vectors that are dominated by a convex combination of other $s$-vectors is presented in reference 9.

Overall, the DM should either make a final choice from the set of $s$-vectors or decide to continue our algorithm for forming the next $s$-vectors.

## NUMERICAL EXAMPLE

Let us consider the numerical example used by Zionts:[10]

$$\begin{aligned} 3x_1 + x_2 + 2x_3 + x_4 &\geq 66, \\ x_1 - x_2 + 2x_3 + 4x_4 &\geq 80, \\ -x_1 + 5x_2 + x_3 + 2x_4 &\geq 75, \\ 2x_1 + x_2 + 4x_3 + 3x_4 &\leq 60, \\ 3x_1 + 4x_2 + x_3 + 2x_4 &\leq 60. \end{aligned} \tag{3.1}$$

The dual problem primary is:

$$\begin{aligned} -3y_1 - y_2 + y_3 + 2y_4 + 3y_5 &\geq 0, \\ -y_1 + y_2 - 5y_3 + y_4 + 4y_5 &\geq 0, \\ -2y_1 - 2y_2 - y_3 + 4y_4 + y_5 &\geq 0, \\ -y_1 - 4y_2 - 2y_3 + 3y_4 + 2y_5 &\geq 0. \end{aligned} \tag{3.2}$$

The constraint $y_1 + y_2 + y_3 + y_4 + y_5 = 1$ is added:

$$\min\,(-66y_1 - 80y_2 - 75y_3 + 60y_4 + 60y_5). \tag{3.3}$$

The intervals of permissible alterations of constraints in the model of equation 3.1 are as follows:

$$20\text{–}70,\ 30\text{–}80,\ 25\text{–}90,\ 40\text{–}70,\ 40\text{–}70. \tag{3.4}$$

Uniform distributions for forming 500 random computer-formed vectors of constraints are used.

The optimal solution $y$ after solving the LP problem (equation 2.19) on step 4 is

$$0.352,\ 0.095,\ 0.146,\ 0.212,\ 0.194. \tag{3.5}$$

We considered 427 random computer-formed vectors $d \in D_{\bar{s}}$ that satisfied the inequality (equation 2.8) as $\bar{s}$-vectors.

The matrix $\Delta$ for finding $s$- and $\bar{s}$-vectors from the 73 remaining vectors is examined. We found that 36 vectors from this set are $s$-vectors ($\Delta_{tj} \geq 0$ for these $t$). The special rule of introducing a variable $y_8$ into the basis is used. The solution $y_2^*$ is

$$0.247,\ 0.227,\ 0.085,\ 0.441,\ 0.0. \tag{3.6}$$

The second decision rule $f_2 = d_t y_2^*$ and the matrix $\Delta$ for this basis are considered. We receive 14 $\bar{s}$-vectors and 11 $s$-vectors. As a result, 47 $s$-vectors and 441 $\bar{s}$-vectors were formed. [We considered 12 vectors out of 500 (or 2.4%) to be the values of errors.] These 47 vectors are preferable to DM. Analysis of such vectors as the beginning of the interactive process is considered.

In addition, the other intervals ($\underline{d}$, $\bar{d}$) or the probability laws may be examined. Intervals that are more narrow (limited) are considered:

$$30\text{–}50, 45\text{–}55, 43\text{–}55, 58\text{–}62, 58\text{–}62. \quad \textbf{(3.7)}$$

We considered 500 random computer-formed vectors using uniform distributions in these intervals. The first decision rule $d_t y_1^*$ is examined. As a result, 109 $s$-vectors and 391 $\bar{s}$-vectors are obtained. The vectors in this set are not of interest to DM. Therefore, DM decides to not only change the intervals, but also the probability laws.

Other intervals,

$$40\text{–}60, 40\text{–}60, 40\text{–}60, 50\text{–}70, 50\text{–}70, \quad \textbf{(3.8)}$$

using normal distributions are considered. In this case, only one vector

$$38.0, 51.5, 44.2, 63.5, 58.7$$

is an $s$-vector. This means that 499 random vectors are $\bar{s}$-vectors.

The normal distribution for other intervals,

$$30\text{–}70, 30\text{–}70, 30\text{–}70, 40\text{–}80, 40\text{–}80, \quad \textbf{(3.9)}$$

is also used. Five hundred random vectors were formed. As a result, 25 vectors are $s$-vectors and 474 are $\bar{s}$-vectors, with one vector as the value of error.

There is another choice suggested that is more preferable to DM. This consists of a vector $d$ whose two first components are equal to 50, whose third component is in the interval 30–70, and whose other two components are in the intervals 50–70 using normal law. As a result of calculations, all 500 vectors are $s$-vectors.

After solving in reference 10, two goal programming (GP)[11] problems are proposed with objectives minimizing two kinds of sums of deviations from the goal (66, 80, 75); the two right-hand vectors are

$$\begin{array}{ccccc} 24 & 66 & 66 & 60 & 60, \\ 66 & 30 & -12 & 60 & 60. \end{array}$$

After minimizing the maximum deviation from each goal and using preemptive priorities, two other similar vectors are obtained:[10]

$$\begin{array}{ccccc} 36.6 & 50.6 & 45.6 & 60 & 60, \\ 50 & 50 & 13.65 & 60 & 60. \end{array}$$

With the help of the procedure suggested above, more than 100 $s$-vectors are received after solving only one LP problem in equations 3.2 and 3.3 and after one iteration of simplex method. The most interesting of them are proposed in TABLE 1.

As a result of the investigation by using our algorithm, DM fully obtained information on how to find available resources and objectives that are of interest to the proper functioning.

**TABLE 1.** $s$-Vectors

| $s$-Vectors | | | | | | |
|---|---|---|---|---|---|---|
| 1 | 2 | 3 | 4 | 5 | 6 | 7 |
| 50 | 50 | 50 | 49.0 | 46.6 | 45.3 | 39.0 |
| 50 | 50 | 50 | 54.2 | 44.0 | 57.8 | 45.7 |
| 30.4 | 32.6 | 32.7 | 44.5 | 37.2 | 42.7 | 44.9 |
| 64.1 | 69.8 | 68.85 | 71.8 | 67.6 | 65.9 | 61.9 |
| 69.9 | 66.9 | 68.75 | 72.8 | 62.9 | 72.8 | 61.0 |

Let us consider the numerical example used in reference 5:

$$\underset{\sim}{\max} \quad x_1 + x_2,$$

$$-x_1 + 3x_2 \lessapprox 21,$$

$$x_1 + 3x_2 \lessapprox 27,$$

$$4x_1 + 3x_2 \lessapprox 45,$$

$$3x_1 + x_2 \lessapprox 30. \qquad \textbf{(3.10)}$$

Intervals of permissible alterations of objectives and constraints in the model of equation 3.10 are as follows: 13.5–14.5, 17–25, 23–30, 40–50, 25–30. A uniform distribution for forming 500 random computer-formed vectors of constraints is used. As a result, 11 $s$-vectors and 450 $\bar{s}$-vectors are obtained. At the next step, 8 $s$-vectors and 31 $\bar{s}$-vectors are received. As a result, 19 $s$-vectors and 481 $\bar{s}$-vectors are formed. Five vectors of them are preferable to DM. Analysis of such vectors is considered to be the beginning of the interactive process.

The vectors are:

| | | | | |
|---|---|---|---|---|
| 13.73 | 13.62 | 14.25 | 13.77 | 13.72 |
| 17.48 | 18.20 | 14.47 | 12.82 | 9.52 |
| 29.33 | 29.54 | 28.61 | 27.06 | 25.34 |
| 47.10 | 46.54 | 49.83 | 48.42 | 49.07 |
| 25.58 | 24.96 | 28.39 | 28.00 | 29.54. |

After solving (in reference 6) the LP problem $13 \times 7$, only one right-hand vector

13.75 17.25 29.25 47.25 25.75

is proposed.

## CONCLUDING REMARKS

In the above examples, it was seen how a DM can obtain a set of $s$-vectors. The advantage of the approach given in this paper is that it is able to form the set of $s$-vectors using computing timesaving procedures.

The DM obtains not only one solution to select the level of resources or objectives; by changing intervals and (or) probability laws, DM is given the possibility of analyzing the set of $s$-vectors before doing such serious actions as the decision-making process in the FPP.

As demonstrated, it is possible to use our approach for solving GP and FLP problems. This approach may be considered only as a first step for solving MOLP, GP, and FLP problems.

The information obtained may be examined as input for the next investigation. It is well known that DM can analyze the small number of alternatives. We must thus aggregate the information given to DM.

## REFERENCES

1. ZELENY, M. 1980. De Novo Programming: A case study in multiobjective design. *In* Multicriteria Analysis in Practice. Nijkamp & J. Spronk, Eds. Nijhoff. The Hague.
2. NEGOITA, C. 1979. Management Applications Of System Theory. Birkhäuser. Basel.
3. ZIMMERMAN, H. J. 1974. Optimization in fuzzy environment. Paper presented at ORSA TIMS Meeting, San Juan, Puerto Rico.
4. CHERNIKOVA, N. 1965. Algorithm for finding the general formula of nonnegative solutions of a system of linear unequalities. J. Vopr. Magmat. Metamorf. **5:** 2. Moscow.
5. ZIMMERMAN, H. J. 1978. Fuzzy programming and linear programming with several objective functions. Fuzzy Sets Systems **1:** 45–55.
6. HAMACHER, H., H. LEBERING & H. J. ZIMMERMAN. 1978. Sensitivity analysis in fuzzy linear programming. Fuzzy Sets Systems.
7. GALE, D. 1960. The Theory of Linear Economic Models. McGraw–Hill. New York.
8. POSNER, M. & C-T. WU. 1981. Linear max-min programming. Math. Programming **20:** 166–172.
9. ZIONTS, S. & J. WALLENIUS. 1980. Identifying efficient vectors: some theory and computational results. Oper. Res. 3.
10. ZIONTS, S. 1980. Multiple criteria decision making: an overview and several approaches. Working Paper no. 454 at SUNY at Buffalo, New York.
11. STEUER, R. 1979. Goal programming sensitivity analysis using interval penalty weights. Math. Programming **17:** 16–31.

# Index of Contributors